T0135020

Intelligent Systems Reference Library

Volume 77

Series editors

Janusz Kacprzyk, Polish Academy of Sciences, Warsaw, Poland
e-mail: kacprzyk@ibspan.waw.pl

Lakhmi C. Jain, University of Canberra, Canberra, Australia, and
University of South Australia, Adelaide, Australia
e-mail: Lakhmi.Jain@unisa.edu.au

About this Series

The aim of this series is to publish a Reference Library, including novel advances and developments in all aspects of Intelligent Systems in an easily accessible and well structured form. The series includes reference works, handbooks, compendia, textbooks, well-structured monographs, dictionaries, and encyclopedias. It contains well integrated knowledge and current information in the field of Intelligent Systems. The series covers the theory, applications, and design methods of Intelligent Systems. Virtually all disciplines such as engineering, computer science, avionics, business, e-commerce, environment, healthcare, physics and life science are included.

More information about this series at http://www.springer.com/series/8578

Lech Polkowski · Piotr Artiemjew

Granular Computing in Decision Approximation

An Application of Rough Mereology

 Springer

Lech Polkowski
Department of Computer Science
Polish–Japanese Institute of Information
  Technology
Warsaw
Poland

and

Department of Mathematics and Computer
  Science
University of Warmia and Mazury
Olsztyn
Poland

Piotr Artiemjew
Department of Mathematics and Computer
  Science
University of Warmia and Mazury
Olsztyn
Poland

ISSN 1868-4394   ISSN 1868-4408 (electronic)
Intelligent Systems Reference Library
ISBN 978-3-319-36621-0   ISBN 978-3-319-12880-1 (eBook)
DOI 10.1007/978-3-319-12880-1

Printed on acid-free paper

Springer International Publishing AG Switzerland is part of Springer Science+Business Media
(www.springer.com)

To Professor Lotfi Asker Zadeh

Preface

Would You tell me, please, which way I ought to go from here?
That depends a good deal on where You want to get to, said the
Cat. I don't much care where—said Alice. Then it doesn't matter
which way You go, said the Cat.

[Lewis Carroll. Alice's Adventures
in Wonderland]

In this Work when it shall be found that much is omitted, let it not
be forgotten that much likewise is performed.

[Samuel Johnson]

This book may be regarded as a sequel to "Approximate Reasoning by Parts" in this Series. The topic of granular computing in classifier synthesis, mentioned in the former book, among other themes, is fully developed in this book. The idea came to the first author in the early 1990s when, in cooperation with Prof. Andrzej Skowron, he proposed to modify the proposition put forth 30 years earlier and converted into Fuzzy Set Theory by Prof. Lotfi Asker Zadeh of the partial membership in a concept, by discussing a partial containment in its most general features.

When attempting at formalization of partial containment, our attention was turned to the theory of mereology created by Stanislaw Leśniewski, an eminent member of the famous Warsaw Logical School; the basic notion of a part in that theory was extended to the notion of a part to a degree subjected to some axiomatic postulates.

In this way the theoretical proposition of Rough Mereology came into effect, and along with it expectations that it may open up a new venue for approximate reasoning under uncertainty. The primitive construct in it, called a *rough inclusion* (1994) does reflect most general and universally true properties of partial inclusion on concepts and its relations to the fuzzy idea were clearly recognized, cf., Prof. Achille Varzi's statement'...*which results in a fuzzification of parthood that parallels in many ways to the fuzzification of membership in Zadeh's (1965) set theory, and it is this sort of intuition that also led to the development of such formal theories as Polkowsky and Skowron's (1994) rough mereology...* (*Stanford Encyclopedia of Philosophy.* http://plato.stanford.edu/entries/mereology).

The first author recognized the potential of the new theory not only in applications to classical already topics of mereotopology and mereogeometry, but also in a more recent area of granular computing, introduced by Prof. Lotfi A. Zadeh, see *Outline of a new approach to the analysis of complex systems and decision processes, IEEE Trans. on Systems, Man, and Cybernetics, SMC-3, 28–44, 1973*.

The idea of granular computing, extended later by Prof. L.A. Zadeh to the idea of *computing with words* see L.A. Zadeh *Computing with Words. Principal Concepts and Ideas. Springer 2012* turned out to become an attractive methodology in intelligent algorithm theory, see the list of principal monographs in reference to Chap. 1. Granules of knowledge are defined as aggregates (or, *clumps*) of objects drawn together by a similarity relation, a good example can be an indiscernibility class in a decision system, see Prof. Z. Pawlak's *Rough Sets: Theoretical Aspects of Reasoning about Data, Kluwer 1991*, or, a fiber of a fuzzy membership function, see L.A. Zadeh *Fuzzy sets, Information and Control 8, 338–353, 1965* in both theories one computes not with single objects but with their classes, i.e., granules.

Rough Mereology is especially suited toward granular theory as its precise assumptions allow for a formal theory of granules. The aggregate (or, formally the mereological class in the sense of Leśniewski) of all things included in the given thing (the *granule center*) to a degree of at least r forms a granule in the sense of Prof. L.A. Zadeh, i.e., a collection of things drawn together by a similarity. A topological character of granules which resemble neighborhoods allows for applications in not only classifier synthesis but, e.g., in behavioral robotics.

The notion of a granule is developed in Chap. 1 and a similarity relation underlying it is modeled on a tolerance relation, or a weaker form of it, called a weak tolerance relation. As such it does encompass some strategies for intelligent algorithms construction like, e.g., indiscernibility relations in information/decision systems.

One of the upshots of rough mereological granular concept theory was presented by the first author at GrC conferences in Beijing (1995) and Atlanta (1996) which consisted of the idea of granular classification. It was proposed to preprocess data (a decision system, a data table) by forming for a given real number $r \in [0, 1]$ the set of granules to this degree about data and representing each granule by a new data averaged in a sense over the granule. From the collection of granules, a covering of the universe of objects is selected by some strategy and obtained in this way new data set called a *granular reflection* of the original data set undergoes a classification process with some classification algorithms. Expectations that this method of data treatment should reduce noise, decrease ambiguity in data, and in effect increase quality of classification measured, e.g., by standard factors like accuracy and coverage were taken to test with real data.

At this stage new concepts were introduced, in an essential part by the second author, like ideas of concept-dependent granulation, layered, or multi-stage granulation, and applications to the missing values problems. Concept-dependent granulation consists in forming granules within the decision class of the center of the granule which yields a greater number of smaller but more compact with respect to similarity in question granules. Layered granulation consists in the repeated granulation process up to the stable data set. The idea for a treatment of missing values consists in treating the missing value as a value of an attribute on its own and forming granules in order to assign in place of the missing value a value assigned by some averaging process over the granule.

These ideas were put to test in many research and conference papers with some real data but it is now that the authors undertake the task of as complete as possible verification of those ideas with numerous real data available in the University of California at Irvine Data Mining Repository. As classifying algorithms a kNN classifier as well as the Naive Bayes classifier have been applied due to the well-known least error property of the latter and the asymptotic convergence to it of the former.

Complexity issues were also instrumental in the choice of algorithms for classification: a full rough set exhaustive classifier turned out to be too time-consuming on some parameterized variants of granular classification.

A most striking feature of the granular approach seems to be the fact that granular reflections yield classification results as good or in some cases better than in the non-granulated case with much smaller sizes of data sets and appropriately smaller numbers of classifying rules. It seems that these facts predestine this approach to such applications like tele-medicine or other emergency events where the simplicity and compactness of diagnostic systems are of vital importance.

The book consists of 11 chapters and an appendix. The first chapters provide an introductory background knowledge. Similarity relations which serve as the means for granule formation are formally discussed as tolerance or weak tolerance relations in Chap. 1 along with a formal idea of a tolerance granule. Chapter 2 is devoted to mereology and rough mereology. Basics of machine learning are discussed in Chap. 3 with emphasis on Bayesian and kNN classifiers and asymptotic properties of classifiers. Classification problems are discussed in detail in Chap. 4 along with a validation scheme known as cross-validation (CV).

The experimental part extends over Chaps. 5–9. In Chap. 5, a study of granular coverings is presented aimed at estimating the impact of a covering strategy on the quality of classification. Sixteen covering strategies are examined on all 13 data sets and best strategies are applied in the following chapters. Chapter 6 brings forth results of classifying in granular and non-granular cases by means of kNN classifier applied to the multi-layer granulation, whereas Chap. 7 shows analogous results for Naive Bayes classifier. In Chap. 8, the problem of missing values is studied by perturbing randomly in 5 or 10 % all 13 data sets and recovering the data by four strategies for missing value treatment.

Chapter 9 is devoted to classifiers employing additional parameters of ε and catch radius, based on weak rough inclusions which diffuse the indiscernibility relations. In Chap. 10, a study of effects of granulation on entropy of data and noise in them is presented. This study leads to parameters maximizing the noise reduction, reduction in size between the first and second granulation layers as well as reduction in number of pairs of indiscernibility classes with a given mereological distance. Chapter 11 called Conclusions brings an analysis of relations among optimal granulation radius and radii defined in Chap. 10. It is shown that radii of maximal noise reduction and maximal size decrease between first and second layers of granulation are fairly close approximators to the optimal granulation radius. In the Appendix that follows, main parameters concerning the distribution of values of attributes are collected and visualized like the central class, intersections with other classes, etc., which parameters bear on characteristics of granular reflections in terms of granule size and distribution.

The applied algorithms are illustrated with simple hand examples with the intention that the book may serve also as a text for students or other people wanting to acquaint themselves with these techniques.

The authors express the hope that this work will be useful for researchers in granular computing in that it will illuminate the techniques based on rough inclusions and will, by this, stimulate further progress in this area.

The authors wish to express their gratitude to Profs. Lakhmi Jain and Janusz Kacprzyk, the Series Editors, for their kind support and encouragement. Their thanks go also to Dr. Thomas Ditzinger of Springer Verlag for His kind help with publication.

This book is dedicated to Prof. Lotfi Asker Zadeh, whose insight brought into the realm of approximate reasoning the idea of partial membership.

The authors would like also to recall on this occasion the memory of Prof. Helena Rasiowa (1917–1994), an eminent logician and one of the founders of Theoretical Computer Science in Poland, on the 20th anniversary of her departure.

Warsaw, Poland   Lech Polkowski
Olsztyn, Poland   Piotr Artiemjew

Contents

Chapter 1
Similarity and Granulation

You can please some of the people all of the time, you can please all of the people some of the time, but you can't please all of the people all of the time

[John Lydgate]

This chapter introduces into topics of similarity and granulation. We define similarity relations as tolerance and weak tolerance relations and give basic information on their structure. In preparation for theme of granulation, we extend notions of tolerance to graded tolerance and weak tolerance. We outline approaches to granulation and the notion of a granule.

1.1 Introduction

Granulation is an operation performed on our knowledge. This knowledge, in its raw form, consists of percepts, records, various data collected through observation, servation, polling, etc. The raw knowledge is rectified in order to exhibit things (objects) of interest, their properties of interest, and relations among those properties, cf., Bocheński [2]. The choice of things depends on the resolution of our perception mechanisms, a good metaphor here may be that of a microscope: depending on its resolution we may perceive discrete atomic structures or continuously looking but magnified amalgams of atoms. Each of those amalgams consists of many atoms collected together by the fact that their pairwise distances are less than the resolution threshold of the microscope.

An abstract form was given to such observations by Zadeh [44] who used the term *granule* for an amalgam of things collected together by a common property; this property is usually called a *similarity*; hence, a granule is a collection of similar things with respect to a chosen similarity relations. Therefore, we begin with a review of similarity relations.

© Springer International Publishing Switzerland 2015
L. Polkowski and P. Artiemjew, *Granular Computing in Decision Approximation*,
Intelligent Systems Reference Library 77, DOI 10.1007/978-3-319-12880-1_1

1.2 Similarity

The most radical similarity is an *equivalence relation* of which we know perfectly well that it is a relation *eq* which is

1. *Reflexive*, i.e., $eq(x, x)$ for each thing x;
2. *Symmetric*, i.e., $eq(x, y)$ implies $eq(y, x)$;
3. *Transitive*, i.e., $eq(x, y)$ and $eq(y, z)$ imply $eq(x, z)$.

Letting $[x]_{eq} = \{y : eq(x, y)\}$ for each thing x, defines *equivalence classes* which partition the collection of things into pairwise disjoint sub-collections. In this case, similarity means classification into disjoint categories.

A less radical notion of similarity is offered by a *tolerance relation*, cf., Poincaré [21], Schreider [37], Zeeman [45], which is a relation $tlr(x, y)$ which is merely 1. *reflexive*; 2. *symmetric*, but need not be transitive. A good example of a tolerance relation was given in Poincaré [21]: consider a metric $\rho(x, y)$ on, e.g., a real line along with a relation $dist_\rho$ defined as

$$dist_\rho(x, y) \text{ if and only if } \rho(x, y) < \delta$$

for some small positive value of δ. As $2\delta > \delta$, it may happen that $dist_\rho(x, y)$ and $dist_\rho(y, z)$ hold whereas $dist_\rho(x, z)$ holds not. Another manifest example is the relation

$$int_\cap(X, Y) \text{ if and only if } X \cap Y \neq \emptyset$$

of a non-empty intersection on a collection of non-empty subsets of a set. The relational system $(V, int_\cap)$ provides the *universal tolerance relation*.

The structure of a tolerance relation *tlr* is more complicated than that induced by equivalence. Mimicking the procedure for equivalences outlined before, one may define *tolerance sets*; a tolerance set $tlr - set$ is a collection of things such that

$$\text{if } y, z \in tlr - set \text{ then } tlr(y, z). \tag{1.1}$$

A tolerance set which is maximal with respect to the property (1.1) is called a *tolerance class*. Due to reflexivity of tolerance, tolerance classes form a *covering* of the universe U. Tolerance classes allow for a characterization of tolerance relations in terms of an embedding into the universal tolerance relation.

Proposition 1 (cf., Schreider [37]) *For each tolerance relation tlr on a set U, there exists a set V endowed with the tolerance relation $int_\cap$ and there exists an embedding $\iota : (U, tlr) \rightarrow (V, int_\cap)$ of the relational system (U, tlr) into the relational system $(V, int_\cap)$, i.e., the equivalence holds: $tlr(x, y)$ if and only if $\iota(x) \cap \iota(y) \neq \emptyset$.*

Proof Consider the set V of all tolerance classes of *tlr*, and, for each $x \in U$ let

$$\iota(x) = \{A : A \text{ is a tolerance class and } x \in A\}. \tag{1.2}$$

Assume that $A \in \iota(x) \cap \iota(y) \neq \emptyset$; for the tolerance class A it does hold that $x, y \in A$, hence, by the definition of the class, $tlr(x, y)$. Conversely, let $tlr(x, y)$, hence, $B = \{x, y\}$ is a tolerance set and the set B does extend to a tolerance class A, as the property defining the class is of a finite character (by, e.g., the Teichmueller principle, cf., Kuratowski and Mostowski [7]). Clearly, $A \in \iota(x) \cap \iota(y)$ □

Remark In the case when the relation *tlr* satisfies the *extensionality property*:

For each pair $x, y \in U$, $x = y$ if and only if for each $z \in U$, $tlr(x, z)$ if and only if $tlr(y, z)$, the embedding ι is injective, i.e., $x \neq y$ if and only if $\iota(x) \neq \iota(y)$.

A weaker yet similarity is reflexive only; we call such a similarity relation a *weak tolerance*, denoted $w - tlr$. It may arise, e.g., in the context where things are characterized by a set of Boolean features, so each thing either possesses or does not possess a given feature. We let in this context

$$w - tlr(x, y) \text{ if and only if } F(x) \subseteq F(y), \tag{1.3}$$

where $F(x)$ is the set of features possessed by x; this relation is transitive which need not be true in the general case.

1.2.1 Graded Similarity

The last example can be generalized, viz., consider things x, y such that

$$\frac{|F(x) \cap F(y)|}{|F(x)|} = r \in [0, 1], \tag{1.4}$$

where $|A|$ denotes the cardinality of the set A. Then we say that x is similar to y to a degree of r. The relation

$$part - w - tlr(x, y, r) \text{ if and only if } \frac{|F(x) \cap F(y)|}{|F(x)|} \geq r \tag{1.5}$$

is an example of a *graded weak tolerance relation*.

In general, we will call a *graded (weak) similarity* a relation $\tau(x, y, r)$ if and only if the conditions

$$\tau(x, y, 1) \text{ is a (weak) similarity} \tag{1.6}$$

$$\tau(x, y, 1) \text{ and } \tau(z, x, r) \text{ imply } \tau(z, y, r) \tag{1.7}$$

$$\text{If } s \leq r \text{ then } \tau(x, y, r) \text{ implies } \tau(x, y, s) \tag{1.8}$$

are fulfilled. When similarity in (1.6) is, respectively, an equivalence, tolerance, weak tolerance, then we call the relation $\tau(x, y, r)$, respectively, a graded equivalence, tolerance, weak tolerance.

A well-known example of a graded similarity is the *T-fuzzy similarity relation*, introduced in Zadeh [43] which in the most general formulation may be defined as a fuzzy set E_T on the square $X \times X$ satisfying the following requirements, where χ_{E_T} is the fuzzy membership function of E_T,

$$\chi_{E_T}(x, x) = 1 \tag{1.9}$$

$$\chi_{E_T}(x, y) = \chi_{E_T}(y, x) \tag{1.10}$$

$$\chi_{E_T}(x, z) \geq T(\chi_{E_T}(x, y), \chi_{E_T}(y, z)), \tag{1.11}$$

where T is a t-norm. As $T(1, r) = r$ for each t-norm T, the property (1.11) is a partial case of the property (1.7), and the property (1.9) conforms to the property (1.6); the only postulate lacking in our scheme of the graded similarity is the postulate of symmetry (1.10). For our purposes, it would go too far, nevertheless, some important rough inclusions defined in the next chapter, will be symmetric as well.

Particular cases are *similarity relations*, cf., Zadeh [43], defined with $T(x, y) = min(x, y)$, *probability similarity relations* defined with $T(x, y) = P(x, y) = x \times y$, i.e., with the *Menger t-norm*, cf., Menger [17], *likeness relations*, in Ruspini [36] defined with $T(x, y) = L(x, y) = max\{0, x+y-1\}$, i.e., with the *Łukasiewicz t-norm*.

Our definition of a graded similarity suits well the case important for our discourse, i.e., it is tailored to fit rough inclusions of Chap. 2. In literature, there are many proposals for similarity measures. Many of them are modeled and are variants of the classical measure for finite sets modeled on a probability of a finite implication defined in Łukasiewicz [16], viz.,

$$\mu(x, y, r) \text{ if and only if } \frac{|x \cap y|}{|x|} \geq r \tag{1.12}$$

Clearly, μ is a graded weak tolerance relation. Some modifications are symmetric, e.g., the *Kulczyński measure*

$$\mu_K(x, y, r) \text{ if and only if } \frac{1}{2}(\frac{|x \cap y|}{|x|} + \frac{|x \cap y|}{|y|}) \geq r, \tag{1.13}$$

which is a graded tolerance relation, or, the *Jaccard measure*

$$\mu_J(x, y, r) \text{ if and only if } \frac{|x \cap y|}{|x|} + \frac{|x \cap y|}{|y|} - 1 \geq r, \tag{1.14}$$

which also is a graded tolerance relation. Reviews of similarity measures can be found in Albatineh et al. [1] or Gomolińska and Wolski [4].

1.3 Granulation

According to the definition posed by Zadeh [44], granules are formed with respect to a similarity relation. Be this similarity an equivalence eq, natural candidates for granules are sets of equivalent things, i.e., equivalence classes of the form $[x]_{eq}$.

They may be called *elementary granules*, minimal in the sense of containment. More coarse granules may be obtained by means of coarsening the equivalence relation eq to a relation $eq*$ such that $eq \subset eq*$. New granules of the form $[x]_{eq*}$ are of the form

$$[x]_{eq*} = \bigcup \{[y]_{eq} : [y]_{eq} \subseteq [x]_{eq*}\}. \tag{1.15}$$

Examples of this mechanism can be perceived in Pawlak's information systems [20]: given an information system $I = (U, A)$, where U is a universe of things and A a finite non-empty set of attributes, each of them mapping U into a set V of values, the *indiscernibility relation* $Ind(B)$ for a set of attributes $B \subseteq A$ is an equivalence relation defined via the condition

$$Ind(B)(x, y) \text{ if and only if } a(x) = a(y) \text{ for each } a \in B \tag{1.16}$$

Equivalence classes $[x]_B$ of $Ind(B)$ are *elementary B-granules*. Coarsening of granules in this context is effected by means of a choice of a subset C of attributes such that $C \subset B$. Then, $[x]_C = \bigcup \{[y]_B : [y]_B \subseteq [x]_C\}$.

On similar lines goes elementary granulation in fuzzy set theory [42]; given a fuzzy membership function $\chi : D \to [0, 1]$ on a domain D, we induce an equivalence relation eq_χ defined as

$$eq_\chi(x, y) \text{ if and only if } \chi(x) = \chi(y) \tag{1.17}$$

Equivalence classes of eq_χ form χ-*elementary granules*.

Granulation by means of a tolerance relation tlr, follows not on such regular lines. Here, we are faced with tolerance classes, $[x]_{tlr}$ which form a covering of the universe U. To form minimal granules, let us, for each thing $x \in U$, consider the sets

$$K(x) = \bigcap \{C : C \text{ is a tolerance class and } x \in C\} \tag{1.18}$$

called *kernels* (of x in this case).

Properties of kernels are highly interesting; the main of them are listed in

Proposition 2 1. *For each thing x: $x \in K(x)$. 2. For each pair x, z: $z \in K(x)$ if and only if for each u, if $tlr(x, u)$ then $tlr(z, u)$. 3. If tlr is extensional then for each pair $x \neq z$: $z \notin K(x)$ or $x \notin K(z)$. 4. For each triple x, y, z, if $z \in K(x) \cap K(y)$ then $K(x) \cap K(y) \subseteq K(z)$.*

Proof Property 1 is obvious by reflexivity of tlr. Assume that $z \in K(x)$; let $tlr(x, u)$. Then the set $\{x, u\}$ extends to a tolerance class C. As $z \in C$, it follows that $tlr(z, u)$. The converse is obvious: if C is a tolerance class and $x \in C$ then $C \cup \{z\}$ is a tolerance set, hence, $z \in C$. This proves Property 2. For Property 3, let $x \neq z$, hence, by extensionality e.g. there exists u such that $tlr(x, u)$ but not $tlr(z, u)$. As the set $\{x, u\}$ extends to a tolerance class C, it follows that $z \notin C$ hence $z \notin K(x)$.

By symmetry, if there exists v such that $tlr(z, v)$ but not $tlr(x, v)$ then $x \notin K(z)$. By extensionality of tlr one at least of the two cases must happen. For Property 4, assume that $z \in K(x) \cap K(y)$. Consider a tolerance class C with $z \in C$. Clearly, $K(x) \cap K(y) \subseteq C$ hence the thesis 4 follows   □

Properties 3 and 4 imply

Proposition 3 *In the case when for each pair x, y there exists a pair u, v such that $tlr(x, u)$, $tlr(y, v)$ but neither $tlr(x, v)$ nor $tlr(y, u)$ then $K(x) = \{x\}$ for each thing x.*

The consequence of Property 4 is

Proposition 4 *The family $K(x) : x \in U$ is an open subbase for a topology on the universe U, a fortiori, open sets in U in the topology induced by this sub-base are unions of finite intersections of kernels.*

Kernels may be justly called *elementary tolerance granules*.

Observe that in case of a weak tolerance relation $w - tlr$, one may define a tolerance relation tlr by means of

$$tlr(x, y) \text{ if and only if } w - tlr(x, y) \text{ and } w - tlr(y, x). \tag{1.19}$$

Then, classes and kernels of the relation $w - tlr$ can be defined as those of the relation tlr.

Let us observe that in case of a graded similarity $\tau(x, y, r)$, one has

Proposition 5 *Denoting with the symbol $C_\tau(x, r)$ a class of $\tau(x, y, r)$ containing x and with $K_\tau(x, r)$ the respective kernel, we have*

(i) $C_\tau(x, r)$ *extends to a class $C_\tau(x, s)$ whenever $s < r$;*
(ii) $K_\tau(x, r)$ *extends to the kernel $K_\tau(x, s)$ whenever $s < r$.*

Summary Summing up what was said in this chapter, one has the following prototypes for granules

- kernels;
- classes;
- sets $\rho^{\rightarrow}(x) = \{y : \rho(x, y)\}$, $\rho^{\leftarrow}(x) = \{y : \rho(y, x)\}$, cf., Lin [14], Yao [40].

Propositions 1–3 order granule prototypes in the increasing order of cardinality. A judicious choice of the granulation mechanism should take into consideration the anticipated context of usage of granules. However, Proposition 3 renders usage of kernels doubtful as in rich tolerance structures kernel will be identical to things i.e. no effective granulation would follow. We now survey briefly some approaches to granulation bypassing the rough mereological approach discussion of which we reserve for forthcoming chapters.

1.4 On Selected Approaches to Granulation

In Lin [8, 9], topological character of granules was recognized and the basic notion of a neighborhood system as the meaning of the collection of granules on the universe of objects was brought forth, Lin [9] recognized the import of tolerance relations in rough set theory, see Nieminen [19], cf., Polkowski et al. [22], by discussing tolerance induced neighborhoods.

In all hybrid approaches involving fuzzy or rough sets along with neural networks, genetic algorithms, etc., one is therefore bound to compute with granules; this fact testifies to the importance of granular structures.

In search of adequate similarity relations, various forms of granules were proposed and considered as well as experimentally verified as to their effectiveness. In information systems, indiscernibility classes were proposed as granules, or, more generally, *templates* have served that purpose, i.e., meanings of generalized descriptors of the form $(a \in W_a)$ where $W_a \subseteq V_a$ with the meaning $[[(a \in W_a)]] = \{u \in U : a(u) \in W_a\}$, see Nguyen [18]; clearly, templates are aggregates, in ontological sense, of descriptors, i.e., they form "big" granules. Their usage is motivated by their potentially greater descriptive force than that of descriptors; a judicious choice of sets W_a should allow for constructing of a similarity relation that would reflect satisfactorily well the dependence of decision on conditional attributes.

As means for granule construction, rough inclusions have been considered and applied, in Polkowski [23–34]. The idea of granule formation and analysis rests on usage of the mereological class operator in the framework of mereological reasoning.

Granules formed by rough inclusions are used in models of fusion of knowledge, rough-neural computing and in building many-valued logics reflecting the rough set ideology in reasoning.

Granulation of knowledge can be considered from a few angles:

1. General purpose of granulation;
2. Granules from binary relations;
3. Granules in information systems from indiscernibility;
4. Granules from generalized descriptors;
5. Granules from rough inclusions, i.e., the mereological approach.

We briefly examine facets 1–4 of granulation; for the facet 5, please see Chaps. 2 and 3.

Granulation of knowledge comes into existence for a few reasons; the principal one is founded on the underlying assumption of basically all paradigms, viz., that reality exhibits a fundamental continuity, i.e., objects with identical descriptions in the given paradigm should exhibit the same properties with respect to classification or decision making, in general, in their behavior towards the external world.

For instance, fuzzy set theory assumes that objects with identical membership descriptions should behave identically and rough set theory assumes that objects indiscernible with respect to a group of attributes should behave identically, in particular they should fall into the same decision class.

Hence, granulation is forced by assumptions of the respective paradigm and it is unavoidable once the paradigm is accepted and applied. Granules induced in the given paradigm form the first level of granulation.

In the search for similarities better with respect to applications like classification or decision making, more complex granules are constructed, e.g., as unions of granules of the first level, or more complex functions of them, resulting, e.g., in fusion of various granules from distinct sources.

Among granules of the first two levels, some kinds of them can be exhibited by means of various operators, e.g., the class operator associated with a rough inclusion.

1.4.1 Granules from Binary Relations

Granulation on the basis of binary general relations as well as their specializations to, e.g., tolerance relations, has been studied by Lin [10–15], in particular as an important notion of a neighborhood system; see also Yao [40, 41]. This approach extends the special case of the approach based on indiscernibility. A general form of this approach according to Yao [40, 41], exploits the classical notion of *Galois connection*: two mappings $f : X \rightarrow Y, g : Y \rightarrow X$ form a Galois connection between ordered sets $(X, <)$ and $(Y, \prec)$ if and only if the equivalence $x < g(y) \Leftrightarrow f(x) \prec y$ holds.

In an information system (U, A), for a binary relation R on U, one considers the sets $xR = \{y \in U : xRy\}$ and $Rx = \{y \in U : yRx\}$, called, respectively, the *successor neighborhood* and the *predecessor neighborhood* of x. These sets are considered as granules formed by a specific relation of being affine to x in the sense of the relation R. Other forms of granulation can be obtained by comparing objects with identical neighborhoods, e.g., $x \equiv y \Leftrightarrow xR = yR$.

Saturation of sets of objects X, Y with respect to the relation R leads to sets X^*, Y^* such that $X * R = Y* \Leftrightarrow RY^* = X^*$, forming a Galois connection. This approach is closely related to the Formal Concept Analysis of Wille [39].

1.4.2 Granules in Information Systems from Indiscernibility

Granules based on indiscernibility in information/decision systems, are constructed as indiscernibility classes: given an information system (U, A) and the collection $IND = \{IND(B) : B \subseteq A\}$ of indiscernibility relations, each elementary granule is of the form of an indiscernibility relation $[u]_B = \{v \in U : (u, v) \in IND(B)\}$ for some B. Among those granules, there are minimal ones: granules of the form $[u]_A$ induced from the set A of all attributes.

Granules $[u]_A$ form the finest partition of the universe U which can be obtained by means of indiscernibility; given the class $[u]_B$ and the class $[u]_{A \setminus B}$, we have

$[u]_A = [u]_B \cap [u]_{A \setminus B}$ hence $[u]_B = \bigcup_{v \in [u]_B \cap DIS_{A \setminus B}(u)} [v]_A$, where $v \in DIS_{A \setminus B}(u)$ if and only if there exists an attribute $a \in A \setminus B$ such that $a(u) \neq a(v)$. It is manifest that granules $[u]_B$ can be arranged into a tree with the root $[u]_\emptyset = U$ and leaves of the form $[u]_A$.

Granules based on indiscernibility form a complete Boolean algebra generated by atoms of the form $[u]_A$: unions of these atomic granules are closed on intersections and complements and these operations induce into granules the structure of a field of sets.

Atomic granules are made into some important unions by approximation operators: given a concept $X \subseteq U$, the lower approximation $\underline{B}X$ to X over the set B of attributes is defined as the union $\bigcup \{[u]_B : [u]_B \subseteq X\}$; the operator L_B : $Concepts \rightarrow Granules$ sending X to $\underline{B}X$ is monotone increasing and idempotent: $X \subseteq Y$ implies $LX \subseteq LY$ and $L \circ L = L$. Similarly, the upper approximation $\overline{B}X = \bigcup \{[u]_B : [u]_B \cap X \neq \emptyset\}$ to X over B, makes some elementary granules into the union; the operator U^B sending concepts into upper approximations is also monotone increasing and idempotent.

1.4.3 Granules from Generalized Descriptors

Some authors have made use of generalized descriptors called *templates*, see Nguyen [18]; a template is a formula $T : (a \in W_a)$, where $W_a \subseteq V_a$, with the meaning $g(T) : \{u \in U : a(u) \in W_a\}$. The granule $g(T)$ can be represented as the union $\bigcup_{u \in W_a} [u]_a$; granules of the form $[u]_a$ are also called *blocks*, see Grzymala–Busse [5], Grzymala-Busse and Ming [6].

1.5 A General Approach to Similarity Based Granules

We begin with a technical lemma.

Lemma 1 *Assume a universe U and a family Σ of tolerance relations. Given sets $A \subseteq U$, $B \subseteq U$ with A a tolerance set for a tolerance $\sigma \in \Sigma$ and B a tolerance set for a tolerance $\tau \in \Sigma$. there exists the smallest tolerance $\delta_{\sigma,\tau}^{A,B}$ with the property that $\sigma \cup \tau \cup \{(a,b), (b,a) : a \in A, b \in B\} \subseteq \delta_{\sigma,\tau}^{A,B}$.*

Proof of Lemma 1 *The tolerance relation $\sigma \cup \tau \cup U \times U$ contains the set $\sigma \cup \tau \cup \{(a,b), (b,a) : a \in A, b \in B\}$, i.e., the family of tolerances $\Delta_{\sigma,\tau}^{A,B}$: $\delta \supseteq \sigma \cup \tau \cup \{(a,b), (b,a) : a \in A, b \in B\}$ is non-empty. The intersection $\bigcap \Delta_{\sigma,\tau}^{A,B}$ is the required tolerance relation.*

Example 1 Consider the Euclidean plane R^2 and tolerance relations: $\sigma((a_1, b_1), (a_2, b_2))$ if and only if $|a_1 - b_1| < r_1$, $\tau((a_1, b_1), (a_2, b_2))$ if and only if $|a_2 - b_2| < r_2$ for some fixed positive real numbers r_1, r_2. Then $\delta_{\sigma,\tau}^{\emptyset,\emptyset} = \sigma \cup \tau$.

Let $\sigma^* = (\sigma_0 \subseteq \sigma_1 \subseteq \sigma_2 \subseteq \ldots \sigma_n)$ be a sequence of tolerance relations in Σ. A (σ^*, g_0)—*granule* g^* is the union $g^* = \bigcup_{i=0}^{n} g_i$ where

(i) g_0, called the *granule seed* is a tolerance class with respect to σ_0.
(ii) g_{i+1} is the tolerance class containing the set g_i with respect to the tolerance σ_{i+1}, for $i = 0, 1, \ldots, n - 1$.

The granule g^* therefore contains objects of decreasing but still deemed relevant similarity to g_0. Complexity of a granule g^* computation can be of order $|U|^2$, hence, it can be useful to define a simplified version of it, which, by the way, is the one exploited by us in our experiments.

A simple granule g is a g^* granule in case the sequence σ^* is of the form (' $='$, σ). The structure of the granule g is thus as follows:

(i) g_0 is a singleton u_0 called in this case *the center* of the granule.
(ii) g_1 is the set of the tolerance σ containing u_0.

Complexity of any simple granule computation is linear in $|U|$.

1.5.1 Operations on Granules

In case of two granules, g^* induced by a sequence σ^* and h^* induced by a sequence τ^*, with σ^* and τ^* of the same length n, the *fusion* $g^* + h^*$ is defined as the union $\bigcup_{i=1}^{n}(g+h)_i$ where

(i) $(g+h)_0$ is a tolerance class of $\delta_{\sigma_0, \tau_0}^{g_0^*, h_0^*}$;
(ii) $(g+h)_{i+1}$ is a tolerance class of $\delta_{\sigma_{i+1}, \tau_{i+1}}^{(g_i^* \cup h_i^* \cup (g+h)_i}$.

Fusion in case of unequal length of sequences of tolerances, e.g., $|\sigma^*| = m > n = |\tau^*|$ can proceed as in (i), (ii) up to the step n and then the resulting granule $(g+h)_n$ is fused subsequently with layers $g_{n+1}^*, \ldots, g_m^*$.

Fusion of simple granules should regain the form of a simple granule, hence, the above recipe is not adequate to this case. Instead, we may adopt heuristics known, e.g., from path compression, i.e., of attaching one granule to the other. The criterion of selecting the leading granule can be taken as size of the granule (the smaller attached to the bigger), a preference given to one of the centers for some reasons, etc. In some cases, fusion can lead to emergence of a new center which absorbs centers of fused granules. A good example of this phenomenon is the process of assembling/design.

1.5.2 An Example of Granule Fusion: Assembling

In order to introduce the relevant tolerance relations, we resort to mereology based on the connection relation, cf., Polkowski [35]. The primitive notion is here the relation

C of *being connected*, defined on the universe U of things, which is bound to satisfy the conditions

(i) $C(u, u)$ for each $u \in U$.
(ii) If $C(u, v)$ then $C(v, u)$ for each pair, u, v in U.

The notion of an *ingredient ingr* is defined as $ingr(u, v)$ if and only if $C(u, z)$ implies $C(v, z)$ for each thing z in U. Two things u, v *overlap*, in symbols $Ov(u, v)$, if and only if there is a thing z in U such that $ingr(z, u), ingr(z, v)$.

Things u, v are *externally connected*, in symbols $EC(u, v)$ if and only if $C(u, v)$, $\neg Ov(u, v)$. A thing u is a *tangential part* of a thing v, in symbols $TP(u, v)$, if and only if $ingr(u, v)$ and there exists a thing z such that $EC(z, u), EC(z, v)$. A thing u is a *non-tangential part* of a thing v, in symbols $NTP(u, v)$, if and only if $ingr(u, v)$, $\neg TP(u, v)$.

A mereological algebra due to Tarski [38], cf., Polkowski [35], defines sums $+$ and products $\times$ by means of recipes:

$$ingr(z, u + v) \text{ if and only if } ingr(z, u) \text{ or } ingr(z, v), \tag{1.20}$$

respectively,

$$ingr(z, u \times v) \text{ if and only if } ingr(z, u) \text{ and } ingr(z, v). \tag{1.21}$$

One can define now the *complement* z to a thing u in a thing v, in symbols $comp(z, u, v)$, if and only if $ingr(z, v), ingr(u, v), \neg Ov(z, u), v = z + u$.

Predicates related to shape of thing can be introduced now, see Casati and Varzi [3]. A thing u is a *hole* in a thing v, in symbols $hole(u, v)$, see Fig. 1.1, if and only if there exists a thing z such that $NTP(u, z), ingr(v, z), comp(u, v, z)$.

A *dent* in a thing v, in symbols $dent(u, v)$, is a thing u such that there exists a thing z such that $TP(u, z), comp(u, v, z)$, see Fig. 1.2. Predicates relevant to assembling proper can be defined now. One says that a thing u *fills* a thing v in a dent w, in symbols *fill*(u, v, w), if and only if $dent(w, v)$ and $w = u \times v$. A thing u *meets* a thing

Fig. 1.1 A hole

Fig. 1.2 A dent

Fig. 1.3 u meets v

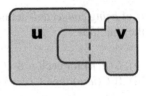

Fig. 1.4 z meets u and v

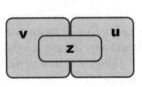

v, in symbols $meet(u, v)$, if and only if there exists a thing w such that $dent(w, v)$, $fill(u, v, w)$, see Fig. 1.3. The final predicate of *joining*, in symbols $join(u, v, z)$, holds if and only if there exists a thing w such that $w = u + v + z$, $meet(z, u)$, $meet(z, v)$, see Fig. 1.4. Given two things u, v, we say that u, v are *in an assembly tolerance*, in symbols $asmbl(u, v)$, if and only if $u = v$ or $join(u, v, z)$ for some thing z. In case $asmbl(u, v)$, we can take u as the granule center g_0 and then g_1 is the granule composed of u and v joined together, which in turn is capable of being extended by means of another join. A thing u for which there is no thing v such that $asmbl(u, v)$ is an *artefact* over $(U, asmbl)$.

References

1. Albatineh, A.N., Niewiadomska-Bugaj, M., Mihalko, D.: On similarity indices and correction for chance agreement. J. Classif. **23**, 301–313 (2006)
2. Bocheński, I.M.: Die Zeitgönossischen Denkmethoden. A. Francke AG, Bern (1954)
3. Casati, R., Varzi, A.C.: Parts and Places: The Structures of Spatial Representations. MIT Press, Cambridge (1999)
4. Gomolińska, A., Wolski, M.: Rough inclusion functions and similarity indices. In: Proceedings of CSP 2013 (22nd International Workshop on Concurrency, Specification and Programming), Warsaw University, September 2013. Białystok University of Technology, pp. 145–156 (2013). ISBN 978-83-62582-42-6
5. Grzymala-Busse, J.W.: Data with missing attribute values: generalization of indiscernibility relation and rule induction. Transactions on Rough Sets I. Lecture Notes in Computer Science, vol. 3100, pp. 78–95. Springer, Berlin (2004)
6. Grzymala-Busse, J.W., Ming, H.: A comparison of several approaches to missing attribute values in data mining. Lecture Notes in Artificial Intelligence, vol. 2005, pp. 378–385. Springer, Berlin (2004)
7. Kuratowski, C., Mostowski, A.: Set Theory. Polish Scientific Publishers, Warszawa (1968)
8. Lin, T.Y.: Neighborhood systems and relational database. Abstract. In: Proceedings of CSC'88, p. 725 (1988)
9. Lin, T.Y.: Neighborhood systems and approximation in database and knowledge based systems. In: Proceedings of the 4th International Symposium on Methodologies for Intelligent Systems (ISMIS), pp. 75–86 (1989)

10. Lin, T.Y.: Topological and fuzzy rough sets. In: Słowiński, R. (ed.) Intelligent Decision Support. Handbook of Applications and Advances of the Rough Sets Theory, pp. 287–304. Kluwer, Dordrecht (1992)
11. Lin, T.Y.: From rough sets and neighborhood systems to information granulation and computing with words. In: Proceedings of the European Congress on Intelligent Techniques and Soft Computing, pp. 1602–1606 (1994)
12. Lin, T.Y.: Granular computing: fuzzy logic and rough sets. In: Zadeh, L.A., Kacprzyk, J. (eds.) Computing with Words in Information/Intelligent Systems, vol. 1, pp. 183–200. Physica Verlag, Heidelberg (1999)
13. Lin, T.Y.: Granular Computing. Lecture Notes in Computer Science, vol. 2639, pp. 16–24. Springer, Berlin (2003)
14. Lin, T.Y.: Granular computing: examples, intuitions, and modeling. In: Proceedings of IEEE 2005 Conference on Granular Computing GrC05, pp. 40–44. IEEE Press, Beijing (2005)
15. Lin, T.Y.: A roadmap from rough set theory to granular computing. In: Proceedings RSKT 2006, 1st International Conference on Rough Sets and Knowledge Technology, Chongqing, China. Lecture Notes in Artificial Intelligence, vol. 4062, pp. 33–41. Springer, Berlin (2006)
16. Łukasiewicz, J.: Die Logischen Grundlagen der Wahrscheinlichtkeitsrechnung. Cracow (1913)
17. Menger, K.: Statistical metrics. Proc. Natl. Acad. Sci. USA **28**, 535–537 (1942)
18. Nguyen, S.H.: Regularity analysis and its applications in data mining. In: Polkowski, L., Tsumoto, S., Lin, T.Y. (eds.) Rough Set Methods and Applications. New Developments in Knowledge Discovery in Information Systems, pp. 289–378. Physica Verlag, Heidelberg (2000)
19. Nieminen, J.: Rough tolerance equality and tolerance black boxes. Fundamenta Informaticae **11**, 289–296 (1988)
20. Pawlak, Z.: Rough Sets: Theoretical Aspects of Reasoning About Data. Kluwer, Dordrecht (1991)
21. Poincaré, H.: La Science et l'Hypothèse. Flammarion, Paris (1902)
22. Polkowski, L., Skowron, A, Żytkow, J.: Tolerance based rough sets. In: Lin, T.Y., Wildberger, M. (eds.) Soft Computing: Rough Sets, Fuzzy Logic, Neural Networks, Uncertainty Management, Knowledge Discovery, pp. 55–58. Simulation Councils Inc., San Diego (1994)
23. Polkowski, L.: A rough set paradigm for unifying rough set theory and fuzzy set theory. Fundamenta Informaticae **54**, 67–88; also. In: Proceedings RSFDGrC03, Chongqing, China, 2003. Lecture Notes in Artificial Intelligence, vol. 2639, pp. 70–78. Springer, Berlin (2003)
24. Polkowski, L.: A note on 3-valued rough logic accepting decision rules. Fundamenta Informaticae **61**, 37–45 (2004)
25. Polkowski, L.: Toward rough set foundations. Mereological approach. In: Proceedings RSCTC04, Uppsala, Sweden. Lecture Notes in Artificial Intelligence, vol. 3066, pp. 8–25. Springer, Berlin (2004)
26. Polkowski, L.: A rough-neural computation model based on rough mereology. In: Pal, S.K., Polkowski, L., Skowron, A. (eds.) Rough-Neural Computing. Techniques for Computing with Words, pp. 85–108. Springer, Berlin (2004)
27. Polkowski, L.: Formal granular calculi based on rough inclusions. In: Proceedings of IEEE 2005 Conference on Granular Computing GrC05, pp. 57–62. IEEE Press, Beijing (2005)
28. Polkowski, L.: Rough-fuzzy-neurocomputing based on rough mereological calculus of granules. Int. J. Hybrid Intell. Syst. **2**, 91–108 (2005)
29. Polkowski, L.: A model of granular computing with applications. In: Proceedings of IEEE 2006 Conference on Granular Computing GrC06, pp. 9–16. IEEE Press, Atlanta (2006)
30. Polkowski, L.: Granulation of knowledge in decision systems: the approach based on rough inclusions. The method and its applications. In: Proceedings RSEiSP'07 (in memory of Z. Pawlak). Lecture Notes in Artificial Intelligence, vol. 4585, pp. 69–79 (2007)
31. Polkowski, L.: The paradigm of granular rough computing. In: Proceedings ICCI'07. 6th IEEE International Conference on Cognitive Informatics, pp. 145–163. IEEE Computer Society, Los Alamitos (2007)
32. Polkowski, L.: Rough mereology in analysis of vagueness. In: Proceedings RSKT 2008. Lecture Notes in Artificial Intelligence, vol. 5009, pp. 197–205. Springer, Berlin (2008)

33. Polkowski, L.: A unified approach to granulation of knowledge and granular computing based on rough mereology: a survey. In: Pedrycz, W., Skowron, A., Kreinovich, V. (eds.) Handbook of Granular Computing, pp. 375–400. Wiley, Chichester (2008)

34. Polkowski, L.: Granulation of knowledge: similarity based approach in information and decision systems. In: Meyers, R.A. (ed.) Springer Encyclopedia of Complexity and System Sciences. Springer, Berlin, article 00 788 (2009)

35. Polkowski, L.: Approximate Reasoning by Parts. An Introduction to Rough Mereology. ISRL, vol. 20. Springer, Berlin (2011)

36. Ruspini, E.H.: On the semantics of fuzzy logic. Int. J. Approx. Reason. **5**, 45–88 (1991)

37. Schreider, Yu.: Equality, Resemblance, and Order. Mir Publishers, Moscow (1975)

38. Tarski, A.: Zur Grundlegung der Booleschen Algebra. I. Fundamenta Mathematicae **24**, 177–198 (1935)

39. Wille, R.: Restructuring lattice theory: an approach based on hierarchies of concepts. In: Rival, I. (ed.) Ordered Sets, pp. 445–470. D. Reidel, Dordrecht (1982)

40. Yao, Y.Y.: Granular computing: basic issues and possible solutions. In: Proceedings of the 5th Joint Conference on Information Sciences, vol. 1, pp. 186–189. Association for Intelligent Machinery, Atlantic (2000)

41. Yao, Y.Y.: Perspectives of granular computing. In: Proceedings of IEEE 2005 Conference on Granular Computing GrC05, pp. 85–90. IEEE Press, Beijing (2005)

42. Zadeh, L.A.: Fuzzy sets. Inf. Control **8**, 338–353 (1965)

43. Zadeh, L.A.: Similarity relations and fuzzy orderings. Inf. Sci. **3**, 177–200 (1971)

44. Zadeh, L.A.: Fuzzy sets and information granularity. In: Gupta, M., Ragade, R., Yager, R.R. (eds.) Advances in Fuzzy Set Theory and Applications, pp. 3–18. North-Holland, Amsterdam (1979)

45. Zeeman, E.C.: The topology of the brain and the visual perception. In: Fort, K.M. (ed.) Topology of 3-manifolds and Selected Topics, pp. 240–256. Prentice Hall, Englewood Cliffs (1965)

Further Reading

46. Bargiela, A., Pedrycz, W.: Granular Computing: An Introduction. Kluwer Academic Publishers, Dordrecht (2003)

47. Kacprzyk, J., Nurmi, H., Zadrożny, S.: Towards a comprehensive similarity analysis of voting procedures using rough sets and similarity measures. In: Skowron, A., Suraj, Z. (eds.) Rough Sets and Intelligent Systems—Professor Zdzisław Pawlak in Memoriam, vol. 1. Springer, Heidelberg, ISRL vol. 42, pp. 359–380 (2013)

48. Pal, S.K., Polkowski, L., Skowron, A. (eds.): Rough-Neural Computing: Techniques for Computing with Words. Springer, Heidelberg (2004)

49. Pedrycz, W. (ed.): Granular Computing: An Emerging Paradigm. Physica Verlag, Heidelberg (2001)

50. Pedrycz, W., Skowron, A., Kreinovich, V. (eds.): Handbook of Granular Computing. Wiley, Chichester (2008)

51. Pawlak, Z.: Elementary rough set granules: Toward a rough set processor. In: Pal, S.K., Polkowski, L., Skowron, A. (eds.) Rough Neuro-Computing: Techniques for Computing with Words, pp. 5–14. Springer, Heidelberg (2013)

52. Polkowski, L., Skowron, A.: Grammar systems for distributed synthesis of of approximate solutions extracted from experience. In: Paun, Gh., Salomaa, A. (eds.) Grammatical Models of Multi-Agent Systems, pp. 316–333. Gordon and Breach, Amsterdam (1999)

53. Polkowski, L., Skowron, A.: Towards adaptive calculus of granules. In: Zadeh, L.A., Kacprzyk, J. (eds.) Computing with Words in Information/Intelligent Systems. Foundations, vol. 1. Physica Verlag, Heidelberg (1999)

54. Polkowski, L., Skowron, A.: Rough mereological calculi of granules: a rough set approach to computation. Comput. Intell. Int. J. **17**(3), 472–492 (2001)
55. Polkowski, L., Semeniuk-Polkowska, M.: On rough set logics based on similarity relations. Fundamenta Informaticae **64**, 379–390 (2005)
56. Skowron, A., Stepaniuk, J.: Information granules: towards foundations of granular computing. Int. J. Intell. Syst. **16**(1), 57–86 (2001)
57. Skowron, A., Suraj, Z. (eds.): Rough Sets and Intelligent Systems. Professor Zdzisław Pawlak in Memoriam, vol. 1. Springer, ISRL, vol. 42. Heidelberg (2013)
58. Stepaniuk, J. (ed.): Rough-Granular Computing in Knowledge Discovery and Data Mining. Springer, Heidelberg (2008)

Chapter 2
Mereology and Rough Mereology: Rough Mereological Granulation

Ex pluribus unum
[Saint Augustine. Confessions]

In this chapter, we embark on a more specific granulation theory, stemming from the mereological theory of things. This mechanism provides us with tolerance and weak tolerance relations forming a graded similarity in the sense of Chap. 1 and with resulting therefrom granules. To put the necessary notions in a proper order, we are going to discuss mereology, rough mereology and the mereological granulation.

2.1 Mereology

Mereology does address things in terms of parts, cf., [1]. A formal theory of Mereology due to Leśniewski [5] axiomatically defines the notion of a part.

The reader may be aware of the existence of a vast literature on philosophical and ontological aspects of mereology which cannot be mentioned nor discussed here, and, we advise them to consult, e.g., Simons [27], Luschei [11] or Casati and Varzi [2] for discussions of those aspects.

2.1.1 Mereology of Leśniewski

Mereology due to Leśniewski arose from attempts at reconciling antinomies of naïve set theory, see Leśniewski [5, 6, 8], Surma et. al. [9], Sobociński [29, 30]. Leśniewski [5] was the first presentation of the foundations of this attempt as well as the first formally complete exposition of mereology.

© Springer International Publishing Switzerland 2015
L. Polkowski and P. Artiemjew, *Granular Computing in Decision Approximation*,
Intelligent Systems Reference Library 77, DOI 10.1007/978-3-319-12880-1_2

2.1.1.1 On the Notion of Part

The primitive notion of mereology in this formalism is the notion of a *part*. Given some category of things, a relation of a part is a binary relation π which is required to be

M1 *Irreflexive: For each thing x, it is not true that* $\pi(x, x)$

M2 *Transitive: For each triple x, y, z of things, if* $\pi(x, y)$ *and* $\pi(y, z)$*, then* $\pi(x, z)$

Remark In the original scheme of Leśniewski, the relation of parts is applied to *individual things* as defined in Ontology of Leśniewski, see Leśniewski [7], Iwanuś [4], Słupecki [28]. Ontology due to Leśniewski is based on the Ontology Axiom:

(AO) $x\varepsilon y$ if and only if $(x\varepsilon x)$ and there is z such that $z\varepsilon y$ and for each z if $z\varepsilon x$

then $z\varepsilon y$,

which singles out x as an *individual* (characterized by the formula $x\varepsilon x$) and y as a *collective thing*, with the copula ε read as *'is'*.

The relation of *part* induces the relation of an *ingredient ingr*, defined as

$$ingr(x, y) \text{ if and only if } \pi(x, y) \ \vee \ x = y. \tag{2.1}$$

The relation of ingredient is a partial order on things, i.e.,

1. $ingr(x, x)$;
2. If $ingr(x, y)$ and $ingr(y, x)$ then $(x = y)$;
3. If $ingr(x, y)$ and $ingr(y, z)$ then $ingr(x, z)$.

We formulate the third axiom with a help from the notion of an ingredient, see Polkowski [24], Chap. 5,

M3 (*Inference*) For each pair of things x, y, if the property

$I(x, y) :$ if $ingr(t, x)$, then exist w, z with $ingr(w, t), ingr(w, z), ingr(z, y)$

is satisfied, then $ingr(x, y)$.

The predicate of *overlap*, Ov in symbols, is defined by means of

$$Ov(x, y) \text{ if and only if there is } z \text{ such that } ingr(z, x) \text{ and } ingr(z, y). \tag{2.2}$$

Using the overlap predicate, one can write $I(x, y)$ down in the form

$I_{Ov}(x, y) :$ if $ingr(t, x)$, then there is z such that $ingr(z, y)$ and $Ov(t, z)$.

2.1.1.2 On the Notion of a Class

The notion of a *mereological class* follows; for a non-vacuous property Φ of things, the *class of* Φ, denoted $Cls\Phi$ is defined by the conditions

C1 If $\Phi(x)$, then $ingr(x, Cls\Phi)$

C2 If $ingr(x, Cls\Phi)$, then there is z such that $\Phi(z)$ and $I_{Ov}(x, z)$

In plain language, the class of Φ collects in an individual thing all things satisfying the property Φ. The existence of classes is guaranteed by an axiom

M4 For each non-vacuous property Φ there exists a class $Cls\Phi$

The uniqueness of the class follows by M3. M3 implies also that, for the non-vacuous property Φ, if for each thing z such that $\Phi(z)$ it holds that $ingr(z, x)$, then $ingr(Cls\Phi, x)$.

The notion of an overlap allows for a succinct characterization of a class: for each non-vacuous property Φ and each thing x, it happens that $ingr(x, Cls\Phi)$ if and only if for each ingredient w of x, there exists a thing z such that $Ov(w, z)$ and $\Phi(z)$.

Remark Uniqueness of the class along with its existence is an axiom in the Leśniewski [5] scheme, from which M3 is derived. Similarly, it is an axiom in the Tarski [32–34] scheme.

Consider two examples,

1. The strict inclusion $\subset$ on sets is a part relation. The corresponding ingredient relation is the inclusion $\subseteq$. The overlap relation is the non-empty intersection. For a non-vacuous family F of sets, the class $ClsF$ is the union $\bigcup F$.
2. For reals in the interval [0, 1], the strict order $<$ is a part relation and the corresponding ingredient relation is the weak order $\leq$. Any two reals overlap; for a set $F \subseteq [0, 1]$, the class of F is $supF$.

2.1.1.3 Notions of Element, Subset

The notion of an element is defined as follows

$$el(x, y) \text{ if and only if for a property } \Phi \ y = Cls\Phi \text{ and } \Phi(x). \qquad (2.3)$$

In plain words, $el(x, y)$ means that y is a class of some property and x responds to that property. To establish some properties of the notion of an element, we begin with the property $INGR(x) = \{y : ingr(y, x)\}$, for which the identity $x = ClsINGR(x)$ holds by M3. Hence, $el(x, y)$ is equivalent to $ingr(x, y)$. Thus, each thing x is its own element. This is one of means of expressing the impossibility of the Russell paradox within the mereology, cf., Leśniewski [5], Thms. XXVI, XXVII, see also Sobociński [29].

We observe the extensionality of overlap: *For each pair x, y of things, $x = y$ if and only if for each thing z, the equivalence $Ov(z, x) \Leftrightarrow Ov(z, y)$ holds.* Indeed, assume the equivalence $Ov(z, x) \Leftrightarrow Ov(z, y)$ to hold for each z. If $ingr(t, x)$ then $Ov(t, x)$ and $Ov(t, y)$ hence by axiom M3 $ingr(t, y)$ and with $t = x$ we get $ingr(x, y)$. By symmetry, $ingr(y, x)$, hence $x = y$.

The notion of a subset follows,

$$sub(x, y) \text{ if and only if for each z if } ingr(z, x) \text{ then } ingr(z, y). \tag{2.4}$$

It is manifest that for each pair x, y of things, $sub(x, y)$ holds if and only if $el(x, y)$ holds if and only if $ingr(x, y)$ holds.

For the property $Ind(x) \Leftrightarrow ingr(x, x)$, one calls the class $ClsInd$, *the universe*, in symbols V.

2.1.1.4 The Universe of Things, Things Exterior, Complement

It follows by definition of the universe that

1. The universe V is unique;
2. $ingr(x, V)$ holds for each thing x;
3. For each non-vacuous property Φ, it is true that $ingr(Cls\Phi, V)$.

The notion of an *exterior* thing x to a thing y, $extr(x, y)$, is the following

$$extr(x, y) \text{ if and only if } \neg Ov(x, y). \tag{2.5}$$

In plain words, x is exterior to y when no thing is an ingredient both to x and y.

Clearly, the operator of exterior has properties

1. No thing is exterior to itself;
2. $extr(x, y)$ implies $extr(y, x)$;
3. If for a non-vacuous property Φ, a thing x is exterior to every thing z such that $\Phi(z)$ holds, then $extr(x, Cls\Phi)$.

The notion of a *complement* to a thing, with respect to another thing, is rendered as a ternary predicate $comp(x, y, z)$, cf., Leśniewski [5], par. 14, Def. IX, to be read: 'x is the complement to y with respect to z', and it is defined by means of the following requirements

1. $x = ClsEXTR(y, z)$;
2. $ingr(y, z)$, where $EXTR(y, z)$ is the property which holds for a thing t if and only if $ingr(t, z)$ and $extr(t, y)$ hold.

This definition implies that the notion of a complement is valid only when there exists an ingredient of z exterior to y. Following are basic properties of complement,

1. If $comp(x, y, z)$, then $extr(x, y)$ and $\pi(x, z)$;
2. If $comp(x, y, z)$, then $comp(y, x, z)$.

We let for a thing x, $-x = ClsEXTR(x, V)$. It follows that

1. $-(-x) = x$ *for each thing* x;
2. $-V$ *does not exist*.

We conclude this paragraph with two properties of classes useful in the following.

$$\text{If } \Phi \Rightarrow \Psi \text{ then } ingr(Cls\Phi, Cls\Psi), \tag{2.6}$$

and, a corollary

$$\text{If } \Phi \Leftrightarrow \Psi \text{ then } Cls\Phi = Cls\Psi. \tag{2.7}$$

2.2 Rough Mereology

A scheme of mereology, introduced into a collection of things, sets an exact hierarchy of things of which some are (exact) parts of others; to ascertain whether a thing is an exact part of some other thing is in practical cases often difficult if possible at all, e.g., a robot sensing the environment by means of a camera or a laser range sensor, cannot exactly perceive obstacles or navigation beacons. Such evaluation can be done approximately only and one can discuss such situations up to a degree of certainty only. Thus, one departs from the exact reasoning scheme given by decomposition into parts to a scheme which approximates the exact scheme but does not observe it exactly.

Such a scheme, albeit its conclusions are expressed in an approximate language, can be more reliable, as its users are aware of uncertainty of its statements and can take appropriate measures to fend off possible consequences.

Introducing some measures of overlapping, in other words, the extent to which one thing is a part to the other, would allow for a more precise description of relative position, and would add an expressional power to the language of mereology. Rough mereology answers these demands by introducing the notion of a *part to a degree* with the degree expressed as a real number in the interval [0, 1]. Any notion of a part by necessity relates to the general idea of *containment*, and thus the notion of a part to a degree is related to the idea of *partial containment* and it should preserve the essential intuitive postulates about the latter.

The predicate of a part to a degree stems ideologically from and has as one of motivations the predicate of an element to a degree introduced by Zadeh as a basis for fuzzy set theory [36]; in this sense, rough mereology is to mereology as the fuzzy set theory is to the naive set theory. To the rough set theory, owes rough mereology the interest in concepts as things for analysis.

The primitive notion of rough mereology is the notion of a *rough inclusion* which is a ternary predicate $\mu(x, y, r)$ where x, y are *things* and $r \in [0, 1]$, read as '*the thing x is a part to degree at least of r to the thing y*'. Any rough inclusion is associated with a mereological scheme based on the notion of a part by postulating that $\mu(x, y, 1)$ is equivalent to $ingr(x, y)$, where the ingredient relation is defined by the adopted mereological scheme. Other postulates about rough inclusions stem from intuitions about the nature of partial containment; these intuitions can be manifold, a fortiori, postulates about rough inclusions may vary. In our scheme for rough mereology, we begin with some basic postulates which would provide a most general framework. When needed, other postulates, narrowing the variety of possible models, can be introduced.

2.2.1 Rough Inclusions

We have already stated that a rough inclusion is a ternary predicate $\mu(x, y, r)$. We assume that a collection of things is given, on which a part relation π is introduced with the associated ingredient relation *ingr*. We thus apply inference schemes of mereology due to Leśniewski, presented above.

Predicates $\mu(x, y, r)$ were introduced in Polkowski and Skowron [25, 26]; they satisfy the following postulates, relative to a given part relation π and the induced by π relation *ingr* of an ingredient, on a set of things:

RINC1 $\mu(x, y, 1)$ if and only if $ingr(x, y)$

This postulate asserts that parts to degree of 1 are ingredients.

RINC2 If $\mu(x, y, 1)$ then $\mu(z, x, r)$ implies $\mu(z, y, r)$ for every z

This postulate does express a feature of partial containment that a 'bigger' thing contains a given thing 'more' than a 'smaller' thing. It can be called a *monotonicity condition* for rough inclusions.

RINC3 If $\mu(x, y, r)$ and $s < r$ then $\mu(x, y, s)$

This postulate specifies the meaning of the phrase 'a part to a degree at least of r'. From postulates RINC1–RINC3, and known properties of ingredients some consequences follow

1. $\mu(x, x, 1)$;
2. If $\mu(x, y, 1)$ and $\mu(y, z, 1)$ then $\mu(x, z, 1)$;
3. $\mu(x, y, 1)$ and $\mu(y, x, 1)$ if and only if $x = y$;
4. If $x \neq y$ then either $\neg\mu(x, y, 1)$ or $\neg\mu(y, x, 1)$;
5. If, for each z, r, $[\mu(z, x, r)$ if and only if $\mu(z, y, r)]$ then $x = y$.

Property 5 may be regarded as an *extensionality postulate* in rough mereology.

It follows that rough inclusions are in general graded weak tolerance relations in the sense of Chap. 1, par. 2.1.

By a *model* for rough mereology, we mean a quadruplex

$$M = (V_M, \pi_M, ingr_M, \mu_M)$$

where V_M is a set with a part relation $\pi_M \subseteq V_M \times V_M$, the associated ingredient relation $ingr_M \subseteq V_M \times V_M$, and a relation $\mu_M \subseteq V_M \times V_M \times [0, 1]$ which satisfies RINC1–RINC3.

We now describe some models for rough mereology which at the same time give us methods by which we can define rough inclusions, see Polkowski [14–17, 20–22], a detailed discussion may be found in Polkowski [24].

2.2.1.1 Rough Inclusions from T-norms

We resort to *continuous t-norms* which are continuous functions $T : [0, 1]^2 \to [0, 1]$ which are 1. symmetric; 2. associative; 3. increasing in each coordinate; 4. satisfying boundary conditions $T(x, 0) = 0, T(x, 1) = x$, cf., Polkowski [24], Chaps. 4, 6, Hájek [3], Chap. 2. Classical examples of continuous t-norms are

1. $L(x, y) = max\{0, x + y - 1\}$ (the *Łukasiewicz t-norm*);
2. $P(x, y) = x \cdot y$ (the *product t-norm*);
3. $M(x, y) = min\{x, y\}$ (the *minimum t-norm*).

The *residual implication* $\Rightarrow_T$ induced by a continuous t-norm T is defined as

$$x \Rightarrow_T y = max\{z : T(x, z) \le y\}. \tag{2.8}$$

One proves that $\mu_T(x, y, r) \Leftrightarrow x \Rightarrow_T y \ge r$ is a rough inclusion; particular cases are

1. $\mu_L(x, y, r) \Leftrightarrow min\{1, 1 - x + y \ge r\}$ (the *Łukasiewicz implication*);
2. $\mu_P(x, y, r) \Leftrightarrow \frac{y}{x} \ge r$ when $x > 0$, $\mu_P(x, y, 1)$ when $x = 0$ (the *Goguen implication*);
3. $\mu_M(x, y, r) \Leftrightarrow y \ge r$ when $x > 0$, $\mu_M(x, y, 1)$ when $x = 0$ (the *Gödel implication*).

A particular case of continuous t-norms are *Archimedean t-norms* which satisfy the inequality $T(x, x) < x$ for each $x \in (0, 1)$. It is well–known, see Ling [10], that each archimedean t-norm T admits a representation

$$T(x, y) = g_T(f_T(x) + f_T(y)), \tag{2.9}$$

where, the function $f_T : [0, 1] \to R$ is continuous decreasing with $f_T(1) = 0$, and $g_T : R \to [0, 1]$ is the *pseudo–inverse* to f_T, i.e., $g \circ f = id$. It is known, cf., e.g., Hájek [3], that up to an isomorphism there are two Archimedean t-norms: L and P. Their representations are

$$f_L(x) = 1 - x; \ g_L(y) = 1 - y, \tag{2.10}$$

and,

$$f_P(x) = exp(-x); \ g_P(y) = -ln \ y. \tag{2.11}$$

For an Archimedean t-norm T, we define the rough inclusion μ^T on the interval $[0, 1]$ by means of

$$(ari) \ \mu^T(x, y, r) \Leftrightarrow g_T(|x - y|) \ge r, \tag{2.12}$$

equivalently,

$$\mu^T(x, y, r) \Leftrightarrow |x - y| \le f_T(r). \tag{2.13}$$

It follows from (2.13), that μ^T satisfies conditions RINC1–RINC3 with *ingr* as identity =.

To give a hint of proof: for RINC1: $\mu^T(x, y, 1)$ if and only if $|x - y| \le f_T(1) = 0$, hence, if and only if $x = y$. This implies RINC2. In case $s < r$, and $|x - y| \le f_T(r)$, one has $f_T(r) \le f_T(s)$ and $|x - y| \le f_T(s)$.

Specific recipes are

$$\mu^L(x, y, r) \Leftrightarrow |x - y| \le 1 - r, \tag{2.14}$$

and,

$$\mu^P(x, y, r) \Leftrightarrow |x - y| \le -\ln r. \tag{2.15}$$

Both residual and archimedean rough inclusions satisfy the *transitivity condition*, cf., Polkowski [24], Chap. 6,

(Trans) *If $\mu(x, y, r)$ and $\mu(y, z, s)$, then $\mu(x, z, T(r, s))$.*

To recall the proof, assume, e.g., $\mu^T(x, y, r)$ and $\mu^T(y, z, s)$, i.e., $|x - y| \le f_T(r)$ and $|y - z| \le f_T(s)$. Hence, $|x - z| \le |x - y| + |y - z| \le f_T(r) + f_T(s)$, hence, $g_T(|x - z|) \ge g_T(f_T(r) + f_T(s)) = T(r, s)$, i.e., $\mu^T(x, z, T(r, s))$. Other cases go along the same lines. Let us observe that rough inclusions of the form (ari) are also *symmetric*, hence they are graded tolerance relations, whereas residual rough inclusions are graded weak tolerance relations needing not be symmetric.

2.2.1.2 Rough Inclusions in Information Systems (Data Tables)

An important domain where rough inclusions will play a dominant role in our analysis of reasoning by means of parts is the realm of *information systems* of Pawlak [13], cf., Polkowski [24], Chap. 6. We will define information rough inclusions denoted with a generic symbol μ^I.

We recall that an *information system* (a *data table*) is represented as a pair (U, A) where U is a finite set of things and A is a finite set of *attributes*; each attribute $a : U \to V$ maps the set U into the *value set V*. For an attribute a and a thing v, $a(v)$ is the value of a on v.

For things u, v the *discernibility set DIS(u, v)* is defined as

$$DIS(u, v) = \{a \in A : a(u) \ne a(v)\}. \tag{2.16}$$

For an (ari) μ_T, we define a rough inclusion μ_T^I by means of

$$(airi) \ \mu_T^I(u, v, r) \Leftrightarrow g_T(\frac{|DIS(u, v)|}{|A|}) \geq r. \qquad (2.17)$$

Then, μ_T^I is a rough inclusion with the associated ingredient relation of identity and the part relation empty. These relations are graded tolerance relations.

For the Łukasiewicz t-norm, the *airi* μ_L^I is given by means of the formula

$$\mu_L^I(u, v, r) \Leftrightarrow 1 - \frac{|DIS(u, v)|}{|A|} \geq r. \qquad (2.18)$$

We introduce the set $IND(u, v) = A \setminus DIS(u, v)$. With its help, we obtain a new form of (2.18)

$$\mu_L^I(u, v, r) \Leftrightarrow \frac{|IND(u, v)|}{|A|} \geq r. \qquad (2.19)$$

The formula (2.19) witnesses that the reasoning based on the rough inclusion μ_L^I is the probabilistic one which goes back to Łukasiewicz [12]. Each (airi)–type rough inclusion μ_T^I satisfies the transitivity condition (Trans) and is symmetric.

2.2.1.3 Rough Inclusions on Sets and Measurable Sets

Formula (2.19) can be abstracted for set and geometric domains. For finite sets A, B,

$$\mu^S(A, B, r) \Leftrightarrow \frac{|A \cap B|}{|A|} \geq r \qquad (2.20)$$

defines a rough inclusion μ^S. For bounded measurable sets X, Y in an Euclidean space E^n,

$$\mu^G(A, B, r) \Leftrightarrow \frac{||A \cap B||}{||A||} \geq r, \qquad (2.21)$$

where, $||A||$ denotes the area (the Lebesgue measure) of the region A, defines a rough inclusion μ^G. Both μ^S, μ^G are neither symmetric nor transitive, hence, they are graded weak tolerance relations.

Other rough inclusions and their weaker variants will be defined in later chapters.

2.3 Granules from Rough Inclusions

The idea of mereological granulation of knowledge, see Polkowski [16–19], cf., surveys Polkowski [21, 22], presented here finds an effective application in problems of synthesis of classifiers from data tables. This application consists in granulation of data at preprocessing stage in the process of synthesis: after granulation, a new data set is constructed, called a *granular reflection*, to which various strategies for rule synthesis can be applied. This application can be regarded as a process of *filtration* of data, aimed at reducing noise immanent to data. Application of rough inclusions leads to a formal theory of granules of various *radii* allowing for various choices of coarseness degree in data.

Granules are formed here as simple granules in the sense of Chap. 1, with tolerance or weak tolerance induced by a rough inclusion. Assume that a rough inclusion μ is given along with the associated ingredient relation *ingr*, as in postulate RINC1.

The *granule* $g_\mu(u, r)$ of the radius r about the center u is defined as the class of property $\Phi_{u,r}^\mu$, i.e.,

$$\Phi_{u,r}^\mu(v) \text{ if and only if } \mu(v, u, r). \tag{2.22}$$

The granule $g_\mu(u, r)$ is defined by means of

$$g_\mu(u, r) = Cls\Phi_{u,r}^\mu. \tag{2.23}$$

Properties of granules depend, obviously, on the type of rough inclusion used in their definitions. We consider separate cases, as some features revealed by granules differ from a rough inclusion to a rough inclusion. The reader is asked to refer to the axiom M3 for the tool for mereological reasoning, which is going to be used in what follows.

In case of Archimedean t-norm–induced rough inclusions (ari), or (airi)–type rough inclusions, by their transitivity, and symmetry, the important property holds, see Polkowski [21, 24].

Proposition 6 *In case of a symmetric and transitive rough inclusion μ, for each pair u, v of objects, and $r \in [0, 1]$, ingr$(v, g_\mu(u, r))$ if and only if $\mu(v, u, r)$ holds. In effect, the granule $g_\mu(u, r)$ can be represented as the set $\{v : \mu(v, u, r)\}$.*

Proof (op.cit., op.cit.) Assume that *ingr*$(v, g_\mu(u, r))$ holds. Thus, there exists z such that $Ov(z, v)$ and $\mu(z, u, r)$. There is x with *ingr*(x, v), *ingr*(x, z), hence, by transitivity of μ, also $\mu(x, u, r)$ holds. By symmetry of μ, *ingr*(v, x), hence, $\mu(v, x, r)$ holds also $\square$

In case of rough inclusions in information systems, induced by residual implications generated by continuous t-norms, we have a positive case, for the minimum t-norm M, see Polkowski [24].

Proposition 7 *For the rough inclusion μ induced by the residual implication $\Rightarrow_M$, due to the minimum t-norm M, and $r < 1$, the relation ingr$(v, g_\mu(u, r))$ holds if and only if $\mu(v, u, r)$ holds.*

Proof (loc.cit.) We recall the proof. The rough inclusion μ has the form $\mu(v, u, r)$ if and only if $\frac{|IND(v,s)|}{|A|} \Rightarrow_M \frac{|IND(u,s)|}{|A|} \geq r$. If *ingr*$(v, g_\mu(u, r))$ holds, then by the class definition, there exists z such that $Ov(v, z)$ and $\mu(z, u, r)$ hold. Thus, we have

w with $ingr(w, v)$ and $\mu(w, u, r)$ by transitivity of μ and the fact that $ingr(w, z)$. By definition of μ, $ingr(w, v)$ means that $|IND(w, s)| \leq |IND(v, s)|$. As $\mu(w, u, r)$ with $r < 1$ means that $|IND(u, s)| \geq r$ because of $|IND(w, s)| \geq |IND(u, s)|$, the condition $|IND(w, s)| \leq |IND(v, s)|$ implies that $\mu(v, u, r)$ holds as well   □

The case of the rough inclusion μ induced either by the product t-norm $P(x, y) = x \cdot y$, or by the Łukasiewicz t-norm L, is a bit more intricate. To obtain in this case some positive result, we exploit the averaged t-norm $\vartheta(\mu)$ defined for the rough inclusion μ, induced by a t-norm T, by means of the formula, see Polkowski [24], Chap. 7, par. 7.3, from which this result is taken,

$$\vartheta(\mu)(v, u, r) \Leftrightarrow \forall z. \exists a, b. \mu(z, v, a), \mu(z, u, b), a \Rightarrow_T b \geq r. \tag{2.24}$$

Our proposition for the case of the t-norm P is, op.cit.

Proposition 8 *For $r < 1$, $ingr(v, g_{\vartheta(\mu)}(u, r))$ holds if $\mu(v, u, a \cdot r)$, where $\mu(v, t, a)$ holds for t which obeys conditions $ingr(t, v)$ and $\vartheta(\mu)(t, u, r)$.*

Proof $ingr(v, g_{\vartheta(\mu)}(u, r))$ implies that there is w such that $Ov(v, w)$ and $\vartheta(\mu)(w, u, r)$, so we can find t with properties, $ingr(t, w)$, $ingr(t, v)$, hence, by transitivity of $\vartheta(\mu)$ also $\vartheta(\mu)(t, u, r)$.

By definition of $\vartheta(\mu)$, there are a, b such that $\mu(v, t, a)$, $\mu(v, u, b)$, and $a \Rightarrow_P b \geq r$, i.e., $\frac{b}{a} \geq r$. Thus, $\mu(v, u, b)$ implies $\mu(v, u, a \cdot r)$ □

An analogous reasoning brings forth in case of the rough inclusion μ induced by residual implication due to the Łukasiewicz implication L, the result that, op.cit.

Proposition 9 *For $r < 1$, $ingr(v, g_{\vartheta(\mu)}(u, r))$ holds if and only if $\mu(v, u, r + a - 1)$ holds, where $\mu(v, t, a)$ holds for t such that $ingr(t, v)$ and $\vartheta(\mu)(t, u, r)$.*

The two last propositions can be recorded jointly in the form

Proposition 10 *For $r < 1$, and μ induced by residual implications either $\Rightarrow_P$ or $\Rightarrow_L$, $ingr(v, g_{\vartheta(\mu)}(u, r))$ holds if and only if $\mu(v, u, T(r, a))$ holds, where $\mu(v, t, a)$ holds for t such that $ingr(t, v)$ and $\vartheta(\mu)(t, u, r)$.*

Granules as collective concepts can be objects for rough mereological calculi.

2.3.1 Rough Inclusions on Granules

Due to the feature of mereology that it operates (due to the class operator) only on level of individuals, one can extend rough inclusions from objects to granules; the formula for extending a rough inclusion μ to a rough inclusion $\overline{\mu}$ on granules is a modification of mereological axiom M3, see Polkowski [24], Chap. 7, par. 7.4:

$\overline{\mu}(g, h, r)$ if and only if for each z if $ingr(z, g)$ then there is w such that

$$ingr(w, h) \text{ and } \mu(z, w, r).$$

Proposition 11 *The predicate* $\overline{\mu}(g, h, r)$ *is a rough inclusion on granules.*

Proof To recall the proof, see that $\mu(g, h, 1)$ means that for each object z with $ingr(z, g)$ there exists an object w with $ingr(w, h)$ such that $\mu(z, w, 1)$, i.e., $ingr(z, w)$, which, by the inference rule implies that $ingr(g, h)$. This proves RINC1. For RINC2, assume that $\mu(g, h, 1)$ and $\mu(k, g, r)$ so for each $ingr(x, k)$ there is $ingr(y, g)$ with $\mu(x, y, r)$. For y there is z such that $ingr(z, h)$ and $\mu(y, z, 1)$, hence, $\mu(x, z, r)$ by property RINC2 of μ. Thus, $\mu(k, h, r)$. RINC2 follows and RINC3 is obviously satisfied.   □

We now examine rough mereological granules with respect to their properties.

2.4 General Properties of Rough Mereological Granules

In this exposition, we follow the results presented in Polkowski [24] with references given therein. The basic properties are collected in

Proposition 12 *The following constitute a set of basic properties of rough mereo-logical granules*

1. *If* $ingr(y, x)$ *then* $ingr(y, g_{\mu}(x, r))$;
2. *If* $ingr(y, g_{\mu}(x, r))$ *and* $ingr(z, y)$ *then* $ingr(z, g_{\mu}(x, r))$;
3. *If* $\mu(y, x, r)$ *then* $ingr(y, g_{\mu}(x, r))$;
4. *If* $s < r$ *then* $ingr(g_{\mu}(x, r), g_{\mu}(x, s))$,

which follow straightforwardly from properties RINC1–RINC3 of rough inclusions and the fact that *ingr* is a partial order, in particular it is transitive, regardless of the type of the rough inclusion μ.

For T–transitive rough inclusions, we can be more specific, and prove

Proposition 13 *For each T-transitive rough inclusion* μ,

1. *If* $ingr(y, g_{\mu}(x, r))$ *then, for each s,* $ingr(g_{\mu}(y, s), g_{\mu}(x, T(r, s)))$;
2. *If* $\mu(y, x, s)$ *with* $1 > s > r$, *then there exists* $\alpha < 1$ *with the property that* $ingr(g_{\mu}(y, \alpha), g_{\mu}(x, r))$.

Proof Property 1 follows by transitivity of μ with the t-norm T. Property 2 results from the fact that the inequality $T(s, \alpha) \geq r$ has a solution in α, e.g., for $T = P$, $\alpha \geq \frac{r}{s}$, and, for $T = L$, $\alpha \geq 1 - s + r$ □

It is natural to regard granule system $\{g_r^{\mu_t}(x) : x \in U; r \in (0, 1)\}$ as a neighborhood system for a topology on U that may be called the *granular topology*.

In order to make this idea explicit, we define classes of the form

$$N^T(x, r) = Cls(\psi_{r,x}^{\mu_T}),   \tag{2.25}$$

where,

$$\psi_{r,x}^{\mu_T}(y) \text{ if and only if there is } s > r \text{ such that } \mu_T(y, x, s). \tag{2.26}$$

We declare the system $\{N^T(x, r) : x \in U; r \in (0, 1)\}$ to be a neighborhood basis for a topology θ_μ. This is justified by the following

Proposition 14 *Properties of the system* $\{N^T(x, r) : x \in U, r \in (0, 1)\}$ *are as follows:*

1. *If* $ingr(y, N^T(x, r))$ *then there is* $\delta > 0$ *such that* $ingr(N^T(y, \delta), N^T(x, r))$;
2. *If* $s > r$ *then* $ingr(N^T(x, s), N^T(x, r))$;
3. *If* $ingr(z, N^T(x, r))$ *and* $ingr(z, N^T(y, s))$ *then there is* $\delta > 0$ *such that* $ingr(N^T(z, \delta), N^T(x, r))$ *and* $ingr(N^T(z, \delta), N^T(y, s))$.

Proof For Property 1, $ingr(y, N^t(x, r))$ implies that there exists an $s > r$ such that $\mu_t(y, x, s)$. Let $\delta < 1$ be such that $t(u, s) > r$ whenever $u > \delta$; δ exists by continuity of t and the identity $t(1, s) = s$. Thus, if $ingr(z, N^t(y, \delta))$, then $\mu_t(z, y, \eta)$ with $\eta > \delta$ and $\mu_t(z, x, t(\eta, s))$ hence $ingr(z, N^t(x, r))$.

Property 2 follows by RINC3 and Property 3 is a corollary to properties 1 and 2. This concludes the argument.   □

2.5 Ramifications of Rough Inclusions

In problems of classification, it turns out important to be able to characterize locally the distribution of values in data. The idea that metrics used in classifier construction should depend locally on the training set is, e.g., present in classifiers based on the idea of nearest neighbor, see, e.g., a survey in Polkowski [23]: for nominal values, the metric Value Difference Metric (VDM) in Stanfill and Waltz [31] takes into account conditional probabilities $P(d = v | a_i = v_i)$ of decision value given the attribute value, estimated over the training set *Trn*, and on this basis constructs in the value set V_i of the attribute a_i a metric $\rho_i(v_i, v_i') = \sum_{v \in V_d} |P(d = v | a_i = v_i) - P(d = v | a_i = v_i')|$. The global metric is obtained by combining metrics ρ_i for all attributes $a_i \in A$ according to one of many-dimensional metrics.

This idea was also applied to numerical attributes in Wilson and Martinez [35] in metrics *IVDM* (Interpolated VDM) and *WVDM* (Windowed VDM). A modification of the *WVDM* metric based again on the idea of using probability densities in determining the window size was proposed as *DBVDM* metric.

In order to construct a measure of similarity based on distribution of attribute values among objects, we resort to residual implications, of the form μ_T; this rough inclusion can be transferred to the universe U of an information system; to this end, first, for given objects u, v, and $\varepsilon \in [0, 1]$, factors

$$dis_\varepsilon(u, v) = \frac{|\{a \in A : |a(u) - a(v)| \geq \varepsilon\}|}{|A|}, \tag{2.27}$$

and,

$$ind_\varepsilon(u, v) = \frac{|\{a \in A : |a(u) - a(v)| < \varepsilon\}|}{|A|}, \tag{2.28}$$

are introduced. The weak variant of rough inclusion $\mu_{\to T}$ is defined, see Polkowski [20], as

$$\mu_T^*(u, v, r) \text{ if and only if } dis_\varepsilon(u, v) \to_T ind_\varepsilon(u, v) \geq r. \tag{2.29}$$

Particular cases of this similarity measure induced by, respectively, t-norm min, t-norm $P(x, y)$, and t-norm L are,

1. For $T = M(x, y) = min(x, y)$, $x \Rightarrow_{min} y$ is y in case $x > y$ and 1 otherwise, hence, $\mu_{min}^*(u, v, r)$ if and only if $dis_\varepsilon(u, v) > ind_\varepsilon(u, v) \geq r$ with $r < 1$ and 1 otherwise.
2. For $t = P$, with $P(x, y) = x \cdot y$, $x \Rightarrow_P y = \frac{y}{x}$ when $x > y$ and 1 when $x \leq y$, hence, $\mu_P^*(u, v, r)$ if and only if $\frac{ind_\varepsilon(u,v)}{dis_\varepsilon(u,v)} \geq r$ with $r < 1$ and 1 otherwise.
3. For $t = L$, $x \Rightarrow_L y = min\{1, 1 - x + y\}$, hence, $\mu_L^*(u, v, r)$ if and only if $1 - dis_\varepsilon(u, v) + ind_\varepsilon(u, v) \geq r$ with $r < 1$ and 1 otherwise.

References

1. Aristotle. Lawson-Tancred, H. (transl.): Metaphysics. Penguin Classics. The Penguin Group, London, Book Delta 1203 b (2004)
2. Casati, R., Varzi, A.C.: Parts and Places. The Structures of Spatial Representations. MIT Press, Cambridge (1999)
3. Hájek, P.: Metamathematics of Fuzzy Logic. Kluwer, Dordrecht (1998)
4. Iwanuś, B.: On Leśniewski's elementary ontology. Stud. Log. **XXXI**, 73–119 (1973)
5. Leśniewski, S.: Podstawy Ogólnej Teoryi Mnogości, I (Foundations of General Set Theory, I, in Polish). Prace Polskiego Koła Naukowego w Moskwie, Sekcya Matematyczno-przyrodnicza, No. 2, Moscow (1916)
6. Leśniewski, S.: O podstawach matematyki (On foundations of mathematics, in Polish). Przegląd Filozoficzny, XXX, pp. 164–206; Przegląd Filozoficzny, XXXI, pp. 261–291; Przegląd Filozoficzny, XXXII, pp. 60–101; Przegląd Filozoficzny, XXXIII, pp. 77–105; Przegląd Filozoficzny, XXXIV, pp. 142–170 (1927–1931)
7. Leśniewski, S.: Über die Grundlagen der Ontologie. Comptes rendus Social Science Letters de Varsovie Cl. III, 23 Anneé, pp. 111–132 (1930)
8. Leśniewski, S.: On the foundations of mathematics. Topoi 2, pp. 7–52 [[5] transl. E. Luschei] (1982)
9. Leśniewski, S., Srzednicki, J., Surma, S.J., Barnett, D., Rickey, V.F. (eds.): Collected Works of Stanisław Leśniewski. Kluwer, Dordrecht (1992)
10. Ling, C.-H.: Representations of associative functions. Publ. Math. Debr. **12**, 189–212 (1965)
11. Luschei, E.C.: The Logical Systems of Leśniewski. North Holland, Amsterdam (1962)
12. Łukasiewicz, J.: Die Logischen Grundlagen der Wahrscheinlichkeitsrechnung. Cracow (1913); cf., [Engl. transl.] In: Borkowski L. (ed.) Jan Łukasiewicz. Selected Works. North Holland, pp. 16–63. Polish Scientific Publishers, Amsterdam-Warsaw (1970)

13. Pawlak, Z.: Rough Sets: Theoretical Aspects of Reasoning about Data. Kluwer, Dordrecht (1991)
14. Polkowski, L.: Rough Sets: Mathematical Foundations. Physica Verlag, Heidelberg (2002)
15. Polkowski, L.: A rough set paradigm for unifying rough set theory and fuzzy set theory. Fundamenta Informaticae **54**, 67–88. In: Proceedings RSFDGrC03, Chongqing, China, 2003. Lecture Notes in Artificial Intelligence, vol. 2639, pp. 70–78. Springer, Berlin (2003)
16. Polkowski, L.: Toward rough set foundations. Mereological approach. In: Proceedings RSCTC04, Uppsala, Sweden. Lecture Notes in Artificial Intelligence, vol. 3066, pp. 8–25. Springer, Berlin (2004)
17. Polkowski, L.: Formal granular calculi based on rough inclusions. In: Proceedings of IEEE 2005 Conference on Granular Computing GrC05, pp. 57–62. IEEE Press, Beijing (2005)
18. Polkowski, L.: Rough-fuzzy-neurocomputing based on rough mereological calculus of granules. Int. J. Hybrid Intell. Syst. **2**, 91–108 (2005)
19. Polkowski, L.: A model of granular computing with applications. In: Proceedings of IEEE 2006 Conference on Granular Computing GrC06, pp. 9–16. IEEE Press, Atlanta (2006)
20. Polkowski L.: Granulation of knowledge in decision systems: the approach based on rough inclusions. The method and its applications. In: Proceedings RSEISP 07, Warsaw, Poland, June 2007. Lecture Notes in Artificial Intelligence, vol. 4585, pp. 271–279. Springer, Berlin (2007)
21. Polkowski, L.: A unified approach to granulation of knowledge and granular computing based on rough mereology: a survey. In: Pedrycz W., Skowron A., Kreinovich V. (eds.) Handbook of Granular Computing, pp. 375–400. Wiley, Chichester (2008)
22. Polkowski, L.: Granulation of knowledge: similarity based approach in information and decision systems. In: Meyers, R.A. (ed.) Springer Encyclopedia of Complexity and System Sciences. Springer, Berlin, article 00 788 (2009)
23. Polkowski, L.: Data-mining and knowledge discovery: case based reasoning, nearest neighbor and rough sets. In: Meyers, R.A. (ed.) Encyclopedia of Complexity and System Sciences. Springer, Berlin, Article 00 391 (2009)
24. Polkowski, L.: Approximate Reasoning by Parts. An Introduction to Rough Mereology. Springer, Berlin (2011)
25. Polkowski, L., Skowron, A.: Rough mereology. In: Proceedings of ISMIS'94. Lecture Notes in Artificial Intelligence, vol. 869, pp. 85–94. Springer, Berlin (1994)
26. Polkowski, L., Skowron, A.: Rough mereology: a new paradigm for approximate reasoning. Int. J. Approx. Reason. **15**(4), 333–365 (1997)
27. Simons, P.: Parts: A Study in Ontology, 2nd edn. Clarendon Press, Oxford (2003)
28. Słupecki, J.: S. Leśniewski's calculus of names. Stud. Log. **III**, 7–72 (1955)
29. Sobociński, B.: L'analyse de l'antinomie Russellienne par Leśniewski. Methodos I, pp. 94–107, 220–228, 308–316; Methodos II, pp. 237–257 (1949–1950)
30. Sobociński, B.: Studies in Leśniewski's mereology. Yearbook for 1954–55 of the Polish Society of Art and Sciences Abroad, vol. V, pp. 34–43. London (1955)
31. Stanfill, C., Waltz, D.: Toward memory-based reasoning. Commun. ACM **29**, 1213–1228 (1986)
32. Tarski, A.: Les fondements de la géométrie des corps. Supplement to Annales de la Société Polonaise de Mathématique **7**, 29–33 (1929)
33. Tarski, A.: Zur Grundlegung der Booleschen Algebra. I. Fundamenta Mathematicae **24**, 177–198 (1935)
34. Tarski, A.: Appendix E. In: Woodger, J.H. (ed.) The Axiomatic Method in Biology, p. 160. Cambridge University Press, Cambridge (1937)
35. Wilson, D.R., Martinez, T.R.: Improved heterogeneous distance functions. J. Artif. Intell. Res. **6**, 1–34 (1997)
36. Zadeh, L.A.: Fuzzy sets. Inf. Control **8**, 338–353 (1965)

Chapter 3
Learning Data Classification: Classifiers in General and in Decision Systems

Comparisons do ofttime great grievance
[John Lydgate. Fall of Princes.]

In this chapter, we recall basic facts about data classifiers in machine learning. Our survey is focused on Bayes and kNN classifiers which are employed in our experiments. Some basic facts form Computational Learning Theory are followed by an account of classifiers in real decision systems, mostly elaborated within Rough Set Theory.

3.1 Learning by Machines: A Concise Introduction

Studied in many areas of Computer Science like Machine Learning, Pattern Recognition, Rough Set Theory, Fuzzy Set Theory, Kernel Methods, Cognitive Methods, etc., the subject of automated learning of concepts does encompass many paradigms and specific techniques. Though distinct in their assumptions and methods, yet they share a general setting of the problem of machine learning which may be described as follows.

Learning of a concept proceeds by examining a finite set of *training examples* called often the *training set* or the *training sample*; when presented with the training set, one faces a dichotomy: either training examples are *labeled* with labels indicating their proper concept to which they belong or they are *unlabeled*. In the former case, concepts to which examples answer are often called *categories* or *decision classes* and the task is to recognize the scheme according to which concepts are assigned to things on the basis of their features; in the latter case, the task is to aggregate examples into a number of *clusters* which are to represent categories, the number of clusters not set in advance but resulting in the process. The labeled case goes under the name of the *supervised learning* and the other is then the *unsupervised learning*.

© Springer International Publishing Switzerland 2015
L. Polkowski and P. Artiemjew, *Granular Computing in Decision Approximation*,
Intelligent Systems Reference Library 77, DOI 10.1007/978-3-319-12880-1_3

We are following the path of the supervised learning and we formalize this paradigm. We assume a number of *categories* or *decision classes*, $C_1, C_2, \ldots, C_m$ into which things in our scope may fall. The training set is then a finite set of ordered pairs of the form (x, y) where x is an example and y is its category index, viz., $y = i$ if and only if $x \in C_i$. One would like to have categories C_i disjoint but it is not always the case due to noise in data, i.e., it may happen that things which are regarded as identical belong in distinct categories as is often the case in decision tables.

The training set Trn_n comes as the finite set of n indexed pairs,

$$Trn_n = \{(x_1, y_1), (x_2, y_2), \ldots, (x_n, y_n)\}. \tag{3.1}$$

The problem of machine learning may be posed as follows: On the basis of Trn_n find a mapping, called a *classifier*, h, which would map the set of things under consideration X into the set of categories Y. The classifier h comes from a class of mappings H called *the hypotheses space*.

One assumes that the space H contains the true classifier h_0 which assigns to each thing $x \in X$ its correct category index $h_0(x)$. The difficulty in finding h equal to h_0 lies in the fact that our knowledge of the set X of things is provided by the 'window' Trn_n and besides it our knowledge is nil (*the closed world assumption*). Nevertheless, one may consider a virtual case when some knowledge of X is assumed in order to create some virtual classifiers. An important example is the *Bayes classifier*.

3.1.1 Bayes Classifier

It is a classifier defined within Statistical Learning Theory under the assumption that the probability distribution $P[y|x]$ on the product $X \times Y$ is known. Then, it is possible to determine probabilities

$$P[y_i|x] = p_i(x), \tag{3.2}$$

where y_i is the index for the category C_i.

The Bayes classifier assigns to the thing $x \in X$ the category C_{i_0} with the property that

$$h_{Bayes}(x) = i_0 = argmax_i p_i(x). \tag{3.3}$$

Thus, the Bayes classifier assigns to each thing x its most probable category (though the assignment may be non-deterministic in case of two or more most probable choices resolved randomly), hence, it may be chosen and serve as the 'reference classifier' to which 'real' classifiers can be compared and against it judged.

The comparison is measured in terms of errors made in classification; the global error of the classifier h, $Err(h)$, is defined as the expected value of individual errors on things with respect to the distribution P. For a thing x, the eventual error $Err_h(x)$ of the classification of x by h is defined as

$$Err_h(x) = 1 \text{ when } h(x) \neq h_0(x) \text{ and } 0 \text{ otherwise.} \tag{3.4}$$

The global error is then

$$Err(h) = E[Err_h(x)]. \tag{3.5}$$

Agnosticism about the real probability P calls for making assumptions about P redundant by requiring that the comparison should be made simultaneously for all distributions of probability on the set $X \times Y$. This leads to the notion of *consistency*. In order to define it, one adopts a standard principle of 'solving a problem in finite case by going to infinity' (according to Stanislaw Ulam), hence, one considers an infinite sequence $(Trn_n)_{n=1}^{\infty}$ of training sets with $Trn_n \subset Trn_{n+1}$ for each n independently and identically drawn from X under a distribution P. The symbol h_n refers to the classifier obtained by a classifying algorithm on the training set Trn_n.

One says then that the classifying algorithm is *P(Bayes)-consistent* if and only if

$$lim_{n \to \infty} P[Err(h_n) - Err(h_{Bayes}] > \varepsilon = 0 \tag{3.6}$$

for each positive ε.

The classifying algorithm is *(Bayes) consistent* if and only if it is P(Bayes)-consistent for each probability distribution P on X.

The idea of Bayes classifier can be expressed more clearly in case of the binary classification into two categories which can be then represented as opposite to each other with indices $+1$ and -1; the Bayes classifier in this case acts according to the recipe,

$$h_{Bayes}(x) = +1 \text{ if and only if } P[y = +1|x] \geq \frac{1}{2} \tag{3.7}$$

i.e., in majority of cases the category pointed to for x is the one indexed by $+1$.

This observation is the basis of the classifier called the *majority voting*, MV in short. MV acts on a set of *voters* on a given issue by allotting each voter one choice of the issue category and the winning category is that chosen by the majority of voters, with ties resolved randomly or by the runoff voting.

In search of the real classifier most close to the Bayesian one, one modifies MV to the *nearest neighbor* voting.

3.1.2 Nearest Neighbor Classifier: Asymptotic Properties

Nearest neighbor method goes back to Fix and Hodges [25, 26], Skellam [61] and Clark and Evans [21], who used the idea in statistical tests of significance in nearest neighbor statistics in order to discern between random patterns and clusters. The idea is to assign a class value to a tested object on the basis of the class value of the nearest with respect to a chosen metric already classified object. For a generalization to to k-nearest neighbors, see Patrick and Fischer [45].

The nearest neighbor algorithm requires a metric d on the set $X \times X$. The classifier is faced with the training set $Trn_n = \{(x_1, y_1), (x_2, y_2), \ldots, (x_n, y_n)\}$ and a new thing $x \in X \backslash Trn_n$. In order to assign the category index to x, the classifier performs the majority voting on a set $Near(x)$ of training examples nearest in d to x. Denoting the classifier chosen function with the symbol h^{NN}, one may write the formula down

$$h^{NN}(x) = MV(Near(x)). \qquad (3.8)$$

The set $Near(x)$ is determined by its size: in case $card(Near(x)) = k$ for a natural number k, the nearest neighbor algorithm is termed the *k-nearest neighbor algorithm*, abbrev., *kNN*. In case $k = 1$, the 1NN classifier is simply denoted NN.

Theoretical justification of this methodology is provided by a theorem due to Stone [65], cf. Devroye et al. [24].

Theorem 1 (Stone [65]) *Given a training set* $Trn = \{(x_1, y_1), (x_2, y_2), \ldots, (x_n, y_n)\}$ *and* x *in the space of test things, assume that* x_i *is the i-th farthest from* x *thing among* $x_1, x_2, \ldots, x_n$ *with respect to some arbitrary norm-induced metric on* R^d. *Assuming that* x *is distributed as* x_1, *and* f *is any integrable function, if* $n \to \infty$ *and* $\frac{k}{n} \to 0$, *then*

$$\frac{1}{k} \sum_{i=1}^{k} E[|f(x) - f(x_i)|] \to 0. \qquad (3.9)$$

The most essential properties of kNN algorithm are expressed in theorems due, respectively, to Cover and Hart [23] and Stone [65].

The theorem of Cover and Hart relates the error of NN to the Bayesian error. In case of the binary classification, it reads as follows.

Theorem 2 (Cover and Hart [23]) *Assume that the NN classifier acts on a space* $X \times Y$ *endowed with a probability distribution* P *and* X *is equipped with a metric* d. *Let* $Err[h^{NN}]_n$ *denote the error of NN classifier on the sample of size of* n. *Then*

$$lim_{n \to \infty}[Err[h^{NN}]_n] = 2 \cdot E[P[y = 1|x] \cdot P[y = -1|x]], \qquad (3.10)$$

moreover,

$$Err[h_{Bayes}] \leq Err[h^{NN}] \leq 2E[P[y = 1|x] \cdot P[y = -1|x]]$$
$$\leq 2 \cdot Err[h_{Bayes}] \cdot (1 - Err[h_{Bayes}]) \leq 2 \cdot Err[h_{Bayes}].$$

It follows that the limiting error of the NN classifier is sandwiched between the Bayesian error and twice the Bayesian error times 1 minus the Bayesian error.

A similar result is proven in Cover and Hart (op.cit.) for *kNN* classifiers.

The Bayes consistency of the *kNN* was established in Stone [65], viz.,

Theorem 3 (Stone [65]) *Consider the kNN algorithm acting on the increasing sequence of training sets* $(Trn_n)_{n=1}^{\infty}$ *with* h_n^{NN} *denoting the classifier obtained from the nth training set. If* $k \to \infty$ *and* $\frac{k}{n} \to 0$ *as* $n \to \infty$, *then* $Err[h_n^{NN}] \to_{n \to \infty} Err[h_{Bayes}]$ *for each probability distribution P.*

It follows that *kNN* classifier is, under restrictions stated above, Bayes-consistent.

3.1.3 Metrics for kNN

Metrics used in nearest neighbor-based techniques can be of varied form; the basic distance function is the Euclidean metric in a d-space

$$\rho_E(x, y) = \left[\sum_{i=1}^{d} (x_i - y_i)^2 \right]^{\frac{1}{2}} \tag{3.11}$$

and its generalization to the class of *Minkowski's metrics*

$$L_p(x, y) = \left[\sum_{i=1}^{d} |x_i - y_i|^p \right]^{\frac{1}{p}} \tag{3.12}$$

for $p \geq 1$ with limiting cases of the *Manhattan metric*

$$L_1 = \sum_{i=1}^{d} |x_i - y_i|, \tag{3.13}$$

and,

$$L_\infty(x, y) = max\{|x_i - y_i| : i = 1, 2, \ldots, d\}. \tag{3.14}$$

These metrics can be modified by scaling factors (weights) applied to coordinates, e.g.,

$$L_1^w(x, y) = \sum_{i=1}^{d} w_i \cdot |x_i - y_i| \tag{3.15}$$

is the Manhattan metric modified by the non-negative weight vector w, and subject to adaptive training, see Wojna [72].

Metrics like above can be detrimental to nearest neighbor method in the sense that the nearest neighbors are not invariant with respect to transformations like translations, shifts, rotations. A remedy for this difficulty was proposed as the notion of the tangent distance by Simard et al. [60]. The idea consists in replacing each training as well as each test object, represented as a vector x in the feature space R^k, with its invariance manifold, see Hastie et al. [34], consisting of x along with all its images by allowable transformations: translation, scaling of axes, rotation, shear, line thickening; instead of measuring distances among object representing vectors x, y, one can find shortest distances among invariance manifolds induced by x and y; this task can be further simplified by finding for each x the tangent hyperplane at x to its invariance manifold and measuring the shortest distance between these tangents. For a vector x and the matrix T of basic tangent vectors at x to the invariance manifold, the equation of the tangent hyperplane is $H(x) : y = x + Ta$, where a in R^k. The simpler version of the tangent metric method, the "one–sided", assumes that tangent hyperplanes are produced for training objects x whereas for test objects x' no invariance manifold is defined but the nearest to x' tangent hyperplane is found and x' is classified as x that defined the nearest tangent hyperplane. In this case the distance to the tangent hyperplane is given by $arg \ min_a \rho_E(x', x + Ta)$ in case the Euclidean metric is chosen as the underlying distance function. In case of "two–sided" variant, tangent hyperplane at x' is found as well and x' is classified as the training object x for which the distance between tangent hyperplanes $H(x), H(x')$ is the smallest among distances from $H(x')$ to tangent hyperplanes at training objects.

Among problems related to metrics is also the *dimensionality problem*: in high dimensional feature spaces, nearest neighbors can be located at large distances from the test point in question, thus violating the basic principle justifying the method as a window choice; as indicated in [34], the median of the radius R of the sphere about the origin containing the nearest to the origin neighbor from n points uniformly distributed in the cube $[-\frac{1}{2}, \frac{1}{2}]^p$ is equal to

$$\frac{1}{V_p^p} \cdot \left(1 - \frac{1^{\frac{1}{N}}}{2} \right)^{\frac{1}{p}}, \qquad (3.16)$$

where $V_p \cdot r^p$ is the volume of the sphere of radius r in the p-space, and it approaches the value of 0.5 with increase of p for each n, i.e., a nearest neighbor is asymptotically located on the surface of the cube.

A related result in Aggarwal et al. [1, 2] is related to the comparison of the behavior of the distance $dist_{max}$—the maximal distance between the test thing and the farthest training example in a sample of n points and the $dist_{min}$—the distance to the nearest training example.

Theorem 4 (Aggarwal et al. [1]) *For n training examples in the space R^d distributed according to the probability distribution P, with the metric L_k as the distance measure,*

$$C_k \leq lim_{d \to \infty} E \left[\frac{dist_{max}^k - dist_{min}^k}{d^{\frac{1}{k} - \frac{1}{2}}} \right] \leq (n-1) \cdot C_k, \qquad (3.17)$$

where C_k is a constant depending on k.

The conclusion is, e.g., that for metrics L_k with $k \geq 3$ in large dimensions it is useless to select the nearest neighbor as the notion becomes meaningless, whereas for metrics L_1, L_2 this notion is reasonable even at large dimensions. To cope with this problem, a method called *discriminant adaptive nearest neighbor* was proposed by Hastie and Tibshirani [33]; the method consists in adaptive modification of the metric in neighborhoods of test vectors in order to assure that probabilities of decision classes do not vary much; the direction in which the neighborhood is stretched is the direction of the least change in class probabilities.

This idea that metric used in nearest neighbor finding should depend locally on the training set was also used in constructions of several metrics in the realm of decision systems, i.e., objects described in the attribute-value language; for nominal values, the metric Value Difference Metric (*VDM*) in Stanfill and Waltz [64] takes into account conditional probabilities $P(d = v|a_i = v_i)$ of decision value given the attribute value, estimated over the training set *Trn*, and on this basis constructs in the value set V_i of the attribute a_i a metric $\rho_i(v_i, v_i') = \sum_{v \in V_d} |P(d = v|a_i = v_i) - P(d = v|a_i = v_i')|$. The global metric is obtained by combining metrics ρ_i for all attributes $a_i \in A$ according to one of many-dimensional metrics, e.g., Minkowski metrics.

This idea was also applied to numerical attributes in Wilson and Martinez [71] in metrics *IVDM* (Interpolated VDM) and *WVDM* (Windowed VDM). A modification of the *WVDM* metric based again on the idea of using probability densities in determining the window size was proposed as *DBVDM* metric.

3.2 Classifiers: Concept Learnability

In real cases of classification, one is faced with a finite training set of examples which provides the only information about the set X of all relevant things. On the basis of only this set one is to produce a classifier which should be as good as to classify new unknown yet things with as small as possible error. The problem of the opposition 'training set–the whole of X' has been raised and analyzed within the statistical theory. We give a concise summary of basic results in this field.

The training set can be regarded as a sample from the population X, and the average error of a classification on the training set

$$Trn_n = \{(x_1, y_1), (x_2, y_2), \ldots, (x_n, y_n)\}$$

by a function h chosen from a set H of hypotheses can be expressed as

$$errT_{rn_n}(h) = \frac{1}{n} \cdot card\{i : h(x_i) \neq y_i\}. \tag{3.18}$$

We recall that P denotes the probability distribution on the set $X \times Y$ conforming with which examples are drawn identically and independently. The Chernoff bound [20] estimates the probability that the training set error by h differs from the error by h,

Theorem 5 (Chernoff [20]) *The following inequality estimates the probability* $P[errT_{rn_n}(h) - err(h) \geq \varepsilon]$:

$$P[errT_{rn_n}(h) - err(h) \geq \varepsilon] \leq 2 \cdot e^{-2 \cdot n \cdot \varepsilon^2} \tag{3.19}$$

for each positive ε.

The Chernoff bound suggests the way of choosing a classifier h_{opt} from the set H of hypotheses, viz.,

$$h_{opt} = argmin_{h \in H} errT_{rn_n}(h). \tag{3.20}$$

The mapping h_{opt} minimizes the error on the training set, hence, by the Chernoff bound, its training error is close to its error. Hence, the strategy for the classifier synthesis is to choose h_{opt} as the classifier from H (the *ERM (empirical risk minimization) strategy*).

The question is now whether the classifier h_{opt} which minimizes the training error is the best classifier h_{best} in H. This issue depends on the ability of H to discern concepts, formally defined in the Vapnik and Chervonenkis theory of the *Vapnik–Chervonenkis dimension, dim_{VC}*. We return to the problem after a short excursion into the PAC domain.

3.2.1 The VC Dimension and PAC Learning

For a finite set $F = \{x_1, x_2, \ldots, x_n\}$ of examples from X, one says that the set H of hypotheses *shatters* F if and only if the size of the set $res_H(F) = \{[h(x_1), h(x_2), \ldots, h(x_n)] : h \in H\}$ is of the maximal cardinality m^n (recall that $C_1, C_2, \ldots, C_m$ are categories (decision classes) available for classification). The Vapnik–Chervonenkis dimension, see, Vapnik and Chervonenkis [70], Vapnik [69] is defined as

$$dim_{VC}(H) = n \text{ if and only if } n = max\{k : H \text{ shatters a set of size k}\}. \tag{3.21}$$

If H shatters for each natural number n a set of size n, then $dim_{VC}(H) = \infty$. One must be aware that some commonly used H's have infinite dimension, e.g., the nearest neighbor classifier.

Another approach set in probabilistic context is the *probabilistically approximately correct* (PAC) learning due to Valliant [68]; cf. e.g., Blumer et al. [15] on the question of PAC learnability. In this setting concepts are represented as sets in a space X endowed with a probability distribution P. For a given concept class C, samples are defined as sets $S_m(c, x) = \{(x_1, i_1), (x_2, i_2), \ldots, (x_m, i_m)\}$ where i_j is the value of the characteristic function of some $c \in C$ on x_j, i.e., 0 or 1 and $x = \{x_1, x_2, \ldots, x_m\}$. A mapping h on a sample $S_m(c, x)$ *agrees* with C if and only if $h(x_j) = i_j$ for each j. The mapping h is *consistent* if it agrees with C with respect to all samples.

The error $err(h, S_m(c, x))$ of h with respect to a sample $S_m(c, x)$ is defined as the probability $P[c \triangle h(S_m(c, x))]$ where $\triangle$ is the symmetric difference operator.

Given $\varepsilon, \delta \in (0, 1)$, the mapping h is a $(\varepsilon, \delta, P, m = m(\varepsilon, \delta))$-*learning function* for C if for each sample of size m, the probability $P^m[\{x_1, x_2, \ldots, x_m\} : err(h, S_m(c, x) \geq \varepsilon] \leq \delta$. The mapping h is a *universal learning function* for C if it is a $(\varepsilon, \delta, P, m = m(\varepsilon, \delta))$-learning function for C for all ε, δ, P.

Under these assumptions and notions the following statements are true, see Blumer et al. [15, Theorem 2.1].

Theorem 6 (Blumer et al. [15])

1. *A universal learning function for C exists if and only if $\dim_{VC}(C) < \infty$;*
2. *For $m(\varepsilon, \delta) = max\{\frac{4}{\varepsilon} \cdot log\frac{2}{\delta}, \frac{8 \cdot \dim_{VC}(C)}{\varepsilon} \cdot log 13\varepsilon\}$ any mapping h is a universal learning function for C if it is consistent.*

From Theorem 6 one infers, returning to a discussion in the preceding section, and noticing paradigmatic analogy between PAC and the setting in the previous section that

Proposition 15 $P[err[h_{opt} - err[h_{best} > \varepsilon] \to_{n \to \infty} 0$ *for each ε positive and each probability distribution on X if and only if $\dim_{VC}(H) < \infty$.*

A detailed discussion of topics in statistical learning theory can be found in Devroye et al. [24], Hastie et al. [34], Vapnik [69].

3.3 Rough Set Approach to Data: Classifiers in Decision Systems

Introduced in Pawlak [46, 48] rough set theory is based on ideas that go back to Gottlob Frege, Gottfried Wilhelm Leibniz, Jan Łukasiewicz, Stanislaw Leśniewski, to mention a few names of importance. Its characteristics is that they divide notions (concepts) formalized as sets of things into two classes: exact as well as inexact.

The idea for a dividing line between the two comes from Frege [28]: an inexact concept should possess a boundary region into which objects fall which can be classified with certainty neither to the concept nor to its complement, i.e. the decision problem for them fails. This boundary to a concept is constructed in the rough set theory from indiscernibility relations induced by attributes (features) of objects.

Knowledge is assumed in rough set theory to be a *classification* of objects into categories (decision classes). As a frame for representation of knowledge, we adopt following Pawlak [46] *information systems*.

One of languages for knowledge representation is the *attribute-value* language in which notions representing things are described by means of *attributes* (features) and their *values*; information systems are pairs of the form (U, A) where U is a set of objects—representing things—and A is a set of attributes; each attribute a is modeled as a mapping $a : U \rightarrow V_a$ from the set of objects into the *value set* V_a. For an attribute a and its value v, the *descriptor*, see Pawlak [48], Baader et al. [12], $(a = v)$ is a formula interpreted in the set of objects U as $[[(a = v)]] = \{u \in U : a(u) = v\}$.

Descriptor formulas are the smallest set containing all descriptors and closed under sentential connectives $\vee, \wedge, \neg, \Rightarrow$. Meanings of complex formulas are defined recursively

1. $[[\alpha \vee \beta]] = [[\alpha]] \cup [[\beta]]$;
2. $[[\alpha \wedge \beta]] = [[\alpha]] \cap [[\beta]]$;
3. $[[\neg\alpha]] = U \setminus [[\alpha]]$;
4. $[[\alpha \Rightarrow \beta]] = [[\neg\alpha \vee \beta]]$.

In descriptor language each object $u \in U$ can be encoded over a set B of attributes as its information vector $Inf_B(u) = \{(a = a(u)) : a \in B\}$. A formula α is *true* (is a tautology) if and only if its meaning $[[\alpha]]$ equals the set U.

The Leibniz Law (the Principle of Identity of Indiscernibles and Indiscernibility of Identicals), Leibniz [27, 37], affirms that two things are identical if and only if they are indiscernible, i.e., no available operator acting on both of them yields distinct values

$$\text{if } F(x) = F(y) \text{ for each operator F then } x = y. \tag{3.22}$$

In the context of information systems, indiscernibility relations are introduced and interpreted in accordance with the Leibniz Law from sets of attributes: given a set $B \subseteq A$, the *indiscernibility relation* relative to B is defined as

$$Ind(B) = \{(u, u') : a(u) = a(u') \text{ for each } a \in B\}. \tag{3.23}$$

Objects u, u' in the relation $Ind(B)$ are said to be *B-indiscernible* and are regarded as identical with respect to knowledge represented by the information system (U, B). The class $[u]_B = \{u' : (u, u') \in Ind(B)\}$ collects all objects identical to u with respect to B.

Each class $[u]_B$ is *B-definable*, i.e., the decision problem whether $v \in [u]_B$ is decidable: one says that $[u]_B$ is *exact*. More generally, an *exact concept* is a set of objects in the considered universe which can be represented as the union of a collection of indiscernibility classes; otherwise, the set is *inexact* or *rough*. In this case, there exist a boundary about the notion consisting of objects which can be with certainty classified neither into the notion nor into its complement. In order to express the *B-boundary* of a concept X, induced by the set B of attributes, *approximations over B*, see Pawlak, op. cit., op. cit., are introduced, i.e., the *B-lower approximation*,

$$\underline{B}X = \bigcup\{[u]_B : [u]_B \subseteq X\}, \tag{3.24}$$

and, the *B-upper approximation*,

$$\overline{B}X = \bigcup\{[u]_B : [u]_B \cap X \neq \emptyset. \tag{3.25}$$

The difference

$$Bd_B X = \overline{B}X \setminus \underline{B}X \tag{3.26}$$

is the *B-boundary* of X; when non-empty, it does witness that X is B-inexact. The reader has certainly noticed the topological character of the approximations as, respectively, the closure and the interior of a concept in the topology generated by indiscernibility relations.

The inductive character of rough set analysis of uncertainty stems from the fact that the information system (U, A) neither exhausts the world of objects W nor the set A^* of possible attribute of objects. On the basis of induced classification $\Delta : U \to$ *Categories*, one attempts to classify each possible unknown object $x \in W\setminus U$; this classification is subject to uncertainty.

In order to reduce the complexity of classification task, a notion of a *reduct* was introduced, see Pawlak [48]; a reduct B of the set A of attributes is a minimal subset of A with the property that $Ind(B) = Ind(A)$. An algorithm for finding reducts based on Boolean reasoning, see Brown [16], was proposed in Skowron and Rauszer [63]; given input (U, A) with $U = \{u_1, \ldots, u_n\}$ it starts with the *discernibility matrix*

$$M_{U,A} = [c_{i,j} = \{a \in A : a(u_i) \neq a(u_j)\}]_{1 \geq i,j \leq n} \tag{3.27}$$

and the Boolean function

$$f_{U,A} = \bigwedge_{c_{i,j} \neq \emptyset, i < j} \bigvee_{a \in c_{i,j}} \overline{a}, \tag{3.28}$$

where $\overline{a}$ is the Boolean variable associated with the attribute $a \in A$.

The function $f_{U,A}$ is converted to its DNF form

$$f_{U,A}^* : \bigvee_{j \in J} \bigwedge_{k \in K_j} \overline{a_{j,k}}. \tag{3.29}$$

Then, sets of the form $R_j = \{a_{j,k} : k \in K_j\}$ for $j \in J$, corresponding to *prime implicants* $\bigwedge_{k \in K_j} \overline{a_{j,k}}$ of $f_{U,A}^*$ are all reducts of A.

To verify this thesis, it suffices to consider for a subset $B \subseteq A$, the valuation v_B: $v_B(\overline{a}) = 1$ if and only if $a \in B$, and observe that $v_B(f_{U,A}) = 1$ if and only if $Ind(B) = Ind(A)$. On the other hand, as $f_{U,A}$ and $f_{U,A}^*$ are semantically equivalent, we have $v_B(f_{U,A}^*) = 1$ and this happens if and only if $\{\overline{a_{j,k}} : k \in K_j\} \subseteq B$ for some prime implicant $\bigwedge_{k \in K_j} \overline{a_{j,k}}$, so minimal B with $Ind(B) = Ind(A)$ are of the form $B = \{a_{j,k} : k \in K_j\}$ for some prime implicant $\bigwedge_{k \in K_j} \overline{a_{j,k}}$.

For a reduct B, the reduced information system (U, B), secures the same classification as the full system (U, A), hence, it does preserve knowledge.

Moreover, as $Ind(B) \subseteq Ind(C)$ for any subset $C \subseteq A$, one can establish a functional dependence of C on B: as for each object $u \in U$, $[u]_B \subseteq [u]_C$, the assignment $f_{B,C} : Inf_B(u) \rightarrow Inf_C(u)$ is functional, i.e., values of attributes in B determine functionally values of attributes in C on U, object-wise. Thus, any reduct determines functionally the whole system.

3.4 Decision Systems

A *decision system* is a triple (U, A, d) in which d is the *decision attribute, decision*, the attribute not in A, that does express the evaluation of objects by an external oracle, an expert. Attributes in A are called in this case *conditional*, in order to discern them from the decision d. Values v_d of decision d can be regarded as codes for categories into which the decision classifies objects.

Inductive reasoning by means of rough sets aims at finding as faithful as possible description of the concept d in terms of conditional attributes in A in the language of descriptors. This description is effected by means of *decision rules*, i.e., formulas of descriptor logic of the form of an implication

$$\bigwedge_{a \in B} (a = v_a) \Rightarrow (d = v). \tag{3.30}$$

The formula (3.30) is *true* in case

$$[[\bigwedge_{a \in B} (a = v_a)]] = \bigcap_{a \in B} [[(a = v_a)]] \subseteq [[(d = v)]].$$

Otherwise, the formula is *partially true*. An object o which is matching the rule, i.e., $a(o) = v_a$ for $a \in B$ can be classified to the class $[[(d = v)]]$; often a partial match based on a chosen distance measure has to be performed.

The simplest case is when the decision system is *deterministic*, i.e., when $Ind(A) \subseteq Ind(d)$. In this case, the relation between A and d is functional, given by the unique assignment $f_{A,d}$ or in the decision rule form (3.30) as the set of rules: $\bigwedge_{a \in A} (a = a(u)) \Rightarrow (d = f_{A,d}(u))$ for $u \in U$. In place of A any reduct R of A can be substituted leading to shorter rules.

In the contrary case of a non-deterministic system, some classes $[u]_A$ are split into more than one decision classes $[v]_d$ leading to ambiguity in classification. In that case, decision rules are divided into *true* (or, exact, certain) and *possible*; to induce the certain rule set, the notion of a δ-*reduct* was proposed in Skowron and Rauszer [63].

To define δ-reducts, the *generalized decision* δ_B is defined: for $u \in U$,

$$\delta_B(u) = \{v \in V_d : \exists u'.d(u') = v \wedge (u, u') \in Ind(B)\}. \tag{3.31}$$

A subset B of A is a δ-reduct, when it is a minimal subset od A with respect to the property that $\delta_B = \delta_A$. δ-reducts can be obtained from the modified Skowron and Rauszer algorithm as pointed in there; it suffices to modify the entries $c_{i,j}$ of the discernibility matrix to new entries $c'_{i,j}$, by letting

$$c^d_{i,j} = \{a \in A \cup \{d\} : a(u_i) \neq a(u_j)\} \tag{3.32}$$

and

$$c'_{i,j} = \begin{cases} c^d_{i,j} \setminus \{d\} & \text{in case } d(u_i) \neq d(u_j) \\ \emptyset & \text{in case } d(u_i) = d(u_j). \end{cases} \tag{3.33}$$

The algorithm described above input with entries $c'_{i,j}$ forming the matrix $M^\delta_{U,A}$ outputs all δ-reducts to d encoded as prime implicants of the associated Boolean function $f^\delta_{U,A}$.

The problem of finding a reduct of minimal length is NP-hard [63], therefore, one may foresee that no polynomial algorithm is available for computing reducts. Thus, the algorithm based on discernibility matrix has been proposed with stop rules that permit to stop the algorithm and obtain a partial set of reducts, see Bazan [13]. In finding approximations to reducts, methods of artificial intelligence like swarm algorithms are applied, see Yu et al. [74]; in Moshkov et al. [41], an idea of a descriptive set of attributes, computed by a polynomial algorithm is discussed.

In order to precisely discriminate between certain and possible rules, the notion of a *positive region* along with the notion of a *relative reduct*, see Pawlak [48], was studied in Skowron and Rauszer [63].

Positive region $Pos_B(d)$ is the set

$$\{u \in U : [u]_B \subseteq [u]_d\} \tag{3.34}$$

which is equivalent to

$$\bigcup_{v \in V_d} \underline{B}[[(d = v)]]. \tag{3.35}$$

$Pos_B(d)$ is the greatest subset X of U such that (X, B, d) is deterministic; it generates certain rules.

Objects in $U \setminus Pos_B(d)$ are subjected to ambiguity; given such u, and the collection $v_1, \ldots, v_k$ of decision values on the class $[u]_B$, the decision rule describing u can be written down as

$$\bigwedge_{a \in B} (a = a(u)) \Rightarrow \bigvee_{i=1,\ldots,k} (d = v_i). \tag{3.36}$$

Each of the rules $\bigwedge_{a \in B}(a = a(u)) \Rightarrow (d = v_i)$ is possible but not certain, as only for a fraction of objects in the class $[u]_B$ the decision takes the value v_i on.

Relative reducts are minimal sets B of attributes with the property that $Pos_B(d) = Pos_A(d)$; they can also be found by means of discernibility matrix $M^*_{U,A}$ in as pointed to in [63] with entries

$$c^*_{i,j} = \begin{cases} c^d_{i,j} \setminus \{d\} & \text{in case either } d(u_i) \neq d(u_j) \text{ and } u_i, u_j \in Pos_A(d) \\ \text{or} \quad pos(u_i) \neq pos(u_j) \\ \emptyset & \text{otherwise.} \end{cases} \tag{3.37}$$

For a relative reduct B, *certain rules* are induced from the deterministic system $(Pos_B(d), A, d)$, *possible rules* are induced from the non-deterministic system $(U \setminus Pos_B(d), A, d)$. In the last case, one can find δ-reducts and turn the system into a deterministic one $(U \setminus Pos_B(d), A, \delta)$ inducing certain rules of the form $\bigwedge_{a \in B}(a = a(u)) \Rightarrow \bigvee_{v \in \delta(u)}(d = v)$.

3.5 Decision Rules

Forming a decision rule consists in searching in the pool of available semantically non-vacuous descriptors for their combination that describes closely a chosen decision class. The very basic idea of inducing rules consists in considering a set B of attributes: the lower approximation $Pos_B(d)$ allows for rules which are certain, the upper approximation $\bigcup_{v \in V_d} \overline{B}[(d = v)]$ adds rules which are possible.

Let us observe that in the rough set setting, the classifier set H is the set of all mappings on the product $\prod_i V_{a_i}$ of values of attributes into the set of decision classes V_d. In case sets V_{a_i} are all finite, the VC-dimension of H is $card(\prod_i V_{a_i})$, and if at least one set V_{a_i} is infinite, then $dim_{VC}(H) = \infty$. In real cases usually value sets of attributes are intervals in the real line, hence infinite, which calls for other than discussed above methods of rule evaluation.

We write down concisely a decision rule in the form $\phi/B, u \Rightarrow (d = v)$ where $\phi/B, u$ is a descriptor formula $\bigwedge_{a \in B}(a = a(u))$ over B, see Pawlak and Skowron [49]. A method for inducing decision rules in a systematic way of Pawlak and Skowron [49] and Skowron [62] consists in finding the set of all δ-reducts $\mathbf{R} = \{R_1, \ldots, R_m\}$, and defining for each reduct R_j and each object $u \in U$, the rule $\phi/R_j, u \Rightarrow (d = d(u))$. Rules obtained by this method are not minimal usually in the sense of the number of descriptors in the premise ϕ.

A method for obtaining decision rules with minimal number of descriptors, in Pawlak and Skowron [49] and Skowron [62], consists in reducing a given rule $r: \phi/B, u \Rightarrow (d = v)$ by finding a set $R_r \subseteq B$ consisting of irreducible attributes in B only, in the sense that removing any $a \in R_r$ causes the inequality

$$[\phi/R_r, u \Rightarrow (d = v)] \neq [\phi/R_r \setminus \{a\}, u \Rightarrow (d = v)] \tag{3.38}$$

to hold. In case $B = A$, reduced rules $\phi/R_r, u \Rightarrow (d = v)$ are called *optimal basic rules (with minimal number of descriptors)*. The method for finding of all irreducible subsets of the set A in Skowron [62], consists in considering another modification of discernibility matrix: for each object $u_k \in U$, the entry $c'_{i,j}$ into the matrix $M^{\delta}_{U,A}$ for δ-reducts is modified into

$$c^k_{i,j} = \begin{cases} c'_{i,j} & \text{in case } d(u_i) \neq d(u_j) \text{ and } i = k \vee j = k \\ \emptyset & \text{otherwise.} \end{cases} \tag{3.39}$$

Matrices $M^k_{U,A}$ and associated Boolean functions $f^k_{U,A}$ for all $u_k \in U$ allow for finding all irreducible subsets of the set A and in consequence all basic optimal rules (with minimal number of descriptors).

Decision rules can also be found in an *exhaustive way* by selecting a maximal consistent set of decision rules, incrementally adding descriptors object-wise and eliminating contradicting cases, see RSES system [59]. A method for inducting a minimal set of rules is proposed in LERS system of Grzymala–Busse [30].

3.5.1 Exhaustive Rules

The exhaustive method for rule induction consists in a search for combinations of descriptors which are consistent with respect to decision, pair-wise independent, and, maximal in the sense that no new combination can be added to the existing set.

The following description is based on the exposition in Artiemjew [10, 11]. A tool useful in generation of exhaustive rules is *the relative indiscernibility matrix*, here we give an example.

Example 2 We consider a decision system $D = (U, A, d)$ in Table 3.1, with $U = \{u1, u2, \ldots, u8\}, A = \{a_1, a_2, \ldots, a_6\}$.

We recall that the *relative indiscernibility matrix*, cf. Skowron and Rauszer [63] is defined as $[c'_{ij}]$, where,

$$c'_{ij} = \begin{cases} c_{ij} & \text{when } d(u_i) \neq d(u_j) \\ \emptyset & \text{when } d(u_i) = d(u_j) \end{cases}$$

We recall that

$$c_{ij} = \{a \in A : a(u_i) = a(u_j)\}.$$

Table 3.1 Decision system $D = (U, A, d)$

	a_1	a_2	a_3	a_4	a_5	a_6	d
u_1	1	1	1	1	3	1	1
u_2	1	1	1	1	3	2	1
u_3	1	1	1	3	2	1	0
u_4	1	1	1	3	3	2	1
u_5	1	1	2	1	2	1	0
u_6	1	1	2	1	2	2	1
u_7	1	1	2	2	3	1	0
u_8	1	1	2	2	4	1	1

Table 3.2 The relative indiscernibility matrix

	u_1	u_2	u_3	u_4	u_5	u_6	u_7	u_8
u_1	ϕ	ϕ	a_1, a_2, a_3, a_6	ϕ	a_1, a_2, a_4, a_6	ϕ	a_1, a_2, a_5, a_6	ϕ
u_2	ϕ	ϕ	a_1, a_2, a_3	ϕ	a_1, a_2, a_4	ϕ	a_1, a_2, a_5	ϕ
u_3	a_1, a_2, a_3, a_6	a_1, a_2, a_3	ϕ	a_1, a_2, a_3, a_4	ϕ	a_1, a_2, a_5	ϕ	a_1, a_2, a_6
u_4	ϕ	ϕ	a_1, a_2, a_3, a_4	ϕ	a_1, a_2	ϕ	a_1, a_2, a_5	ϕ
u_5	a_1, a_2, a_4, a_6	a_1, a_2, a_4	ϕ	a_1, a_2	ϕ	a_1, a_2, a_3, a_4, a_5	ϕ	a_1, a_2, a_3, a_6
u_6	ϕ	ϕ	a_1, a_2, a_5	ϕ	a_1, a_2, a_3, a_4, a_5	ϕ	a_1, a_2, a_3	ϕ
u_7	a_1, a_2, a_5, a_6	a_1, a_2, a_5	ϕ	a_1, a_2, a_5	ϕ	a_1, a_2, a_3	ϕ	a_1, a_2, a_3, a_4, a_6
u_8	ϕ	ϕ	a_1, a_2, a_6	ϕ	a_1, a_2, a_3, a_6	ϕ	a_1, a_2, a_3, a_4, a_6	ϕ

The relative indiscernibility matrix for the decision system in Table 3.1 is given in Table 3.2.

Now, for the rules of order 1, i.e., containing one descriptor in the premises and pointing to exactly one decision irrespective of the thing, i.e., the line in the matrix, we have two rules, each supported by three things:

1. From u_2: $(a_6 = 2) => (d = 1)$;
2. From u_4: $(a_6 = 2) => (d = 1)$;
3. From u_6: $(a_6 = 2) => (d = 1)$;
4. From u_8: $(a_5 = 4) => (d = 1)$.

Now, we modify Table 3.2, marking descriptors $(a_6 = 2)$, $(a_5 = 4)$ into Table 3.3.

Table 3.3 Relative indiscernibility matrix after elimination of two descriptors

	u_1	u_2	u_3	u_4	u_5	u_6	u_7	u_8
u_1	ϕ	ϕ	a_1, a_2, a_3, a_6	ϕ	a_1, a_2, a_4, a_6	ϕ	a_1, a_2, a_5, a_6	ϕ
u_2	ϕ	ϕ	a_1, a_2, a_3	ϕ	a_1, a_2, a_4	ϕ	a_1, a_2, a_5	ϕ
u_3	a_1, a_2, a_3, a_6	a_1, a_2, a_3	ϕ	a_1, a_2, a_3, a_4	ϕ	a_1, a_2, a_5	ϕ	a_1, a_2, a_6
u_4	ϕ	ϕ	a_1, a_2, a_3, a_4	ϕ	a_1, a_2	ϕ	a_1, a_2, a_5	ϕ
u_5	a_1, a_2, a_4, a_6	a_1, a_2, a_4	ϕ	a_1, a_2	ϕ	a_1, a_2, a_3, a_4, a_5	ϕ	a_1, a_2, a_3, a_6
u_6	ϕ	ϕ	a_1, a_2, a_5	ϕ	a_1, a_2, a_3, a_4, a_5	ϕ	a_1, a_2, a_3	ϕ
u_7	a_1, a_2, a_5, a_6	a_1, a_2, a_5	ϕ	a_1, a_2, a_5	ϕ	a_1, a_2, a_3	ϕ	a_1, a_2, a_3, a_4, a_6
u_8	ϕ	ϕ	a_1, a_2, a_6	ϕ	a_1, a_2, a_3, a_6	ϕ	a_1, a_2, a_3, a_4, a_6	ϕ
neglect	−	a_6	−	a_6	−	a_6	−	a_5

Rules of the second order, i.e., with two descriptors in the premises are

1. From u_1: $(a_3 = 1)$ & $(a_4 = 1) \implies (d = 1)$;
   $(a_3 = 1)$ & $(a_5 = 3) \implies (d = 1)$;
   $(a_4 = 1)$ & $(a_5 = 3) \implies (d = 1)$.
2. From u_2 : $(a_3 = 1)$ & $(a_4 = 1) \implies (d = 1)$;
   $(a_3 = 1)$ & $(a_5 = 3) \implies (d = 1)$;
   $(a_4 = 1)$ & $(a_5 = 3) \implies (d = 1)$.
3. From u_3: $(a_3 = 1)$ & $(a_5 = 2) \implies (d = 0)$;
   $(a_4 = 3)$ & $(a_5 = 2) \implies (d = 0)$;
   $(a_4 = 3)$ & $(a_6 = 1) \implies (d = 0)$;
   $(a_5 = 2)$ & $(a_6 = 1) \implies (d = 0)$.
4. From u_4: $(a_3 = 1)$ & $(a_5 = 3) \implies (d = 1)$;
   $(a_4 = 3)$ & $(a_5 = 3) \implies (d = 1)$.
5. From u_5: $(a_5 = 2)$ & $(a_6 = 1) \implies (d = 0)$.
6. From u_7: $(a_3 = 2)$ & $(a_5 = 3) \implies (d = 0)$;
   $(a_4 = 2)$ & $(a_5 = 3) \implies (d = 0)$.

Continuing with the next relative indiscernibility matrix, we find candidates for rules of order 3.

1. From u_1: $(a_1 = 1)$ & $(a_3 = 1)$ & $(a_4 = 1) \implies (d = 1)$;
   $(a_1 = 1)$ & $(a_3 = 1)$ & $(a_5 = 3) \implies (d = 1)$;
   $(a_1 = 1)$ & $(a_4 = 1)$ & $(a_5 = 3) \implies (d = 1)$;

$(a_2 = 1)$ & $(a_3 = 1)$ & $(a_4 = 1)$ => $(d = 1)$;
$(a_2 = 1)$ & $(a_3 = 1)$ & $(a_5 = 3)$ => $(d = 1)$;
$(a_2 = 1)$ & $(4 = 1)$ & $(a_5 = 1)$ => $(d = 1)$;
$(a_3 = 1)$ & $(a_4 = 1)$ & $(a_5 = 3)$ => $(d = 1)$;
$(a_3 = 1)$ & $(a_4 = 1)$ & $(a_6 = 1)$ => $(d = 1)$;
$(a_3 = 1)$ & $(a_5 = 3)$ & $(a_6 = 1)$ => $(d = 1)$;
$(a_4 = 1)$ & $(a_5 = 3)$ & $(a_6 = 1)$ => $(d = 1)$.

2. From u_2 : $(a_1 = 1)$ & $(a_3 = 1)$ & $(a_4 = 1)$ => $(d = 1)$;
$(a_1 = 1)$ & $(a_3 = 1)$ & $(a_5 = 3)$ => $(d = 1)$;
$(a_1 = 1)$ & $(a_4 = 1)$ & $(a_5 = 3)$ => $(d = 1)$;
$(a_2 = 1)$ & $(a_3 = 1)$ & $(a_4 = 1)$ => $(d = 1)$;
$(a_2 = 1)$ & $(a_3 = 1)$ & $(a_5 = 3)$ => $(d = 1)$;
$(a_2 = 1)$ & $(a_4 = 1)$ & $(a_5 = 3)$ => $(d = 1)$;
$(a_3 = 1)$ & $(a_4 = 1)$ & $(a_5 = 3)$ => $(d = 1)$.

3. From u_3: $(a_1 = 1)$ & $(a_3 = 1)$ & $(a_5 = 2)$ => $(d = 0)$;
$(a_1 = 1)$ & $(a_4 = 3)$ & $(a_5 = 2)$ => $(d = 0)$;
$(a_1 = 1)$ & $(a_4 = 3)$ & $(a_6 = 1)$ => $(d = 0)$;
$(a_1 = 1)$ & $(a_5 = 2)$ & $(a_6 = 1)$ => $(d = 0)$;
$(a_2 = 1)$ & $(a_3 = 1)$ & $(a_5 = 2)$ => $(d = 0)$;
$(a_2 = 1)$ & $(a_4 = 3)$ & $(a_5 = 2)$ => $(d = 0)$;
$(a_2 = 1)$ & $(a_4 = 3)$ & $(a_6 = 1)$ => $(d = 0)$;
$(a_2 = 1)$ & $(a_5 = 2)$ & $(a_6 = 1)$ => $(d = 0)$;
$(a_3 = 1)$ & $(a_4 = 3)$ & $(a_5 = 2)$ => $(d = 0)$;
$(a_3 = 1)$ & $(a_4 = 3)$ & $(a_6 = 1)$ => $(d = 0)$;
$(a_4 = 3)$ & $(a_5 = 2)$ & $(a_6 = 1)$ => $(d = 0)$.

4. From u_4: $(a_1 = 1)$ & $(a_3 = 1)$ & $(a_5 = 3)$ => $(d = 1)$;
$(a_1 = 1)$ & $(a_4 = 3)$ & $(a_5 = 3)$ => $(d = 1)$;
$(a_2 = 1)$ & $(a_3 = 1)$ & $(a_5 = 3)$ => $(d = 1)$;
$(a_2 = 1)$ & $(a_4 = 3)$ & $(a_5 = 3)$ => $(d = 1)$;
$(a_3 = 1)$ & $(a_4 = 3)$ & $(a_5 = 3)$ => $(d = 1)$.

5. From u_5: $(a_1 = 1)$ & $(a_5 = 2)$ & $(a_6 = 1)$ => $(d = 0)$;
$(a_2 = 1)$ & $(a_5 = 2)$ & $(a_6 = 1)$ => $(d = 0)$;
$(a_3 = 2)$ & $(a_4 = 1)$ & $(a_6 = 1)$ => $(d = 0)$;
$(a_3 = 2)$ & $(a_5 = 2)$ & $(a_6 = 1)$ => $(d = 0)$;
$(a_4 = 1)$ & $(a_5 = 2)$ & $(a_6 = 1)$ => $(d = 0)$.

6. From u_7: $(a_1 = 1)$ & $(a_3 = 2)$ & $(a_5 = 3)$ => $(d = 0)$;
$(a_1 = 1)$ & $(a_4 = 2)$ & $(a_5 = 3)$ => $(d = 0)$;
$(a_2 = 1)$ & $(a_3 = 2)$ & $(a_5 = 3)$ => $(d = 0)$;
$(a_2 = 1)$ & $(a_4 = 2)$ & $(a_5 = 3)$ => $(d = 0)$;
$(a_3 = 2)$ & $(a_4 = 2)$ & $(a_5 = 3)$ => $(d = 0)$;
$(a_3 = 2)$ & $(a_5 = 3)$ & $(a_6 = 1)$ => $(d = 0)$;
$(a_4 = 2)$ & $(a_5 = 3)$ & $(a_6 = 1)$ => $(d = 0)$.

Elimination of rules which contain rules of lower orders, leaves only one rule of order 3: $(a_3 = 2)$ & $(a_4 = 1)$ & $(a_6 = 1)$ => $(d = 0)$.

3.5.2 Minimal Sets of Rules: LEM2

On the opposite end of the specter of rule induction algorithms are algorithms producing a minimal number of rules covering the decision system. In addition to the Pawlak–Skowron algorithm already mentioned, we would like to present Algorithm LEM2 due to Grzymala-Busse [19, 30, 31]. LEM2 is based on a general heuristic principle, cf. Michalski's AQ [39], Clark's CN2 [22], or PRISM, Cendrowska [18], by which, the first descriptor is chosen as the best one with respect to some criteria; after the rule is found, all training things covered by the rule are removed and the process stops when the training set is covered by the set of rules (Table 3.4).

Example 3 Consider the decision system $D = (U, A, d)$, with $U = \{u_1, u_2, \ldots, u_7\}$, $A = \{a_1, a_2, \ldots, a_5\}$.

Following the idea of LEM2, we consider the decision class 1, finding the most frequent descriptor

$(a_2 = 1)$ which occurs in u_2, u_3, u_4.

It does not make a rule because of the inconsistency in u_7.

For $(a_2 = 1)$, the next best descriptor for the decision class 1 is

$(a_3 = 1)$ in u_2, u_3, and, we make the premise

$(a_2 = 1) \wedge (a_3 = 1)$ which again is inconsistent with u_7.

Continuing, we find descriptors $(a_1 = 1)$, $(a_4 = 3)$, $(a_5 = 2)$, which finally make the consistent rule,

$(a_2 = 1) \wedge (a_3 = 1) \wedge (a_1 = 1) \wedge (a_4 = 3) \wedge (a_5 = 2) \Rightarrow (d = 1)$.

It is the rule covering u_2, so we consider things u_1, u_3, u_4, and here the best descriptor is

Table 3.4 Decision system for Example 3

	a_1	a_2	a_3	a_4	a_5	d
u_1	2	6	1	2	3	1
u_2	1	1	1	3	2	1
u_3	2	1	1	2	3	1
u_4	4	1	3	1	2	1
u_5	3	5	2	1	3	2
u_6	3	1	3	1	1	2
u_7	1	1	1	3	1	2

$(a_1 = 2)$, which is covering things u_1, u_3, consistently, leading to the rule
$(a_1 = 2) \Rightarrow (d = 1)$.

The thing u_4, yields the consistent rule,
$(a_1 = 4) \Rightarrow (d = 1)$
For the decision class 2, the analogous procedure yields the consistent rules

$(a_1 = 3) \Rightarrow (d = 2)$, which covers things u_5, u_6, and,
$(a_1 = 1) \wedge (a_2 = 1) \wedge (a_3 = 1) \wedge (a_4 = 3) \wedge (a_5 = 1) \Rightarrow (d = 2)$, which covers
the thing u_7.

Summing up, we obtain five decision rules:

rule1 $(a_2 = 1) \wedge (a_3 = 1) \wedge (a_1 = 1) \wedge (a_4 = 3) \wedge (a_5 = 2) \Rightarrow (d = 1)$;
rule2 $(a_1 = 2) \Rightarrow (d = 1)$;
rule3 $(a_1 = 4) \Rightarrow (d = 1)$;
rule4 $(a_1 = 3) \Rightarrow (d = 2)$;
rule5 $(a_1 = 1) \wedge (a_2 = 1) \wedge (a_3 = 1) \wedge (a_4 = 3) \wedge (a_5 = 1) \Rightarrow (d = 2)$.

This *sequential covering idea* for inducing decision rules, see also Mitchell [40], leading to a minimal in a sense set of rules covering all objects, has been realized as covering algorithms in Rough Set Exploration System (RSES) system [59].

Association rules in the sense of Agrawal et al. [3], Agrawal and Srikant [4], induced by means of APRIORI algorithm, loc.cit., i.e., rules with high confidence coefficient can also be induced in information system context by regarding descriptors as boolean features in order to induce certain and possible rules, cf. Kryszkiewicz and Rybiński [36].

3.5.3 Quality Evaluations for Decision Rules

Decision rules are judged by their quality on the basis of the training set, and by quality in classifying new unseen as yet objects, i.e., by their performance on the test set. Quality evaluation is done on the basis of some measures; for a rule $r : \phi \Rightarrow (d = v)$, and an object $u \in U$, one says that u *matches* r in case $u \in [[\phi]]$. *match*(r) is the number of objects matching r.

Support, supp(r), of r is the number of objects in $[[\phi]] \cap [[(d = v)]]$; the fraction

$$cons(r) = \frac{supp(r)}{match(r)} \tag{3.40}$$

is the *consistency degree* of r; $cons(r) = 1$ means that the rule is certain.

Strength, strength(r), of the rule r is defined, Michalski et al. [39], Bazan [14], Grzymala-Busse and Hu [32], as the number of objects correctly classified by the rule in the training phase; *relative strength* is defined as the fraction,

$$rel - strength(r) = \frac{supp(r)}{|[[(d = v)]]|}. \tag{3.41}$$

Specificity of the rule r, $spec(r)$, is the number of descriptors in the premise ϕ of the rule r, cf. op. cit., op. cit.

Simple measures of statistical character are found from the *contingency table*, see Arkin and Colton [5]. This table is built for each decision rule r and a decision value v, by counting the number n_t of training objects, the number n_r of objects satisfying the premise of the rule r (caught by the rule), $n_r(v)$ is the number of objects counted in n_r and with the decision v, and $n_r(\neg v)$ is the number of objects counted in n_r but with decision value distinct from v. To these factors, we add n_v, the number of training objects with decision v and $n_{\neg v}$, the number of remaining objects, i.e., $n_{\neg v} = n_t - n_v$.

For these values, *accuracy of the rule r relative to v* is the quotient

$$acc(r, v) = \frac{n_r(v)}{n_r}, \tag{3.42}$$

and, *coverage of the rule r relative to v* is

$$cov(r, v) = \frac{n_r(v)}{n_v}. \tag{3.43}$$

These values are useful as indicators of a *rule strength* which is taken into account when classification of a test object is under way: to assign the value of decision, a rule pointing to a decision with a maximal value of accuracy, or coverage, or combination of both can be taken; methods for combining accuracy and coverage into a single criterion are discussed, e.g., in Michalski [38]. Accuracy and coverage can, however, be defined in other ways; for a decision algorithm D, trained on a training set Tr, and a test set Tst, the *global accuracy* of D, $acc(D)$, is measured by its efficiency on the test set and it is defined as the quotient

$$acc(D) = \frac{n_{corr}}{n_{caught}}, \tag{3.44}$$

where n_{corr} is the number of test objects correctly classified by D and n_{caught} is the number of test objects classified.

Similarly, *coverage* of D, $cov(D)$, is defined as

$$cov(D) = \frac{n_{caught}}{n_{test}}, \tag{3.45}$$

where n_{test} is the number of test objects. Thus, the product $accuracy(D) \cdot coverage(D)$ gives the measure of the fraction of test objects correctly classified by D.

We have already mentioned that accuracy and coverage are often advised to be combined in order to better express the trade–off between the two: one may have a high accuracy on a relatively small set of caught objects, or a lesser accuracy on a larger set of caught by the classifier objects. Michalski [38] proposes a combination rule of the form

$$MI = \frac{1}{2} \cdot A + \frac{1}{4} \cdot A^2 + \frac{1}{2} \cdot C - \frac{1}{4} \cdot A \cdot C, \tag{3.46}$$

where A stands for accuracy and C for coverage.

Statistical measures of correlation between the rule r and a decision class v are expressed, e.g., by χ^2 statistic,

$$\chi^2 = \frac{n_t \cdot (n_r(v) \cdot n_{\neg r}(\neg v) - n_r(\neg v) \cdot n_{\neg r}(v))^2}{n(v) \cdot n(\neg v) \cdot n_r \cdot n_{\neg r}}, \tag{3.47}$$

where $n_{\neg r}$ is the number of objects not caught by the rule r, see Bruning and Kintz [17].

In the testing phase, rules vie among themselves for object classification when they point to distinct decision classes; in such case, negotiations among rules or their sets are necessary. In these negotiations rules with better characteristics are privileged.

For a given decision class $c : d = v$, and an object u in the test set, the set $Rule(c, u)$ of all rules matched by u and pointing to the decision v, is characterized globally by,

$$Support(Rule(c, u)) = \sum_{r \in Rule(c,u)} strength(r) \cdot spec(r). \tag{3.48}$$

The class c for which $Support(Rule(c, u))$ is the largest wins the competition and the object u is classified into the class $c : d = v$.

It may happen that no rule in the available set of rules is matched by the test object u and partial matching is necessary, i.e., for a rule r, the matching factor $match–fact(r, u)$ is defined as the fraction of descriptors in the premise ϕ of r matched by u to the number $spec(r)$ of descriptors in ϕ. The rule for which the partial support, $Part-Support(Rule(c,u))$, given as

$$\sum_{r \in Rule(c,u)} match - fact(r, u) \cdot strength(r) \cdot spec(r) \tag{3.49}$$

is the largest, wins the competition and it does assign the value of decision to u.

In a similar way, notions based on relative strength can be defined for sets of rules and applied in negotiations among them [32].

3.6 Dependencies

Decision rules are particular cases of dependencies among attributes or their sets; certain rules of the form $\phi/B \Rightarrow (d = v)$ establish functional dependency of decision d on the set B of conditional attributes. Functional dependence of the set B of attributes on the set C, $C \mapsto B$, in an information system (U, A) means that $Ind(C) \subseteq Ind(B)$.

Minimal sets $D \subseteq C$ of attributes such that $D \mapsto B$ can be found from a modified discernibility matrix $M_{U,A}$, see Skowron and Rauszer [63]: letting $\langle B \rangle$ to denote the global attribute representing B: $\langle B \rangle(u) = \langle b_1(u), \ldots, b_m(u) \rangle$ where $B = \{b_1, \ldots, b_m\}$, for objects u_i, u_j, one sets $c_{i,j} = \{a \in C \cup \{\langle B \rangle\} : a(u_i) \neq a(u_j)\}$ and then $c_{i,j}^B = c_{i,j} \setminus \{\langle B \rangle\}$ in case $\langle B \rangle$ is in $c_{i,j}$; otherwise $c_{i,j}^B$ is empty.

The associated Boolean function $f_{U,A}^B$ gives all minimal subsets of C on which B depends functionally; in particular, when $B = \{b\}$, one obtains in this way all subsets of the attribute set A on which b depends functionally, see Skowron and Rauszer [63]. A number of contributions are devoted to this topic in an abstract setting of semi-lattices, see Pawlak [47] and Novotny and Pawlak [44].

Partial dependence of the set B on the set C of attributes takes place when there is no functional dependence $C \mapsto B$; in that case, some measures of a degree to which B depends on C were proposed in Novotny and Pawlak [44]: the degree can be defined, e.g., as the fraction $\gamma_{B,C} = \frac{|Pos_C B|}{|U|}$, where the C-positive region of B is defined in analogy to already discussed positive region for decision, i.e.,

$$Pos_C(B) = \{u \in U : [u]_C \subseteq [u]_B\}.$$

In this case, B depends on C partially to the degree $\gamma_{B,C}$: $C \mapsto_{\gamma_{B,C}} B$.

The relation $C \mapsto_r B$ of partial dependency is transitive in the sense: if $C \mapsto_r B$ and $D \mapsto_s C$, then $D \mapsto_{max\{0,r+s-1\}} B$, where $t(r, s) = max\{0, r + s - 1\}$ is the Łukasiewicz t-norm.

3.7 Granular Processing of Data

Our idea of augmenting existing strategies for rule induction consists in using granules of knowledge. The principal assumption we can make is that the nature acts in a continuous way: if objects are similar with respect to judiciously and correctly chosen attributes, then decisions on them should also be similar. A granule collecting similar objects should then expose the most typical decision value for objects in it while suppressing outlying values of decision, reducing noise in data, hence, leading to a better classifier.

These ideas were developed and proposed in Polkowski [50–53] see also surveys Polkowski [54–56]. In Polkowski and Artiemjew [57, 58] and in Artiemjew [7–10] the theoretical analysis was confirmed as to its application merits.

We apply the scheme for simple granules indicated in Chap. 1 along with tolerance relations or weak tolerance relations induced from rough inclusions as indicated in Chap. 2.

We assume that we are given a decision system (U, A, d) from which a classifier is to be constructed; on the universe U, a rough inclusion μ is given, and a radius $r \in [0, 1]$ is chosen, see Polkowski [50–53].

We can find granules $g_\mu(u, r)$, for a given rough inclusion μ, defined as in Chap. 2, and make them into the set $G(\mu, r)$.

From this set, a covering $Cov(\mu, r)$ of the universe U can be selected by means of a chosen strategy $\mathcal{G}$, i.e.,

$$Cov(\mu, r) = \mathcal{G}(G(\mu, r)). \tag{3.50}$$

We intend that $Cov(\mu, r)$ becomes a new universe of the decision system whose name will be the *granular reflection* of the original decision system. It remains to define new attributes for this decision system.

Each granule g in $Cov(\mu, r)$ is a collection of objects; attributes in the set $A \cup \{d\}$ can be factored through the granule g by means of a chosen strategy $\mathcal{S}$, i.e., for each attribute $q \in A \cup \{d\}$, the new factored attribute $\overline{q}$ is defined by means of the formula

$$\overline{q}(g) = \mathcal{S}(\{a(v) : ingr(v, g_\mu(u, r))\}). \tag{3.51}$$

In effect, a new decision system $(Cov(\mu, r), \{\overline{a} : a \in A\}, \overline{d})$ is defined. The object v with

$$Inf(v) = \{(\overline{a} = \overline{a}(g)) : a \in A\} \tag{3.52}$$

is called the *granular reflection of g*. Granular reflections of granules need not be objects found in data set; yet, the results show that they mediate very well between the training and test sets.

The procedure just described for forming a granular reflection of a decision system can be modified as proposed in Artiemjew [7] with help of the procedure of *concept dependent granulation*. In this procedure, the granule $g_{mu}(u, r)$ is modified to the granule

$$g_\mu^c(u, r) = g_\mu(u, r) \cap [u]_d$$

i.e., it is computed relative to the decision class of u, cf. Chap. 4.

3.8 Validation Methods: CV

Given an algorithm A, and a training set Trn_n, an estimate $h = A(Trn_n)$ of a proper classifier $h_0 \in H$ is produced. To evaluate the error of h, a method called *cross–validation* was proposed by few independent authors, among them, Mosteller and Tukey [42], Stone [66], Geisser [29], cf. Arlot and Celisse [6]. The idea was to split the training set into at least one subset called the *training sample* and its complement in the training set called the *validation sample*; in this case one calls the procedure *validation*, CV in symbols. A more elaborate scheme assumes partitioning of Trn_n into $2 < k \leq n$ subsets $T_1, \ldots, T_k$. In turn, each of sets T_j acts as a validation sample whereas $\bigcup_{i \neq j} T_i$ is a training sample; this procedure is termed the *k-fold cross-validation*, CVk in symbols. In the limiting case $k = n$, the procedure is called the *leave-one-out*, *LOO* in symbols.

For analysis of cross-validation in case of classification by means of granular reflections, we adopt the scheme given in Tibshirani [67], cf. Kohavi [35].

For the sample identically distributed and independently drawn $\{(x_i, y_i) : i = 1, 2, \ldots, n\}$, the decision function dec of the decision system (data table) $D = (U, A, dec)$, and the parameter r_{gran} which is the granulation radius, error in classification is expressed by means of the $0 - 1$-*loss function*

$$L(y, \hat{dec}(x)) = 1 \text{ when } y \neq \hat{dec}(x) \text{ else } 0, \tag{3.53}$$

where $\hat{dec}$ is an approximation to the decision dec.

We consider a K-fold cross-validation with folds $F_1, F_2, \ldots, F_K$ for a radius r_{gran}; we denote with the symbol $dec_{r_{gran}}^{-k}$ the decision induced by treating the kth fold as the test set, and $acc_{r_{gran}}^{-k}$ will denote accuracy of classification by dec^{-k} on the kth fold at r_{gran}. The granulation radius is the parameter of classification.

The CV estimate of the expected test error is

$$CV(r_{gran}) = \frac{1}{n} \cdot \sum_{k=1}^{K} \sum_{(x_i, y_i) \in F_k} L(y_i, dec_{r_{gran}}^{-k}(x_i, r_{gran})). \tag{3.54}$$

The visualization of the parameterized error (3.54) is known as the *CV-error curve*, cf. [67].

Now, let us observe that, by definition of accuracy (3.44),

$$\sum_{(x_i, y_i) \in F_k} L(y_i, dec_{r_{gran}}^{-k}(x_i, r_{gran})) = \frac{n}{K} \cdot (1 - acc_{r_{gran}}^{k}), \tag{3.55}$$

where, we assume that folds are equal cardinality of about $\frac{n}{K}$ and $acc_{r_{gran}}^{k}$ is the accuracy of classification on the kth fold F_k at the radius r_{gran}.

From (3.54) and (3.55), we obtain that

$$CV(r_{gran}) = \frac{1}{n} \cdot \sum_{k=1}^{K} [\frac{n}{K} \cdot (1 - acc^k_{r_{gran}})], \tag{3.56}$$

i.e.,

$$CV(r_{gran}) = 1 - \sum_{k=1}^{K} \frac{1}{K} \cdot acc^k_{r_{gran}} = 1 - \overline{acc}_{r_{gran}}, \tag{3.57}$$

where, $\overline{acc}_{r_{gran}}$ is the average accuracy of classification over all folds at the radius r_{gran}.

We denote by the symbol r^*_{gran} the minimizer of $CV(r_{gran})$, i.e.,

$$r^*_{gran} = argmin_{r_{gran}} CV(r_{gran}). \tag{3.58}$$

The symbol dec^{-k*} denotes the decision induced at r^*_{gran} for each test fold F_k. For a new test item (x_o, y_o), the expected test error is

$$err = L(y_o, dec^{-k*}(x_o)), \tag{3.59}$$

and, bias of estimation is expressed as, cf. [67],

$$bias = err - CV(r_{gran*}) = err - 1 + \overline{acc}_{r_{gran*}}. \tag{3.60}$$

In Tibshirani [67] an estimate for bias is proposed. Assume again for simplicity sake that folds are of equal cardinality of $\frac{n}{K}$. Then the *error curves for folds* are:

$$err^k_{r_{gran}} = \frac{1}{\frac{n}{K}} \sum_{(x_i, y_i) \in F_k} L(y_i, dec^{-k}_{r_{gran}}(x_i)). \tag{3.61}$$

Calculations analogous to those leading to (3.60) give the equality

$$err^k_{r_{gran}} = 1 - acc^k_{r_{gran}}. \tag{3.62}$$

Now, the estimate of bias arises as a result of computing of the average value of the difference between values of $err^k_{r_{gran}}$ at r_{gran*} and r_{gran*_k} which minimizes the value of $err^k_{r_{gran}}$, i.e.

$$\hat{bias} = \frac{1}{K} \cdot \sum_{k=1}^{K} [err^k_{r_{gran*}} - err^k_{r_{gran*_k}}]. \tag{3.63}$$

Again, passing to the form of (3.63) using accuracies, we do express $\hat{bias}$ as

$$\hat{bias} = \frac{1}{K} \cdot \sum_{k=1}^{K} [acc^k_{r_{gran^*_k}} - acc^k_{r_{gran^*}}]. \tag{3.64}$$

Clearly, as $r_{gran^*_k}$ maximizes accuracy $acc^k_{r_{gran}}$ we have, cf. Tibshirani [67], Theorem 1,

$$\hat{bias} \geq 0. \tag{3.65}$$

References

1. Aggarwal, Ch.C., Hinneburg, A., Keim, D.A.: What is the nearest neighbor in high dimensional spaces? In: Proceedings of the 26th VLBD Conference, Cairo, Egypt (2000)
2. Aggarwal, Ch.C., Hinneburg, A., Keim, D.A.: On the surprising behavior of distance metrics in high dimensional space. In: Proceedings of the Eighth International Conference on Database Theory, pp. 420–434. London (2001)
3. Agrawal, R., Imieliński, T., Swami, A.: Mining association rules between sets of items in large databases. In: Proceedings of the ACM Sigmod Conference, pp. 207–216. Washington (1993)
4. Agrawal, R., Srikant, R.: Fast algorithms for mining association rules in large databases. In: Proceedings 20th International Conference on Very Large Data Bases, VLDB, pp. 487–499 (1996)
5. Arkin, H., Colton, R.R.: Statistical Methods. Barnes and Noble, New York (1970)
6. Arlot, S., Celisse, A.: A survey of cross-validation procedures for model selection. Stat. Surv. **4**, 40–79 (2010)
7. Artiemjew, P.: Classifiers from granulated data sets: concept dependent and layered granulation. In: Proceedings RSKD'07. Workshop at ECML/PKDD'07, pp. 1–9. Warsaw University Press, Warsaw (2007)
8. Artiemjew, P.: On Classification of Data by Means of Rough Mereological Granules of Objects and Rules. Lecture Notes in Artificial Intelligence, vol. 5009, pp. 221–228. Springer, Berlin (2008)
9. Artiemjew, P.: Rough Mereological Classifiers Obtained From Weak Rough Set Inclusions. Lecture Notes in Artificial Intelligence, vol. 5009, pp. 229–236. Springer, Berlin (2008)
10. Artiemjew, P.: On Strategies of Knowledge Granulation with Applications to Decision Systems (in Polish). L. Polkowski (supervisor). Ph.D. Dissertation. Polish-Japanese Institute of Information Technology, Warszawa (2009)
11. Artiemjew, P.: Selected Paradigms of Artificial Intelligence (in Polish). PJWSTK Publishers, Warszawa (2013). ISBN 978-83-63103-36-1
12. Baader, F., Calvanese, D., McGuiness, D.L., Nardi, D., Patel-Schneider, P.F. (eds.): The Description Logic Handbook: Theory, Implementation and Applications. Cambridge University Press, Cambridge (2004)
13. Bazan, J.G.: A comparison of dynamic and non-dynamic rough set methods for extracting laws from decision tables. In: Polkowski, L., Skowron, A. (eds.) Rough Sets in Knowledge Discovery, vol. 1, pp. 321–365. Physica Verlag, Heidelberg (1998)
14. Bazan, J.G., Nguyen, H.S., Nguyen, S.H., Synak, P., Wróblewski, J.: Rough set algorithms in classification problems. In: Polkowski, L., Tsumoto, S., Lin, T.Y. (eds.) Rough Set Methods and Applications. New Developments in Knowledge Discovery in Information Systems, pp. 49–88. Physica Verlag, Heidelberg (2000)
15. Blumer, A., Ehrenfeucht, A., Haussler, D., Warmuth, M.K.: Learnability and the Vapnik-Chervonenkis dimension. J. ACM **36**(4), 929–965 (1989)

16. Brown, F.M.: Boolean Reasoning: The Logic of Boolean Equations. Dover, New York (2003)
17. Bruning, J.L., Kintz, B.L.: Computational Handbook of Statistics, 4th edn. Allyn and Bacon, Columbus (1997)
18. Cendrowska, J.: PRISM, an algorithm for inducing modular rules. Int. J. Man-Mach. Stud. **27**, 349–370 (1987)
19. Chan, C.C., Grzymala-Busse, J.W.: On the two local inductive algorithms: PRISM and *LEM2*. Found. Comput. Decis. Sci. **19**(4), 185–204 (1994)
20. Chernoff, H.: A measure of asymptotic efficiency for tests of a hypothesis based on the sum of observations. Ann. Math. Stat. **23**, 493–507 (1952)
21. Clark, P., Evans, F.: Distance to nearest neighbor as a measure of spatial relationships in populations. Ecology **35**, 445–453 (1954)
22. Clark, P., Niblett, T.: The CN2 induction algorithm. Mach. Learn. **3**, 261–283 (1982)
23. Cover, T.M., Hart, P.E.: Nearest neighbor pattern classification. IEEE Trans. Inf. Theory **13**, 21–27 (1967)
24. Devroye, L., Györfi, L., Lugosi, G.: A Probabilistic Theory of Pattern Recognition. Springer, New York (1996)
25. Fix, E., Hodges Jr, J.L.: Discriminatory analysis: nonparametric discrimination: consistency properties. USAF Sch. Aviat. Med. **4**, 261–279 (1951)
26. Fix, E., Hodges Jr, J.L.: Discriminatory analysis: nonparametric discrimination: small sample performance. USAF Sch. Aviat. Med. **11**, 280–322 (1952)
27. Forrest, P.: Leibniz identity of indiscernibles. Stanford Encyclopedia of Philosophy; http://plato.stanford.edu/entries/identity-indiscernible/
28. Frege, G.: Grundgsetzte der Arithmetik II. Verlag Hermann Pohle, Jena (1903)
29. Geisser, S.: The predictive sample reuse method with applications. J. Am. Stat. Asoc. **70**, 320–328 (1975)
30. Grzymala-Busse, J.W.: LERS—a system for learning from examples based on rough sets. In: Słowiński, R. (ed.) Intelligent Decision Support: Handbook of Advances and Applications of the Rough Sets Theory, pp. 3–18. Kluwer Academic Publishers, Dordrecht (1992)
31. Grzymala-Busse, J.W., Lakshmanan, A.: *LEM2* with interval extension: an induction algorithm for numerical attributes. In: Proceedings of the Fourth International Workshop on Rough Sets, Fuzzy Sets, and Machine Discovery (RSFD'96), pp. 67–76. The University of Tokyo (1996)
32. Grzymala-Busse, J.W., Hu, M.: A Comparison of Several Approaches to Missing Attribute Values in Data Mining. Lecture Notes in Artificial Intelligence, vol. 2005, pp. 378–385. Springer, Berlin (2000)
33. Hastie, T., Tibshirani, R.: Discriminant adaptive nearest-neighbor classification. IEEE Pattern Recognit. Mach. Intell. **18**, 607–616 (1996)
34. Hastie, T., Tibshirani, R., Friedman, J.: The Elements of Statistical Learning. Springer, New York (2001)
35. Kohavi R.: A study of cross-validation and bootstrap for accuracy estimation and model selection. In: Proceedings of the Fourteenth International Joint Conference on Artificial Intelligence IJCAI'95, vol. 2(12), pp. 1137–1143. Morgan Kaufmann, San Mateo (1995)
36. Kryszkiewicz, M., Rybiński, H.: Data mining in incomplete information systems from rough set perspective. In: Polkowski, L., Tsumoto, S., Lin, T.Y. (eds.) Rough Set Methods and Applications. New Developments in Knowledge Discovery in Information Systems, pp. 567–580. Physica Verlag, Heidelberg (2000)
37. Leibniz, G.W.: Discourse on Metaphysics. In: Loemker L. (ed.): Philosophical Papers and Letters. 2nd edn., D. Reidel, Dordrecht; The Identity of Indiscernibles. In Stanford Encyclopedia of Philosophy; available at http://plato.stanford.edu/entries/identity-indiscernible/; last entered 01.04.2014 (1996)
38. Michalski, R.: Pattern recognition as rule-guided inductive inference. IEEE Trans. Pattern Anal. Mach. Intell. PAMI **2**(4), 349–361 (1990)
39. Michalski, R.S., Mozetic, I., Hong, J., Lavrac, N.: The multi-purpose incremental learning system AQ15 and its testing to three medical domains. In: Proceedings of AAAI-86, pp. 1041–1045. Morgan Kaufmann, San Mateo (1986)

40. Mitchell, T.: Machine Learning. McGraw-Hill, Englewood Cliffs (1997)
41. Moshkov, M., Skowron, A., Suraj, Z.: Irreducible descriptive sets of attributes for information systems. Transactions on Rough Sets XI. Lecture Notes in Computer Science, vol. 5946. Springer, Berlin (2010)
42. Mosteller, F., Tukey, J.W.: Data analysis, including statistics. In: Lindzey, G., Aronson, E. (eds.) Handbook of Social Psychology, vol. 2. Addison-Wesley, Massachusetts (1968)
43. Nguyen, S.H.: Regularity analysis and its applications in data mining. In: Polkowski, L., Tsumoto, S., Lin, T.Y. (eds.) Rough Set Methods and Applications. New Developments in Knowledge Discovery in Information Systems, pp. 289–378. Physica Verlag, Heidelberg (2000)
44. Novotny, M., Pawlak, Z.: Partial dependency of attributes. Bull. Pol. Acad.: Math. 36, 453–458 (1988)
45. Patrick, E., Fisher, F.: A generalized k-neighbor rule. Inf. Control 16, 128–152 (1970)
46. Pawlak, Z.: Rough sets. Int. J. Comput. Inf. Sci. 11, 341–356 (1982)
47. Pawlak, Z.: On rough dependency of attributes in information systems. Bull. Pol. Acad.: Tech. 33, 551–559 (1985)
48. Pawlak, Z.: Rough Sets: Theoretical Aspects of Reasoning About Data. Kluwer, Dordrecht (1991)
49. Pawlak, Z., Skowron, A.: A rough set approach for decision rules generation. In: Proceedings of IJCAI'93 Workshop W12. The Management of Uncertainty in AI, France; and ICS Research Report 23/93. Warsaw University of Technology, Institute of Computer Science (1993)
50. Polkowski, L.: Formal granular calculi based on rough inclusions (a feature talk). In: Proceedings of IEEE 2005 Conference on Granular Computing, GrC05, Beijing, China, July 2005. IEEE Press, pp. 57–62 (2005)
51. Polkowski, L.: A model of granular computing with applications (a feature talk). In: Proceedings of IEEE 2006 Conference on Granular Computing, GrC06, Atlanta, USA, May 2006. IEEE Press, pp. 9–16 (2006)
52. Polkowski, L.: Granulation of knowledge in decision systems: the approach based on rough inclusions. The method and its applications. In: Proceedings RSEISP 07, Warsaw, Poland, June 2007. Lecture Notes in Artificial Intelligence, vol. 4585, pp. 271–279. Springer, Berlin (2007)
53. Polkowski, L.: On the Idea of Using Granular Rough Mereological Structures in Classification of Data. Lecture Notes in Artificial Intelligence, vol. 5009. Springer, Berlin (2008)
54. Polkowski, L.: A Unified approach to granulation of knowledge and granular computing based on rough mereology: a survey. In: Pedrycz, W., Skowron, A., Kreinovich, V. (eds.) Handbook of Granular Computing. Wiley, Chichester (2008)
55. Polkowski, L.: Granulation of knowledge: similarity based approach in information and decision systems. In: Meyers, R.A. (ed.) Encyclopedia of Complexity and System Sciences. Springer, Berlin (2009). Article 00 788
56. Polkowski, L.: Data-mining and knowledge discovery: case based reasoning, nearest neighbor and rough sets. In: Meyers, R.A. (ed.) Encyclopedia of Complexity and System Sciences. Springer, Berlin (2009). Article 00 391
57. Polkowski, L., Artiemjew, P.: On Granular Rough Computing: Factoring Classifiers Through Granular Structures. Lecture Notes in Artificial Intelligence, vol. 4585, pp. 280–290. Springer, Berlin (2007)
58. Polkowski, L., Artiemjew, P.: On Granular Rough Computing With Missing Values. Lecture Notes in Artificial Intelligence, vol. 4585, pp. 271–279. Springer, Berlin (2007)
59. RSES. available at: http://www.mimuw.edu.pl/logic/rses/; last entered 01.04.2014
60. Simard, P., Le Cun, Y., Denker, J.: Efficient pattern recognition using a new transformation distance. In: Hanson, S.J., Cowan, J.D., Giles, C.L. (eds.) Advances in Neural Information Processing Systems, vol. 5, pp. 50–58. Morgan Kaufmann, San Mateo (1993)
61. Skellam, J.G.: Studies in statistical ecology. I. Spat. Pattern Biom. 39, 346–362 (1952)
62. Skowron, A.: Boolean reasoning for decision rules generation. In: Proceedings of ISMIS'93. Lecture Notes in Artificial Intelligence, vol. 689, pp. 295–305. Springer, Berlin (1993)
63. Skowron, A., Rauszer, C.: The discernibility matrices and functions in decision systems. In: Słowiński, R. (ed.) Intelligent Decision Support. Handbook of Applications and Advances of the Rough Sets Theory, pp. 311–362. Kluwer, Dordrecht (1992)

64. Stanfill, C., Waltz, D.: Toward memory-based reasoning. Commun. ACM **29**, 1213–1228 (1986)
65. Stone, C.J.: Consistent nonparametric regression. Ann. Stat. **5**(4), 595–620 (1977)
66. Stone, M.: Cross-validatory choice and assessment of statistical predictions. J. J. R. Stat. Soc. Ser. B **36**, 111–147 (1974)
67. Tibshirani, R.J., Tibshirani, R.: A bias correction for the minimum error rate in cross-validation. Ann. Appl. Stat. **3**(2), 822–829 (2009)
68. Valliant, L.G.: A theory of the learnable. Commun. ACM **27**(11), 1134–1142 (1984)
69. Vapnik, V.N.: The Nature of Statistical Learning Theory. Springer, New York (2000)
70. Vapnik, V.N., Chervonenkis, A.: On the uniform convergence of relative frequencies of events to their probabilities. Theory Probab. Appl. **16**, 264–280 (1971)
71. Wilson, D.R., Martinez, T.R.: Improved heterogeneous distance functions. J. Artif. Intell. Res. **6**, 1–34 (1997)
72. Wojna, A.: Analogy-based reasoning in classifier construction. Transactions on Rough Sets Vol. IV. Lecture Notes in Computer Science, vol. 3700. Springer, Berlin (2005)
73. Wróblewski, J.: Adaptive aspects of combining approximation spaces. In: Pal, S.K., Polkowski, L., Skowron, A. (eds.) Rough Neural Computing. Techniques for Computing with Words, pp. 139–156. Springer, Berlin (2004)
74. Yu, H., Wang, G., Lan, F.: Solving the attribute reduction problem with ant colony optimization. Transactions on Rough Sets XIII. Lecture Notes in Computer Science, vol. 6499. Springer, Berlin (2001)

Chapter 4
Methodologies for Granular Reflections

There is no uncertain language in which to discuss uncertainty
[Zdzisław Pawlak]

In this chapter, we intend to introduce the reader into the realm of granular reflections induced from decision systems by means of similarity relations being rough inclusions or their weaker variants. The general scheme outlined in Chap. 3, can be ramified into a plethora of variants of which we give a survey. The detailed generic results which are culminating in this comprehensive survey are recorded in Artiemjew [1–5], Polkowski [12–15], and, Polkowski and Artiemjew [16–24].

We begin with the granulation proper which leads from decision systems to their granular reflections.

4.1 Granules: Granular Reflections

We begin with a rough inclusion μ on the universe U of a decision system $D = (U, A, d)$. We introduce the parameter r_{gran}, the *granulation radius* with values $0, \frac{1}{|A|}, \frac{2}{|A|}, \ldots, 1$. For each thing $u \in U$, and $r = r_{gran}$, the *standard granule* $g(u, r, \mu)$, *of radius r about u*, is defined as

$$g(u, r, \mu) \text{ is } \{v \in U : \mu(v, u, r)\}. \tag{4.1}$$

The algorithm GRANULE_FORMATION (D, μ, r) gives a pseudo-code for making granules.
GRANULE_FORMATION (D, μ, r)
Input: a decision system $D = (U, A, d)$, a granulation radius r

© Springer International Publishing Switzerland 2015
L. Polkowski and P. Artiemjew, *Granular Computing in Decision Approximation*,
Intelligent Systems Reference Library 77, DOI 10.1007/978-3-319-12880-1_4

Output: the granule set $\{g(u, r, \mu) : u \in U\}$

1. **for** each $u \in U$ **do**
2.   **for** each $v \in U$ **do**
3.   **if** $\mu(v, u, r)$ **then**
4. $g(u, r, \mu) \leftarrow g(u, r, \mu) \cup \{v\}$
5.   **end if**
6.   **end for**
7. **end for**
8. **return** $\{g(u, r, \mu) : u \in U\}$

The form of μ can be chosen from among a few possibilities outlined already in Chaps. 2 and 3. We recapitulate them here for the purpose of a further discussion.

4.1.1 The Standard Rough Inclusion

It exploits the set

$$IND(u, v) = \{a \in A : a(u) = a(v)\}, \tag{4.2}$$

by means of the formula

$$\mu(v, u, r) \Leftrightarrow \frac{|Ind(u, v)|}{|A|} \geq r \tag{4.3}$$

It follows that this rough inclusion extends the indiscernibility relation to a degree of r.

4.1.2 ε-Modification of the Standard Rough Inclusion

Given a parameter ε valued in the unit interval $[0, 1]$, we define the set

$$Ind_\varepsilon(u, v) = \{a \in A : dist(a(u), a(v)) \leq \varepsilon\}, \tag{4.4}$$

and, we set

$$\mu_\varepsilon(v, u, r) \Leftrightarrow \frac{|Ind_\varepsilon(u, v)|}{|A|} \geq r \tag{4.5}$$

The radius r is called in this case the *catch radius*; later on, it is discerned from the granulation radius.

4.1.3 Residual Rough Inclusions

Given a continuous t-norm T, the residual implication $\rightarrow_T$ induces a rough inclusion μ_T on the unit square $[0, 1] \times [0, 1]$. To transfer it to the realm of decision systems, one defines functions

$$dis_\varepsilon(u, v) = \frac{|\{a \in A : a(u) \neq a(v)\}|}{|A|}, \tag{4.6}$$

and,

$$ind_\varepsilon(u, v) = 1 - dis_\varepsilon(u, v) \tag{4.7}$$

A similarity μ_T is defined as

$$\mu_T(v, u, r) \text{ if and only if } dis_\varepsilon(v, u) \rightarrow_T ind_\varepsilon(v, u). \tag{4.8}$$

4.1.4 Metrics for Rough Inclusions

The parameter $dist$ in formula (4.4) can be filled in a few ways with distinct specific metrics, of which we recall those of common interest in machine learning. All metrics can be normalized by passing for each attribute $a \in A$ from $dist(a(u), a(v))$ to

$$dist'(a(u), a(v)) = \frac{dist(a(u), a(v))}{dist(max_a, min_a)}, \tag{4.9}$$

where max_a, min_a are, respectively, the maximum and the minimum of values of the attribute a.

4.1.4.1 The Hamming Metric

It is defined on things in a decision system as

$$d_H(u, v) = |\{a \in A : a(u) \neq a(v)\}|. \tag{4.10}$$

4.1.4.2 The Epsilon Hamming Metric

A variation on the theme of the Hamming metrics, given ε, it is defined as

$$d_{H,\varepsilon}(u, v) = |\{a \in A : dist(a(u), a(v)) \geq \varepsilon\}|. \tag{4.11}$$

4.1.4.3 The Minkowski Metrics L^p

For $p = 1, 2, \ldots, \infty$, it is defined as

$$L^p(u, v) = \sum_a [dist^p(a(u), a(v))]^{\frac{1}{p}}. \tag{4.12}$$

In case $p = 2$ it is called the *Euclidean metrics*, in case $p = 1$ it is known as the *Manhattan metrics*, and in case $p = \infty$ it is given by the equivalent of the *max metric, Chebyshev metric*

$$L^\infty(u, v) = max_a\{dist(a(u), a(v)) : a \in A\}. \tag{4.13}$$

4.1.4.4 The Canberra Metric

In the normalized form already, it is given as

$$d_C(u, v) = \sum_a \frac{|a(u) - a(v)|}{a(u) + a(v)}, \tag{4.14}$$

where, all values of attributes are normalized to the unit interval.

4.1.4.5 Statistical Metrics

These metrics use statistical parameters of the mean, the standard deviation, and, correlation. For each thing u, one defines the mean $\bar{u} = \frac{1}{m} \cdot \sum_a a(u)$ and then the *Pearson correlation coefficient* is defined as

$$d_P(u, v) = E\left[\frac{\sum_a [a(u) - \bar{u}] \cdot [a(v) - \bar{v}]}{[\sum_a [a(u) - \bar{u}]^2]^{\frac{1}{2}} \cdot [\sum_a [a(v) - \bar{v}]^2]^{\frac{1}{2}}} \right]. \tag{4.15}$$

The χ^2 *(chi square) metric* is a variant on Canberra metric,

$$d_\chi(u, v) = \sum_a \frac{[a(u) - a(v)]^2}{a(u) + a(v)}. \tag{4.16}$$

The *Mahalanobis metric* uses the covariance matrix $Cov = [\sigma_{uv}]_{u,v}$, where $\sigma_{uv} = \frac{1}{m} \sum_a [a(u) - \bar{u}] \cdot [a(v) - \bar{v}]$. Then

$$d_M(u, v) = \{[Inf(u) - Inf(v)]Cov^{-1}[Inf(u) - Inf(v)]^T\}^{\frac{1}{2}}, \tag{4.17}$$

Table 4.1 A ranking of metrics by accuracy; Estimation: Leave One Out (LOO); Classifier: k-NN global; k = 1–10; Wisconsin Breast Cancer (Diagnostic)

Metrics	1	2	3	4	5	6	7	8	9	10	Mean
Canberra	0.956	0.954	0.960	0.960	0.961	0.960	0.963	0.960	0.961	0.958	0.959
Normalized	0.951	0.963	0.972	0.970	0.970	0.965	0.970	0.970	0.970	0.967	0.967
Chisquare	0.937	0.928	0.944	0.935	0.947	0.940	0.946	0.940	0.944	0.944	0.941
Hamming	0.6	0.68	0.657	0.699	0.678	0.710	0.696	0.692	0.657	0.666	0.674
Manhattan	0.930	0.930	0.935	0.933	0.937	0.935	0.935	0.935	0.942	0.935	0.935
Euklidean	0.916	0.923	0.926	0.926	0.933	0.930	0.931	0.930	0.933	0.930	0.928
Cosinus	0.912	0.905	0.921	0.924	0.923	0.921	0.924	0.924	0.928	0.928	0.921
Minkowski p = 3	0.916	0.919	0.923	0.921	0.928	0.924	0.930	0.930	0.931	0.926	0.925
Hamming eps = 0.1	0.951	0.953	0.961	0.958	0.965	0.963	0.961	0.960	0.958	0.956	0.959
Pearson	0.909	0.903	0.921	0.923	0.928	0.924	0.924	0.928	0.930	0.931	0.922
Chebyshev	0.912	0.917	0.916	0.919	0.926	0.921	0.926	0.924	0.928	0.924	0.921

where, $Inf(u) = [a(u) : a \in A]$ is the ordered set called the *information vector* of u (Table 4.1).

4.1.4.6 The Normalized Metric

Again, it is given by

$$d_N(u, v) = \sum_a \frac{|a(u) - a(v)|}{max_a - min_a}. \tag{4.18}$$

4.1.4.7 The Cosinus Metric

For normalized information vectors it is given as the scalar product of them.

4.1.5 A Ranking of Metrics

A ranking of metrics in classification is shown below for the k-NN classifier on the Wisconsin Breast Cancer (Diagnostic) (WBCD) [26] data set. One may realize that this ranking depends possibly on the data set used (Figs. 4.1 and 4.2).

Results of the parallel experiment for normalized data are given in Table 4.2.

In both cases, for normalized as well as for non-normalized data, the Canberra metric and the normalized metric stand out as the best with respect to accuracy of the classifier based on them.

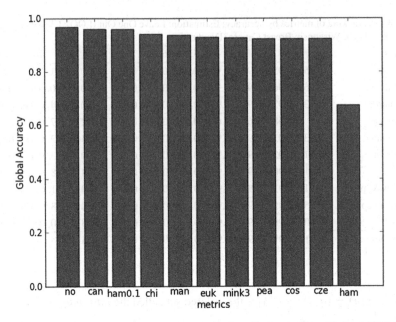

Fig. 4.1 A ranking of metrics by mean accuracy for $k \in [1, 2, \ldots, 10]$, with the k-NN global classifier, data set: Wisconsin Breast Cancer (Diagnostic); can = Canberra, no = normalized, chi = Chisquare, ham = Hamming, man = Manhattan, euk = Euclidean, cos = Cosinus, mik3 = Minkowski p = 3, ham0.1 = Hamming eps = 0.1, pea = Pearson, che = chebyshev

4.2 Granular Coverings

For a decision system $D = (U, A, d)$, a rough (weak) inclusion μ, and a granulation radius r, we can use the procedure GRANULE_FORMATION(D, μ, r) to obtain the collection of granules $GRAN(U, \mu, r) = \{g(u, r, \mu) : u \in U\}$. The complexity of producing it is $O(|U|^2 \cdot |A|)$. The next step is to select a covering $COV(U, \mu, r)$ from granules in $GRAN(U, \mu, r)$.

This choice can be done in a few ways, the detailed analysis and results in terms of the impact on the classifier quality are reported in Chap. 5, here, we only advertise few basic possibilities:

1. Random choice;
2. Order-preserving selection;
3. Granules passing the maximal number of new things;
4. Granules of maximal size.

We show as an example the procedure for a random choice of covering (Figs. 4.3 and 4.4).

RANDOM_COVERING (D, μ, r)

Input: $GRAN(U, \mu, r)$

procedure *random*

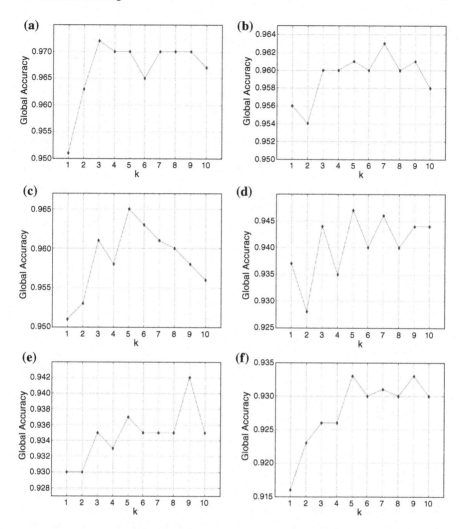

Fig. 4.2 Accuracy for metrics: **a** Normalized, **b** Canberra, **c** Hamming eps = 0.1, **d** Chisquare, **e** Manhattan, **f** Euclidean; global k-NN classifier, result for $k \in [1, 2, \ldots, 10]$, data set: Wisconsin Breast Cancer (Diagnostic)

Output: covering $RANDOM\_COV(U, \mu, r)$
Variables: *cover, temp*

1. $cover \leftarrow \emptyset$
2. $temp \leftarrow GRAN(U, \mu, r)$
3. **loop**
4. $g \leftarrow random(temp)$
5. **if** $g \subseteq cover$ **then**

Table 4.2 A ranking of metrics by accuracy global, for normalized data; validation Leave One Out (LOO); classifier k-NN global, $k = 1, 2, \ldots, 10$; data set: Wisconsin Breast Cancer (Diagnostic)

Metrics k	1	2	3	4	5	6	7	8	9	10	*Mean*
Canberra	0.938	0.940	0.938	0.937	0.938	0.946	0.953	0.942	0.944	0.940	0.942
Normalized	0.961	0.958	0.965	0.963	0.965	0.956	0.961	0.953	0.954	0.947	0.958
Chisquare	0.923	0.919	0.923	0.926	0.929	0.923	0.931	0.923	0.923	0.919	0.924
Hamming	0.420	0.420	0.395	0.395	0.395	0.395	0.395	0.395	0.395	0.395	0.4
Manhattan	0.895	0.909	0.928	0.930	0.932	0.928	0.931	0.926	0.928	0.928	0.924
Euklides	0.912	0.905	0.921	0.924	0.923	0.921	0.924	0.924	0.928	0.928	0.921
Cosinus	0.912	0.905	0.921	0.924	0.923	0.921	0.924	0.924	0.928	0.928	0.921
Minkowski p = 3	0.909	0.910	0.920	0.920	0.924	0.919	0.921	0.926	0.928	0.928	0.921
Hamming eps = 0.1	0.931	0.944	0.944	0.935	0.947	0.953	0.953	0.953	0.954	0.949	0.946
Pearson	0.909	0.903	0.921	0.923	0.928	0.924	0.924	0.928	0.929	0.931	0.922
Czebyshev	0.912	0.912	0.924	0.921	0.926	0.919	0.926	0.926	0.924	0.928	0.922

6.　　　　$temp \leftarrow temp - g$
7.　　**else**
8.　　　　$cover \leftarrow cover \cup g$
9.　　**end if**
10.　　**if** $cover = U$ **then**
11.　　　　**break**
12.　　**end if**
13.　**end loop**

4.3 Granular Reflections

Once the granular covering is selected, the idea is to represent information vectors in each granule by one information vector. The strategy for producing it can be the *majority voting MV*, so for each granule $g \in COV(U, \mu, r)$, the information vector $Inf(g)$ is given by the formula

$$Inf(g) = \{MV(\{a(u) : u \in g\}) : a \in A \cup \{d\}\} \tag{4.19}$$

The granular reflection of the decision system $D = (U, A, d)$ is the decision system $GRANREF(D) = (COV(U, \mu, r), \{Inf(g) : g \in COV(U, \mu, r)\})$.

Example 4 We give a simple example on granular reflections. In Table 4.3, a decision system is given. The granulation radius $r_{gran} \in \{0, 0.25, 0.5, 0.75, 1\}$ (Fig. 4.5) (Table 4.4).

In case $r_{gran} = 0$, the granule is U, for $r_{gran} = 0.25$, granules computed with $\mu(v, u, r) \Leftrightarrow \frac{|IND(v,u)|}{|A|} \geq r$ are

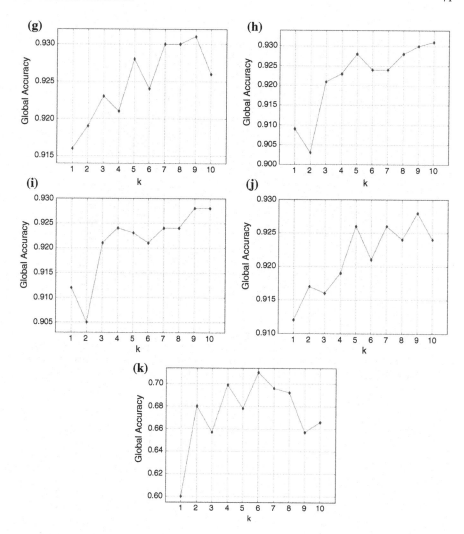

Fig. 4.3 Accuracy for metrics: **g** Minkowski p = 3, **h** Pearson, **i** Cosinus, **j** Czebyshev, **k** Hamming; global k-NN classifier, result for $k \in [1, 2, \ldots, 10]$, data set: Wisconsin Breast Cancer (Diagnostic)

$$g_{0.25}(u_1) = \{u_1, u_6\}$$
$$g_{0.25}(u_2) = \{u_2, u_4\}$$
$$g_{0.25}(u_3) = \{u_3, u_5\}$$
$$g_{0.25}(u_4) = \{u_2, u_4\}$$
$$g_{0.25}(u_5) = \{u_3, u_5, u_6\}$$
$$g_{0.25}(u_6) = \{u_1, u_5, u_6\}$$

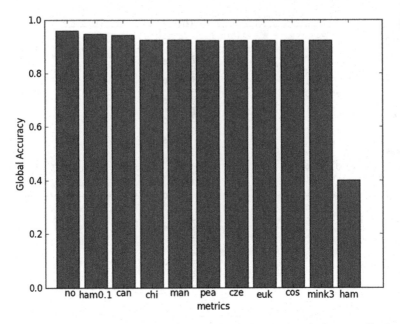

Fig. 4.4 A ranking of metrics by mean accuracy for $k \in [1, 2, \ldots, 10]$ for normalized, with the k-NN global classifier, data set: Wisconsin Breast Cancer (Diagnostic); can = Canberra, no = normalized, chi = Chisquare, ham = Hamming, man = Manhattan, euk = Euclidean, cos = Cosinus, mik3 = Minkowski p = 3, ham0.1 = Hamming eps = 0.1, pea = Pearson, che = chebyshev

Table 4.3 The decision system $D = (U, A, d)$

	a_1	a_2	a_3	a_4	d
u_1	2	1	2	1	1
u_2	3	2	3	3	1
u_3	1	5	1	2	1
u_4	6	2	3	8	2
u_5	4	5	8	6	2
u_6	5	1	8	1	2

The process of granule computing can be visualized with help of the *granular indiscernibility matrix* $[c_{ij}]_{(i,j=1)|U|}$, where

$$
c_{ij} = \begin{cases} 1, & \text{if } \frac{IND(u_i, u_j)}{|A|} \geq r_{gran} \\ 0, & \text{else} \end{cases}
$$

This matrix for $r_{gran} = 0.25$, is given in Table 4.4 (by symmetry we give its triangular form). Reading the matrix line-wise, we read granules off (Fig. 4.6). For $r_{gran} = 0.5$, granules are given in Table 4.5:

$$g_{0.5}(u_1) = \{u_1, u_6\}.$$
$$g_{0.5}(u_2) = \{u_2, u_4\}.$$
$$g_{0.5}(u_3) = \{u_3\}.$$
$$g_{0.5}(u_4) = \{u_2, u_4\}.$$
$$g_{0.5}(u_5) = \{u_5\}.$$
$$g_{0.5}(u_6) = \{u_1, u_6\}.$$

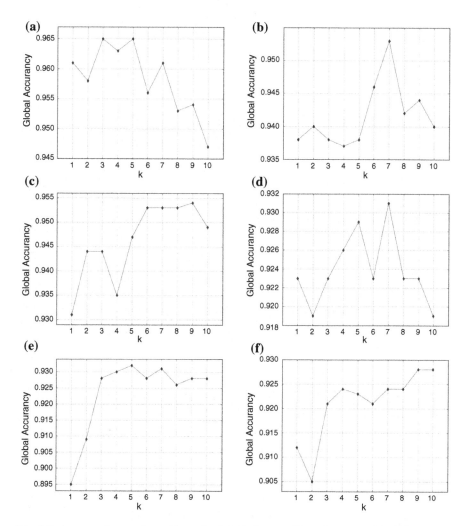

Fig. 4.5 Accuracy for metrics: **a** Normalized, **b** Canberra, **c** Hamming eps = 0.1, **d** Chisquare, **e** Manhattan, **f** Euclidean and Cosinus, **g** Minkowski p = 3, **h** Pearson, **i** Czebyshev, **j** Hamming; global k-NN classifier, result for $k \in [1, 2, \ldots, 10]$, data set: Wisconsin Breast Cancer (Diagnostic)

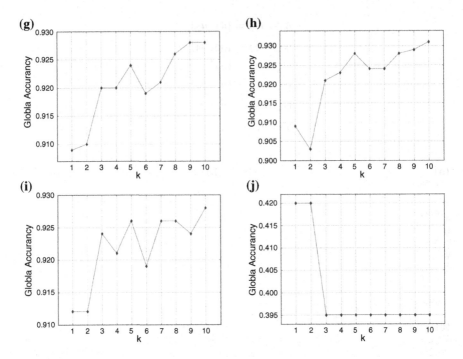

Fig. 4.5 (continued)

Table 4.4 Granular indiscernibility matrix for $r_{gran} = 0.25$		u_1	u_2	u_3	u_4	u_5	u_6
	u_1	1	0	0	0	0	1
	u_2	x	1	0	1	0	0
	u_3	x	x	1	0	1	0
	u_4	x	x	x	1	0	0
	u_5	x	x	x	x	1	1
	u_6	x	x	x	x	x	1

The granular indiscernibility matrix in this case is:

For $r_{gran} = 0.75$ and $r_{gran} = 1$, granules are singletons containing their centers only; granular indiscernibility matrices in those cases are diagonal (Table 4.5).

The granular reflection computed for $r_{gran} = 0.25$ by majority voting with random tie resolution is for $C = COV(U, \mu, 0.25) = \{g_{.25}(u_1), g_{.25}(u_2), g_{.25}(u_3)\}$:

$$(C, \{[2, 1, 2, 1, 1], [3, 2, 3, 3, 1], [1, 5, 1, 2, 1]\}) \qquad (4.20)$$

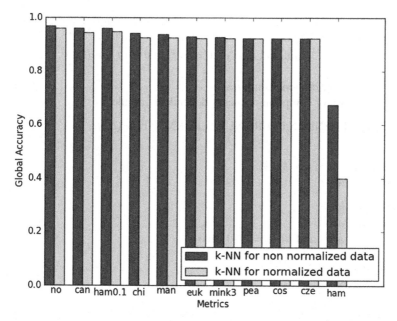

Fig. 4.6 A ranking of metrics before and after normalization of data by mean accuracy; $k \in [1, 2, \ldots, 10]$, classifier k-NN global; data set: Wisconsin Breast Cancer (Diagnostic); can = Canberra, no = normalized, chi = Chisquare, ham = Hamming, man = Manhattan, euk = Euclidean, cos = Cosinus, mik3 = Minkowski p = 3, ham0.1 = Hamming eps = 0.1, pea = Pearson, che = chebyshev

Table 4.5 Granular indiscernibility matrix for $r_{gran} = 0.5$

	u_1	u_2	u_3	u_4	u_5	u_6
u_1	1	0	0	0	0	1
u_2	0	1	0	1	0	0
u_3	0	0	1	0	0	0
u_4	0	1	0	1	0	0
u_5	0	0	0	0	1	0
u_6	1	0	0	0	0	1

4.4 Ramifications of Granulation: Concept-Dependent and Layered

Granules described in the preceding section are unconstrained in the sense that no restrictions are placed upon them. One possible and natural restriction is to enclose granules within decision classes, i.e., within concepts. One obtains then *concept-dependent granules* about which now.

A concept-dependent granule $g_r^{cd}(u)$ of the radius r about u is defined as follows:

$$v \in g_r^{cd}(u) \text{ if and only if } \mu(v, u, r) \text{ and } (d(u) = d(v)) \qquad (4.21)$$

for a given rough (weak) inclusion μ.

The procedure for computing concept-dependent granules is

CONCEPT_DEPENDENT_GRANULE_FORMATION (D, μ, r)

Input:
decision system $D = (U, A, d)$;
granulation radius r

1. **for each** u in U
2.   **for each** v in U
3.   **if** $\mu(v, u, r)$ **and** $d(u) = d(v)$ **then**
4. $g_r^{cd}(u) \leftarrow g_r^{cd}(u) \cup \{v\}$
5.   **end if**
6.   **end for**
7. **end for**
8. **return** $\{g_r^{cd}(u) : u \in U\}$

Example 5 For the decision system in Table 4.6, we find concept-dependent granules (Table 4.7).

For the granulation radius $r_{gran} = 0.25$, the granular concept-dependent indiscernibility matrix (gcdm, for short) is, hence, gcdm's in this case are

$$g_{0.25}^{cd}(u_1) = \{u_1\};$$

Table 4.6 Decision system $D = (U, A, d)$

	a_1	a_2	a_3	a_4	d
u_1	2	1	2	1	1
u_2	3	2	3	3	1
u_3	1	5	1	2	1
u_4	6	2	3	8	2
u_5	4	5	8	6	2
u_6	5	1	8	1	2

Table 4.7 gcdm for $r_{gran} = 0.25$

	u_1	u_2	u_3	u_4	u_5	u_6
u_1	1	0	0	x	x	x
u_2	0	1	0	x	x	x
u_3	0	0	1	x	x	x
u_4	x	x	x	1	0	0
u_5	x	x	x	0	1	1
u_6	x	x	x	0	1	1

$$g_{0.25}^{cd}(u_2) = \{u_2\};$$
$$g_{0.25}^{cd}(u_3) = \{u_3\};$$
$$g_{0.25}^{cd}(u_4) = \{u_4\};$$
$$g_{0.25}^{cd}(u_5) = \{u_5, u_6\};$$
$$g_{0.25}^{cd}(u_6) = \{u_5, u_6\}.$$

For $r_{gran} = 0.5, 0.75, \ldots, 1.0$, granular concept-dependent decision systems are identical to the original decision system in Table 4.6.

Iteration of the granulation process leads to the *layered granulation* carried out until granules stabilize after a finite number of steps.

LAYERED_GRANULE_FORMATION (D, μ, r)
layer loop:
Input:
decision system $D = (U, A, d)$
granulation radius r

1. **for each** u in U
2.   **for each** v in U
3.   **if** $\mu(v, u, r)$ **then**
4. $g_r(u) \leftarrow g_r(u) \cup \{v\}$
5.   **end if**
6.   **end if**
7.   **end for**
8. **end for**
9. **return** $\{g_r(u) : u \in U\}$

4.5 Granular Approximations to Decision Function

Granular reflections serve as vehicles for constructions of granular decision functions which approximate decision functions. We consider a few variations on the theme of how to construct them. We denote descriptors in a concise manner with the formula $a(u)$. Tables $min_{training(rule) set}a, max_{training(rule) set}a$ contain minimal, resp., maximal values of the attribute a over the training set. Outliers are values of attributes in the test set which fall outside the interval $[min_{training set}a, max_{training set}a]$. They are projected onto this interval by the mapping p, defined as follows:

$$\text{if } a(v) < min_{training set}a \text{ then } p(a(v)) = min_{training set}a, \tag{4.22}$$

and,

$$\text{if } a(v) > max_{training set}a \text{ then } p(a(v)) = max_{training set}a \tag{4.23}$$

Symbols $d(v)$, $d(rule)$, $d(g)$, mean, resp., decision value on a thing, a rule, a granule. The concept size $|\{v : d(v)\}|$ is denoted with a shortcut $|d(v)|$.

The quality of approximation is measured by parameters of *accuracy* as well as *coverage*.

We apply the following classification algorithms.

Algorithm_1_v1: Classification by the training set;

Algorithm_1_v2: Classification by the training set with the catch radius;

Algorithm_2_v1: Classification by granules of the training set;

Algorithm_2_v2: Classification by granules of the training set with the catch radius;

Algorthm_3_v1: Classification by decision rules induced from the training set by the *exhaustive method*;

Algorithm_3_v2: Classification by the decision rules *exhaustive* with the catch radius;

Algorithm_4_v1: Classification by the decision rules *exhaustive* from the granular reflection;

Algorithm_4_v2: Classification by the decision rules *exhaustive* from the granular reflection with the catch radius;

Algorithm_5_v1: Classification by things in the decision system weighted by the minimum residual implication;

Algorithm_5_v2: Classification by the decision rules *exhaustive* from the training set weighted by the minimum residual implication;

Algorithm_5_v3: Classification by granules of the training set weighted by the minimum residual implication;

Algorithm_5_v4: Classification by the decision rules *exhaustive* from the granular reflection weighted by the minimum residual implication;

Algorithm_6_v1: Classification by things in the decision system weighted by the product residual implication;

Algorithm_6_v2: Classification by the decision rules *exhaustive* from the training set weighted by the product residual implication;

Algorithm_6_v3: Classification by granules of the training set weighted by the product residual implication;

Algorithm_6_v4: Classification by the decision rules *exhaustive* from the granular reflection weighted by the product residual implication;

Algorithm_7_v1: Classification by things in the decision system weighted by the Łukasiewicz residual implication;

Algorithm_7_v2: Classification by the decision rules *exhaustive* from the training set weighted by the Łukasiewicz residual implication;

Algorithm_7_v3: Classification by granules of the training set weighted by the Łukasiewicz residual implication;

Algorithm_7_v4: Classification by the decision rules *exhaustive* from the granular reflection weighted by the Łukasiewicz residual implication;

Algorithm_8_v1_variants 1,2,3,4,5: Classification is by weighted voting of things in the training set.

The variant to the above called *the ε granulation* rests in replacing the condition $a(u) = a(v)$ with the condition $|a(u) - a(v)| \le \varepsilon$.

The results are validated by means of k-fold cross validation, cf., the discussion in Chap. 3.

We give now a detailed description of those algorithms. In the following procedures, values of $param(d(v))$ are normalized by dividing by cardinality of the decision class $[d(v)]$; the same concerns values of $param(d(r))$ for rules, and, $param(\overline{d}(g))$ for granules, and, in case $\overline{d}(g)$ is not any value of decision on training objects, the default value of $param(\overline{d}(g))$ is 0. The symbol $|r|$ denotes the *length* of the rule r, i.e., the number of attributes from the set A in descriptors in the premise of the rule r. All values of *param* are initialized with the value 0.

Algorithtm_1_v1

Procedure

1. Input the training set and the test set.
2. Create tables $max_{training\ set}a$, $min_{training\ set}a$ for the training set for all $a \in A$.
3. Project outliers in the test set for each attribute a onto $max_{training\ set}a$, $min_{training\ set}a$.
4. Set the value of the parameter $\varepsilon \in \{0, 0.01, \ldots, 1.0\}$.
5. Classify the given test thing u: 1. For each training thing v check the condition $IND_{\varepsilon,a}(u, v)$:

$$\frac{|a(u) - a(v)|}{|max_{training\ set}a - min_{training\ set}a|} \le \varepsilon, \text{ for each } a \in A.$$

2. If v satisfies the condition $IND_{\varepsilon,a}(u, v)$ for a given $a \in A$, then $param(d(v))$++.
3. Assign to u the decision value $d(u) = d(v^*)$, where, $param(d(v^*)) = max\{param(d(v))\}$; in case of a tie choose at random (Fig. 4.7).

Algorithm_1_v2

Procedure

1. Input the training set and the test set.
2. Create tables $max_{training\ set}a$, $min_{training\ set}a$ for the training set for all $a \in A$.
3. Project outliers in the test set for each attribute $a \in A$ onto $max_{training\ set}a$, $min_{training\ set}a$.

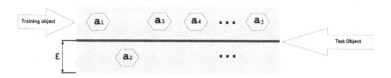

Fig. 4.7 ε-neighborhood of a test thing

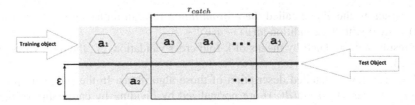

Fig. 4.8 Relative ε, r_{catch}–neighborhood of a test thing

4. Set the value of the parameter $\varepsilon \in \{0, 0.01, \ldots, 1.0\}$.
5. Set the value of the parameter *catch radius* $r_{catch} \in (0, 1)$.
6. Classify the given test thing u: 1. For each decision value w of decision d on training things, and, for each training thing v having decision value w, form the set $IND_\varepsilon(v, u) = \{a \in A : \frac{|a(u)-a(v)|}{|max_{training\ set}a - min_{training\ set}a|} \leq \varepsilon\}$.
   2. Check the condition $\frac{|IND_\varepsilon(v,u)|}{|A|} \geq r_{catch}$.
   3. If v satisfies the condition then $param(d(v)) + +$.
   4. Assign to u the decision value $d(u) = d(v^*)$, where, $param(d(v^*)) = max\{param(d(v))\}$; in case of a tie choose at random (Fig. 4.8).

Algorithm_2_v1

Procedure

1. Input the training set and the test set.
2. Create the granular reflection of the training set with respect to a given radius r_{gran}; in particular, for each granule g in it, collect attribute values for g in the set $Inf(g) = \{\bar{a}(g) : a \in A\}$.
3. Create tables $max_{training\ set}a$, $min_{training\ set}a$ for the training set for all $a \in A$.
4. Project outliers in the test set for each attribute a onto $max_{training\ set}a$, $min_{training\ set}a$.
5. Set the value of the parameter $\varepsilon \in \{0, 0.01, \ldots, 1.0\}$.
6. Classify the given test thing u: 1. for each granule g check the condition $IND_{\varepsilon,a}(u, g)$:

$$\frac{|a(u) - \bar{a}(g)|}{|max_{training\ set}a - min_{training\ set}a|} \leq \varepsilon, \text{ for each } a \in A.$$

2. If the granule g satisfies the condition $IND_{\varepsilon,a}(u, g)$, then $param(\bar{d}(g)) + +$.
3. Assign to u the decision value $d(u) = \bar{d}(g^*)$, where, $param(\bar{d}(g^*)) = max\{param(\bar{d}(g))\}$; in case of a tie choose at random (Fig. 4.9).

Algorithm_2_v2

Procedure

1. Input the training set and the test set.

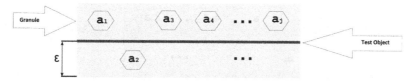

Fig. 4.9 ε-neighborhood in the set of granules

2. Create the granular reflection of the training set with respect to a given radius $r_{gran} \in (0, 1)$; in particular, for each granule g in it, collect attribute values for g in the set $Inf(g) = \{\bar{a}(g) : a \in A\}$.
3. Create tables $max_{training\ set}a$, $min_{training\ set}a$ for the training set for all $a \in A$.
4. Project outliers in the test set for each attribute a onto $max_{training\ set}a$, $min_{training\ set}a$.
5. Set the value of the parameter $\varepsilon \in \{0, 0.01, \ldots, 1.0\}$.
6. Set the value of the parameter $catch\ radius\ r_{catch} \in (0, 1)$.
7. Classify the given test thing u: 1. For each decision value w of d, for each granule g with value of $\bar{d}(g)) = w$, check the condition $IND_{\varepsilon, r_{catch}}(u, g)$:

$$|\{a \in A : \frac{|a(u) - \bar{a}(g)|}{|max_{training\ set}a - min_{training\ set}a|} \leq \varepsilon\}| \geq r_{catch}.$$

2. If the granule g satisfies the condition then $param(\bar{d}(g)) ++$.
3. Assign to u the decision value $d(u) = \bar{d}(g^*)$, where, $param(\bar{d}(g^*)) = max\{param(\bar{d}(g))\}$; in case of a tie choose at random (Fig. 4.10).

Algorithm_3_v1

Procedure

1. Input the training set and the test set.
2. Compute decision rules *exhaustive* from the training set. For each rule r, and each $a \in A$, find values of a in descriptors in the premise of r.
3. Create tables $max_{rule\ set}a$, $min_{rule\ set}a$ for the decision rules in item 2 for all $a \in A$.
4. Project outliers in the test set for each attribute a onto $max_{rule\ set}a$, $min_{rule\ set}a$.
5. Set the value of the parameter $\varepsilon \in \{0, 0.01, \ldots, 1.0\}$.
6. Classify the given test thing u: 1. For each decision value w of d, and, for each decision rule r with $d(r) = w$, check the condition $IND_{\varepsilon,a}(u, r)$:

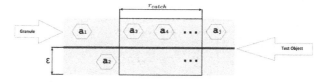

Fig. 4.10 ε, r_{catch}-neighborhood in the granular reflection

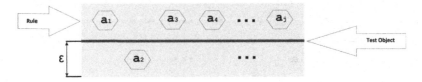

Fig. 4.11 A rule neighborhood of a test thing

$$\frac{|a(u) - a(r)|}{max_{rule\;set}a - min_{rule\;set}a} \leq \varepsilon$$

for each $a \in A$. 2. If the decision rule r satisfies the condition $IND_{\varepsilon,a}(u, r)$ for $|r|$ attributes in A, then $param(d(r)) \leftarrow param(d(r)) + support(r)$. 3. Assign to u the decision value $d(u) = \overline{d}(r^*)$, where, $param(\overline{d}(r^*)) = max\{param(\overline{d}(r)\}$; in case of a tie choose at random (Fig. 4.11).

For a rule r, the *length* $|r|$ of the rule r is the number of descriptors over A in the premise of the rule r.

Algorithm_3_v2

Procedure

1. Input the training set and the test set.
2. Compute decision rules *exhaustive* from the training set.
3. Create tables $max_{rule\;set}a$, $min_{rule\;set}a$ for the decision rules in pt. 1 for all $a \in A$.
4. Project outliers in the test set for each attribute a onto $max_{training\;set}a$, $min_{training\;set}a$.
5. Set the value of the parameter $\varepsilon \in \{0, 0.01, \ldots, 1.0\}$.
6. Set the value of the parameter catch radius r_{catch}.
7. Classify the given test thing u: 1. For each value w of decision d, and, for each decision rule r with $d(r) = w$, check for each $a \in A$ the condition $IND_{\varepsilon,r_{catch},a}(u, r)$:

$$|\{a \in A : \frac{|a(u) - a(r)|}{|max_{training\;set}a - min_{training\;set}a|} \leq \varepsilon\}| \geq r_{catch}.$$

2. If the decision rule r satisfies the condition, then $param(d(r)) \leftarrow param(d(r)) + support(r)$.
4. Assign the value $d(u) = d^*$, where, $d^* = argmax_d param(d)$ (Fig. 4.12).

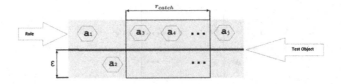

Fig. 4.12 A rule ε–neighborhood of a test thing for r_{catch}

Algorithm_4_v1

Procedure

1. Input the training set and the test set.
2. Compute the granular reflection of the training set for a given radius r_{gran}.
3. Compute decision rules *exhaustive* from the granular reflection of the training set.
4. Create tables $max_{training\ set}a$, $min_{training\ set}a$ for the training set for all $a \in A$.
5. Project outliers in the test set for each attribute a onto $max_{training\ set}a$, $min_{training\ set}a$.
6. Set the value of the parameter $\varepsilon \in \{0, 0.01, \ldots, 1.0\}$.
7. Classify the given test thing u: 1. For each decision rule r determine the set

$$IND_\varepsilon(u, r) = \{a \in A : \frac{|a(u) - a(r)|}{|max_{training\ set}a - min_{training\ set}a|} \leq \varepsilon\}.$$

2. Check if the decision rule r satisfies the condition

$$|IND_\varepsilon(u, r)| = |r|.$$

3. If the decision rule r satisfies the condition, then $param(d(r)) \leftarrow param(d(r)) + support(r)$.
4. Assign the value $d(u) = d^*$ where $d^* = argmax_d param(d)$ (Fig. 4.13).

Algorithm_4_v2

Procedure

1. Input the training set and the test set.
2. Compute the granular reflection of the training set for a given radius r_{gran}.
3. Compute decision rules *exhaustive* from the granular reflection of the training set.
4. Create tables $max_{training\ set}a$, $min_{training\ set}a$ for each $a \in A$.
5. Project outliers in the test set for each attribute a onto $max_{training\ set}a$, $min_{training\ set}a$.
6. Set the value of the parameter $\varepsilon \in \{0, 0.01, \ldots, 1.0\}$.
7. Set the value of the catch radius parameter r_{catch}.

Fig. 4.13 A granular rule ε–neighborhood of a test thing

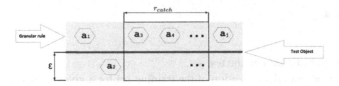

Fig. 4.14 A granular rule ε, r_{catch}-neighborhood of a test thing

8. Classify the given test thing u: 1. For each decision value w of d, and, for each
   decision rule r with $d(r) = w$, determine the set

$$IND_\varepsilon(u, rule) = \{a \in A : \frac{|a(u) - a(r)|}{|max_{training\ set}a - min_{training\ set}a|} \leq \varepsilon\}.$$

2. Check if the decision rule r satisfies the condition

$$\frac{|IND_\varepsilon(u, rule)|}{|A|} \geq r_{catch}.$$

3. If the decision rule r satisfies the condition then $param(d(r)) \leftarrow param$
   $(d(r)) + support(r)$.
4. Assign the value $d(u) = d^*$ where $d^* = argmax_d param(d)$ (Fig. 4.14).

The next collection of algorithms are based on residual implications and weighting
of classifying objects by them. We consider subsequently three classical t-norms: the
minimum, the product and the Łukasiewicz t-norms, see Sect. 2.7.1.1.

The first quadruple of algorithms are based on the minimum residual implication,
i.e., the Goedel implication, Sect. 2.7.1.1:

$$x \Rightarrow_{min} y = 1 \ when \ x \leq y \ y \ when \ x > y. \tag{4.24}$$

Algorithm_5_v1

Procedure

1. Input the training set and the test set.
2. Create tables $max_{training\ set}a$, $min_{training\ set}a$ for the training set for each $a \in A$.
3. Project outliers in the test set for each attribute a onto $max_{training\ set}a$,
   $min_{training\ set}a$.
4. Set the value of the parameter $\varepsilon \in \{0, 0.01, \ldots, 1.0\}$.
5. Classify the given test thing u: 1. For each value w of decision d, and, for
   each training thing v with $d(v) = w$, compute the coefficients $DIS_\varepsilon(u, v)$ and
   $IND_\varepsilon(u, v)$:

$$\frac{|a(u) - a(v)|}{|max_{training\ set}a - min_{training\ set}a|} \geq \varepsilon \rightarrow DIS_\varepsilon(u, v) + + \ each \ a \in A,$$

$$\frac{|a(u) - a(v)|}{|max_{training\ set}a - min_{training\ set}a|} < \varepsilon \rightarrow IND_\varepsilon(u, v) + + \ each \ a \in A.$$

2. Normalize $DIS_\varepsilon(u, v)$ and $IND_\varepsilon(u, v)$:

$$dis_\varepsilon(u, v) = \frac{DIS_\varepsilon(u, v)}{|A|},$$

$$ind_\varepsilon(u, v) = \frac{IND_\varepsilon(u, v)}{|A|}.$$

3. Assign to each training thing v the weight $w(v, u)$ relative to u:

$$dis_\varepsilon(u, v) \leq ind_\varepsilon(u, v) \rightarrow w(u, v) = 1,$$

$$dis_\varepsilon(u, v) > ind_\varepsilon(u, v) \rightarrow w(u, v) = ind_\varepsilon(u, v).$$

4. Normalize weights by dividing each weight by the size of the respective decision class. Let $w(u, v)'$ be the normalized value of $w(u, v)$.
5. For each decision class, sum weights of all training things falling in that class:

$$w(u, d(v)) = \sum \{w(u, v')' : d(v) = d(v')\}.$$

6. Assign the value $d(u) = d(v^*)$ where $v^* = argmax_v w(d(v), u)'$.

Algorithm_5_v2

Procedure

1. Input the training set and the test set.
2. Compute decision rules *exhaustive* from the training set.
3. Create tables $max_{rule\ set}a$, $min_{rule\ set}a$ for each $a \in A$.
4. Project outliers in the test set for each attribute a onto $max_{rule\ set}a$, $min_{rule\ set}a$.
5. Set the value of the parameter $\varepsilon \in \{0, 0.01, \ldots, 1.0\}$.
6. Classify the given test thing u: 1. For each value w of decision d, and, for each rule r with $d(r) = w$, compute the coefficients $DIS_\varepsilon(u, r)$ and $IND_\varepsilon(u, r)$:

$$\frac{|a(u) - a(r)|}{max_{rule\ set}a - min_{rule\ set}a} \geq \varepsilon \rightarrow DIS_\varepsilon(u, r) + + \text{ each } a \in A,$$

$$\frac{|a(u) - a(r)|}{max_{rule\ set}a - min_{rule\ set}a} < \varepsilon \rightarrow IND_\varepsilon(u, r) + + \text{ each } a \in A.$$

2. Normalize $DIS_\varepsilon(u, v)$ and $IND_\varepsilon(u, v)$:

$$dis_\varepsilon(u, r) = \frac{DIS_\varepsilon(u, r)}{|A|},$$

$$ind_\varepsilon(u, r) = \frac{IND_\varepsilon(u, r)}{|A|}.$$

3. Assign to each rule r the weight $w(u, r)$ relative to u:

$$dis_\varepsilon(u, r) \leq ind_\varepsilon(u, r) \rightarrow w(u, r) = 1,$$

$$dis_\varepsilon(u, r) > ind_\varepsilon(u, r) \rightarrow w(u, r) = ind_\varepsilon(u, r).$$

4. Multiply weights by rule supports

$$w(u, r) = w(u, r) * support(r).$$

5. Normalize weights by dividing each weight by the size of the respective decision class. Let $w(u, r)'$ be the normalized value of $w(u, r)$.
6. For each decision class, sum weights of all rules pointing to that class:

$$w(u, d(r)) = \sum \{w(u, r')' : d(r) = d(r')\}.$$

7. Assign the value $d(u) = d(r^*)$ where $r^* = argmax_r w(d(r), u)'$.

Algorithm_5_v3

Procedure

1. Input the training set and the test set.
2. Compute granules from the training set for the granulation radius r_{gran}.
3. Create tables $max_{training\ set}a$, $min_{training\ set}a$ for all $a \in A$.
4. Project outliers in the test set for each attribute a onto $max_{rule\ set}a$, $min_{rule\ set}a$.
5. Set the value of the parameter $\varepsilon \in \{0, 0.01, \ldots, 1.0\}$.
6. Classify the given test thing u: 1. For each value w of decision d, and, for each granule g with $\bar{d}(g) = w$, compute the coefficients $DIS_\varepsilon(u, g)$ and $IND_\varepsilon(u, g)$:

$$\frac{|a(u) - \bar{a}(g)|}{|max_{training\ set}a - min_{training\ set}a|} \geq \varepsilon \rightarrow DIS_\varepsilon(u, g) + + \text{ each } a \in A,$$

$$\frac{|a(u) - a(g)|}{|max_{training\ set}a - min_{training\ set}a|} < \varepsilon \rightarrow IND_\varepsilon(u, g) + + \text{ each } a \in A.$$

2. Normalize $DIS_\varepsilon(u, v)$ and $IND_\varepsilon(u, v)$:

$$dis_\varepsilon(u, g) = \frac{DIS_\varepsilon(u, g)}{|A|},$$

$$ind_\varepsilon(u, g) = \frac{IND_\varepsilon(u, g)}{|A|}.$$

3. Assign to each granule g the weight $w(u, g)$ relative to u:

$$dis_\varepsilon(u, g) \leq ind_\varepsilon(u, g) \rightarrow w(u, g) = 1,$$

$$dis_\varepsilon(u, g) > ind_\varepsilon(u, g) \rightarrow w(u, g) = ind_\varepsilon(u, g).$$

4. Normalize weights by dividing each weight by the size of the respective decision class. Let $w(u, g)'$ be the normalized value of $w(u, g)$.
5. For each decision class, sum weights of all granules pointing to that class:

$$w(u, d(g)) = \sum \{ w(u, g')' : d(g) = d(g') \}.$$

6. Assign the value $d(u) = d(g^*)$ where $g^* = argmax_g w(u, d(g))'$.

Algorithm 5_v4

Procedure

1. Input the training set and the test set.
2. Compute the granular reflection of the training set for the granulation radius r_{gran}.
3. Compute decision rules *exhaustive* from the granular reflection.
4. Create tables $max_{training\ set}a$, $min_{training\ set}a$ for all $a \in A$.
5. Project outliers in the test set for each attribute a onto $max_{training\ set}a$, $min_{training\ set}a$.
6. Set the value of the parameter $\varepsilon \in \{0, 0.01, \dots, 1.0\}$.
7. Classify the given test thing u: 1. for each rule r compute the coefficients $DIS_\varepsilon(u, r)$ and $IND_\varepsilon(u, r)$:

$$\frac{|a(u) - a(r)|}{max_{training\ set}a - min_{training\ set}a} \geq \varepsilon \rightarrow DIS_\varepsilon(u, r) + + \text{ each } a \in A,$$

$$\frac{|a(u) - a(r)|}{max_{training\ set}a - min_{training\ set}a} < \varepsilon \rightarrow IND_\varepsilon(u, r) + + \text{ each } a \in A.$$

2. Normalize $DIS_\varepsilon(u, v)$ and $IND_\varepsilon(u, v)$:

$$dis_\varepsilon(u, r) = \frac{DIS_\varepsilon(u, r)}{|A|},$$

$$ind_\varepsilon(u, r) = \frac{IND_\varepsilon(u, r)}{|A|}.$$

3. Assign to each rule r the weight $w(r, u)$ relative to u:

$$dis_\varepsilon(u, r) \leq ind_\varepsilon(u, r) \rightarrow w(u, r) = 1,$$

$$dis_\varepsilon(u, r) > ind_\varepsilon(u, r) \rightarrow w(u, r) = ind_\varepsilon(u, r).$$

4. Multiply weights by rule supports

$$w(u, r) = w(u, r) * support(r).$$

5. For each decision class, sum weights of all rules pointing to that class:

$$w(d(r), u) = \sum \{w(r', u) : d(r) = d(r')\}.$$

6. Assign the value $d(u) = d(r^*)$ where $r^* = argmax_r w(d(r), u)$.

The next quadruple are algorithms based on the product (Goguen) implication, Sect. 2.7.1.1:

$$x \Rightarrow_{prod} y = 1 \text{ when } x \le y \ \frac{y}{x} \text{ when } x > y. \tag{4.25}$$

Algorithms **Algorithm_6_v1**, **Algorithm_6_v2**, **Algorithm_6_v3**, **Algorithm_6_v4** follow the lines of, respectively, **Algorithm_5_v1**, **Algorithm_5_v2**, **Algorithm_5_v3**, **Algorithm_5_v4** with the exception that

For **Algorithm_6_v1**: replace in the **Algorithm_5_v1** the item 5.3 with the following:

$$dis_\varepsilon(u, v) \le ind_\varepsilon(u, v) \to w(u, v) = 1,$$

$$dis_\varepsilon(u, v) > ind_\varepsilon(u, v) \to w(u, v) = \frac{ind_\varepsilon(u, v)}{dis_\varepsilon(u, v)}.$$

For **Algorithm_6_v2**: replace in the **Algorithm_5_v2** the item 6.3 with the following:

$$dis_\varepsilon(u, r) \le ind_\varepsilon(u, r) \to w(u, r) = 1,$$

$$dis_\varepsilon(u, r) > ind_\varepsilon(u, r) \to w(u, r) = \frac{ind_\varepsilon(u, r)}{dis_\varepsilon(u, r)}.$$

For **Algorithm_6_v3**: replace in the **Algorithm_5_v3** the item 6.3 with the following:

$$dis_\varepsilon(u, g) \le ind_\varepsilon(u, g) \to w(u, g) = 1,$$

$$dis_\varepsilon(u, g) > ind_\varepsilon(u, g) \to w(u, g) = \frac{ind_\varepsilon(u, g)}{dis_\varepsilon(u, g)}.$$

For **Algorithm_6_v4**: replace in the **Algorithm_5_v4** the item 7.3 with the following:

$$dis_\varepsilon(u, r) \le ind_\varepsilon(u, r) \to w(u,) = 1,$$

$$dis_\varepsilon(u, r) > ind_\varepsilon(u, r) \to w(u, r) = \frac{ind_\varepsilon(u, r)}{dis_\varepsilon(u, r)}.$$

The last quadruple of algorithms **Algorithm_7_v1**, **Algorithm_7_v2**, **Algorithm_7_v3**, **Algorithm_7_v4** rest on the Łukasiewicz implication, Sect. 2.7.1.1:

$$x \Rightarrow_L y = min\{1, 1 - x + y\}. \tag{4.26}$$

Algorithms **Algorithm_7_v1**, **Algorithm_7_v2**, **Algorithm_7_v3**, **Algorithm_7_v4** follow the lines of, respectively, **Algorithm_5_v1**, **Algorithm_5_v2**, **Algorithm_5_v3**, **Algorithm_5_v4** with the exception that
For **Algorithm_7_v1**: replace in the **Algorithm_5_v1** the item 5.3 with the following:

$$dis_\varepsilon(u, v) \leq ind_\varepsilon(u, v) \rightarrow w(u, v) = 1,$$

$$dis_\varepsilon(u, v) > ind_\varepsilon(u, v) \rightarrow w(u, v) = min\{1, ind_\varepsilon(u, v) - dis_\varepsilon(u, v) + 1\}.$$

For **Algorithm_7_v2**: replace in the **Algorithm_5_v2** the item 6.3 with the following:

$$dis_\varepsilon(u, r) \leq ind_\varepsilon(u, r) \rightarrow w(u, r) = 1,$$

$$dis_\varepsilon(u, r) > ind_\varepsilon(u, r) \rightarrow w(u, r) = min\{1, ind_\varepsilon(u, r) - dis_\varepsilon(u, r) + 1\}.$$

For **Algorithm_7_v3**: replace in the **Algorithm_5_v3** the item 6.3 with the following:

$$dis_\varepsilon(u, g) \leq ind_\varepsilon(u, g) \rightarrow w(u, g) = 1,$$

$$dis_\varepsilon(u, g) > ind_\varepsilon(u, g) \rightarrow w(u, g) = min\{1, ind_\varepsilon(u, g) - dis_\varepsilon(u, g) + 1\}.$$

For **Algorithm_7_v4**: replace in the **Algorithm_5_v4** the item 7.3 with the following:

$$dis_\varepsilon(u, r) \leq ind_\varepsilon(u, r) \rightarrow w(u, r) = 1,$$

$$dis_\varepsilon(u, r) > ind_\varepsilon(u, r) \rightarrow w(u, r) = min\{1, ind_\varepsilon(u, r) - dis_\varepsilon(u, r) + 1\}.$$

The last couple of algorithms are based on similarity measures which take into account the parameter ε also in weight assignment to training things.

Algorithm_8_v1_variant 1, 2, 3, 4, 5

Procedure

1. Input the training set and the test set.
2. Create tables $max_{training\ set}a$, $min_{training\ set}a$ for the training set for all $a \in A$.
3. Project outliers in the test set for each attribute a onto $max_{training\ set}a - min_{training\ set}a$.
4. Set the value of the parameter $\varepsilon \in \{0, 0.01, \ldots, 1.0\}$ in variant 1, and in $\{0.01, \ldots, 1\}$ in variants 2, 3, 4, 5.

5. Classify the given test thing u:
   1. In variant 1:
      1.1 for each training thing v compute the weight $w(u, v)$. For each $a \in A$:
      $$w(u, v) \leftarrow w(u, v) + \frac{|a(u)-a(v)|}{max_{training\ set}a-min_{training\ set}a}.$$
   2. In variant 2:
      2.1 for each training thing v compute the weight $w(u, v)$. For each $a \in A$:
      $$\frac{|a(u) - a(v)|}{max_{training\ set}a - min_{training\ set}a} \geq \varepsilon \rightarrow$$
      $$w(u, v) \leftarrow w(u, v) + \frac{|a(u) - a(v)|}{*}(1 + \varepsilon).$$
      $$\frac{|a(u) - a(v)|}{max_{training\ set}a - min_{training\ set}a} < \varepsilon \rightarrow$$
      $$w(u, v) \leftarrow w(u, v) + \frac{|a(u) - a(v)|}{max_{training\ set}a - min_{training\ set}a}.$$

      2.2 for each training thing v compute the weight $w(u, v)$. For each $a \in A$:
      $$\frac{|a(u) - a(v)|}{max_{training\ set}a - min_{training\ set}a} \geq \varepsilon \rightarrow$$
      $$w(u, v) \leftarrow w(u, v) + \frac{|a(u) - a(v)|}{|train(a)|} * \frac{(1 + \varepsilon)}{\varepsilon}.$$
      $$\frac{|a(u) - a(v)|}{max_{training\ set}a - min_{training\ set}a} < \varepsilon \rightarrow$$
      $$w(u, v) \leftarrow w(u, v) + \frac{|a(u) - a(v)|}{max_{training\ set}a - min_{training\ set}a} * \frac{1}{\varepsilon}.$$

   3. In variant 3:
      3.1 for each training thing v compute the weight $w(u, v)$. For each $a \in A$:
      $$\frac{|a(u) - a(v)|}{max_{training\ set}a - min_{training\ set}a} \geq \varepsilon \rightarrow$$
      $$w(u, v) \leftarrow w(u, v) + \frac{|a(u) - a(v)|}{max_{training\ set}a - min_{training\ set}a} * (1 + \varepsilon).$$
      $$\frac{|a(u) - a(v)|}{max_{training\ set}a - min_{training\ set}a} < \varepsilon \rightarrow$$

$$w(u, v) \leftarrow w(u, v) + \frac{|a(u) - a(v)|}{max_{training\ set}a - min_{training\ set}a} * \frac{1}{1 + \varepsilon}.$$

4. In variant 4:

4.1 for each training thing v compute the weight $w(u, v)$. For each $a \in A$:

$$\text{if } \frac{|a(u) - a(v)|}{max_{training\ set}a - min_{training\ set}a} \geq \varepsilon \text{ then}$$

$$w(u, v) \leftarrow w(u, v) + \frac{|a(u) - a(v)|}{max_{training\ set}a - min_{training\ set}a}.$$

$$(\varepsilon + \frac{|a(u) - a(v)|}{max_{training\ set}a - min_{training\ set}a})$$

$$\text{if } \frac{|a(u) - a(v)|}{max_{training\ set}a - min_{training\ set}a} < \varepsilon \text{ then}$$

$$w(u, v) \leftarrow w(u, v) + \frac{|a(u) - a(v)|}{max_{training\ set}a - min_{training\ seta} \cdot \varepsilon}$$

5. In variant 5:

5.1 for each training thing v compute the weight $w(u, v)$. For each $a \in A$:

$$\text{if } \frac{|a(u) - a(v)|}{max_{training\ set}a - min_{training\ set}a} \geq \varepsilon \text{ then}$$

$$w(u, v) \leftarrow w(u, v) + \frac{|a(u) - a(v)|}{max_{training\ set}a - min_{training\ set}a}$$

$$\text{if } \frac{|a(u) - a(v)|}{max_{training\ set}a - min_{training\ set}a} < \varepsilon \text{ then}$$

$$w(u, v) \leftarrow w(u, v) + \frac{|a(u) - a(v)|}{(max_{training\ set}a - min_{training\ set}a) \cdot \varepsilon}$$

4.6 Validation of Proposed Algorithms on Real Data Sets

We select a simple real data set **Australian Credit** [26] counting 15 attributes and 690 data items. We apply to this data set our algorithms. For comparison, we reproduce results for this data set obtained by other rough set based methods proposed by various authors, as well as results from Statlog base [10] obtained for this data set by means of other methodologies.

Table 4.8 Australian credit; in [1] the best of all result; in [2] reduction of training things is 17.6 %, in [3], [4], [6] in [7] it is 3.4 %, in [5] it is 3.3 %

No.	Author	Method	Accuracy	Coverage
1.	*Bazan* [6]	*SNAPM*(0.9)	0.870	–
2.	*Statlog* [10]	*Standard (NNANR)*	0.860	–
3.	*Statlog* [10]	*The best dynamic method*	813	–
4.	*Statlog* [10]	*The best RS method*	0.870	–
5.	*literature*	*Cal5 CV − 10*	0.869	–
6.		*k − NN*	0.819	–
7.		*C4.5*	0.845	–
8.	*Nguyen* [11]	*simple.templates*	0.929	0.623
9.	*Nguyen* [11]	*general.templates*	0.886	0.905
10.	*Nguyen* [11]	*closest.simple.templates*	0.821	1.0
11.	*Nguyen* [11]	*closest.gen.templates*	0.855	1.0
12.	*Nguyen* [11]	*tolerance.simple.templ.*	0.842	1.0
13.	*Nguyen* [11]	*tolerance.gen.templ.*	0.875	1.0
14.	*Wróblewski* [27]	*adaptive.classifier*	0.863	–
5-fold Cross-Validation				
14.	*Our result*	$Alg\_8\_v1\_wariant5^1.\varepsilon_{opt} = 0.13$	**0.880**	1.0
15.	*Our result*	$Alg\_2\_v2^2.r_{gran} = 0.714286.$ $r_{catch} = 0.714286.\varepsilon_{opt} = 0.08$	**0.875**	1.0
16.	*Our result*	$Alg\_1\_v2.r_{catch} = 0.785714.\varepsilon_{opt} = 0.18$	**0.872**	1.0
17.	*Our result*	$Alg\_3\_v1.\varepsilon_{opt} = 0.46$	**0.871**	1.0
18.	*Our result*	$Alg\_3\_v2.r_{catch} = 0.142857.\varepsilon_{opt} = 0.35$	**0.868**	1.0
19.	*Our result*	$Alg\_2\_v1^3.r_{gran} = 0.785714.\varepsilon_{opt} = 0.54$	**0.862**	0.996
20.	*Our result*	$Alg\_5\_v2.\varepsilon_{opt} = 0.02$	**0.861**	1.0
21.	*Our result*	$Alg\_1\_v1.\varepsilon_{opt} = 0.83$	**0.859**	0.999
22.	*Our result*	$Alg\_7\_v3^4.r_{gran} = 0.785714.\varepsilon_{opt} = 0.01$	**0.859**	1.0
23.	*Our result*	$Alg\_5\_v3^5.r_{gran} = 0.785714.\varepsilon_{opt} = 0.05$	**0.855**	1.0
24.	*Our result*	$Alg\_6\_v3^6.r_{gran} = 0.785714.\varepsilon_{opt} = 0.01$	**0.852**	1.0
25.	*Our result*	$Alg\_4\_v1^7.r_{gran} = 0.785714, \varepsilon_{opt} = 0.01$	**0.851**	1.0
26.	*Our result*	$Alg\_6\_v2.\varepsilon_{opt} = 0.01$	**0.851**	1.0
27.	*Our result*	$Alg\_5\_v1.\varepsilon_{opt} = 0.04$	**0.848**	1.0
28.	*Our result*	$Alg\_6\_v1.\varepsilon_{opt} = 0.06$	**0.848**	1.0
29.	*Our result*	$Alg\_7\_v1.\varepsilon_{opt} = 0.05$	**0.846**	1.0
30.	*Our result*	$Alg\_7\_v2.\varepsilon_{opt} = 0$	**0.555**	1.0

In Table 4.8 we match results by rough set methods obtained by other authors with results obtained by us for algorithms proposed above.

The best result by algorithms of series 1 to 8 is obtained by Algorithm 8_v1 (line 14) for the optimal value of $\varepsilon = 0.13$ with accuracy 0.880 and coverage of 1.0. Slightly worse accuracy was obtained by Algorithms 2_v2, 1_v2, 3_v1, 3_v2 (lines 15–18) still with coverage $= 1.0$.

Table 4.9 Australian credit; best classification results from Statlog [10]

Source	Method	Accuracy	Coverage
10-fold Cross-Validation			
Statlog [10]	*Cal5*	0.869	–
KG	$K - NN, k = 18, manh, std$	0.864	–
Statlog [10]	*ITrule*	0.863	–
KG	$w - NN, k = 18, manh, simplex, std$	0.862	–
Statlog [10]	*Discrim*	0.859	–
Statlog [10]	*DIPOL92*	0.859	–
Statlog [10]	*C4.5*	0.855 ± 0.007	–
Statlog [10]	*CART*	0.855	–
Statlog [10]	*RBF*	0.855	–
Statlog [10]	*CASTLE*	0.852	–
Statlog [10]	*NaiveBay*	0.849	–
Statlog [10]	*IndCART*	0.848	–
KG	$k - NN.k = 11, std, eucl$	0.848	–
Statlog [10]	*Backprop*	0.846	–
Statlog [10]	*C4.5*	0.845	–
KG	$k - NN, k = 11, fec.sel, eucl, std$	0.844	–
Statlog [10]	*SMART*	0.842	–
Statlog [10]	*Baytree*	0.829	–
Statlog [10]	$k - NN$	0.819	–
Statlog [10]	*NewID*	0.819	–
Statlog [10]	*AC2*	0.819	–
Statlog [10]	*LVQ*	0.803	–
Statlog [10]	*ALLOC80*	0.799	–
Statlog [10]	*CN2*	0.796	–
Statlog [10]	*Quadisc*	0.793	–
Statlog [10]	*Default*	0.760	–

The best accuracy by other methods was obtained by means of the simple(0.929) as well as general (0.886) template approach (lines 8, 9) but with much smaller coverage of, respectively, 0.629 (less than two thirds of data covered) and 0.905.

In Table 4.9, we collect the results for Australian credit obtained by other methods.

4.7 Concept-Dependent and Layered Granulation on Real Data: Granulation as a Compression Tool

We present here results of some tests on some real data sets for concept-dependent as well as layered granulation mechanisms. As the tool for induction of decision rules, we have applied the exhaustive algorithm. As a tool for granulation the rough

inclusion (4.3) has been applied, modified according to the concept-dependent variant. As the strategy for selecting the granular covering a random choice has been
applied and the strategy for decision making has been chosen as the majority voting
with removal of ties by random choice.

We use as a classifier a variant of Algorithm 8_v1_w1 with the use of the rough
inclusion μ^L and a variant of the NN classifier also based on the metric given by (4.3).
The procedure we recall for the convenience of the reader is as follows

Step 1. The training granular decision system $(G^{trn}, \mu^L, r_{gran})$ and the test decision
         system (U_{tst}, A, d) have been input, where A is a set of conditional attributes,
         d a decision attribute, and, r_{gran} a granulation radius.
Step 2. Classification of test objects by means of weighted granules of training
         objects is performed as follows.

For all conditional attributes $a \in A$, training objects $v \in G^{trn}$, and test objects
$u \in U_{tst}$, we compute weights $w(u, v)$ as

$$w(u, v) = max\{r : \mu^L(u, v, r)\} = \frac{|IND(u, v)|}{|A|} \qquad (4.27)$$

The voting procedure of the algorithm in case of 8_v1_w1 Algorithm rests on
values of the parameter

$$Concept\_weight_c(u) = \frac{\sum_{\{v \in G^{trn} : d(v) = c\}} w(u, v)}{|\{v \in G^{trn} : d(v) = c\}|} \qquad (4.28)$$

In the case of the NN classifier, we use

$$Concept\_weight_c(u) = min\{w(u, v)) : v \in \{v' \in G^{trn} : d(v') = c\}\} \qquad (4.29)$$

for all decision classes c.

Finally, the test object u is classified to the class c with a minimal value of
$Concept\_weight_c(u)$.

After all test objects u are classified, the quality parameter of *accuracy*, *acc* for
short, is computed, according to formula

$$acc = \frac{\text{number of correctly classified objects}}{\text{number of classified objects}} \qquad (4.30)$$

Classification after granulation by means of these similarity measures has been
performed for the following data sets from the repository [26]: Adult, Car evaluation,
Hepatitis, Congressional voting records, Mushroom, Nursery, and, SPECT Heart, of
varied size and attribute value type. In Table 6.6, we collect descriptions of those data
sets. The successive columns in Table 6.6 give the name of the data *name*, the type of
attributes *attr type*, the number of attributes *attr no.*, the number of objects *obj no.*,
the number of decision classes *class no.*

Table 4.10 Data sets description

Name	Attr type	Attr no.	Obj no.	Class no.
Adult	categorical, integer	15	20000	2
Car evaluation	categorical	7	1728	4
Hepatitis	categorical, integer, real	20	155	2
Congressional voting records	categorical	17	435	2
Mushroom	categorical	23	8124	2
Nursery	categorical	9	12960	5
SPECT heart	categorical	23	267	2

Results have been validated with 5-fold cross validation CV5, applied five times and accuracy results averaged over five epochs are recorded as $1NN - acc5xCV5$ for NN and as $811 - acc5xCV5$ for 8_v1_w1. The average number of granules is recorded in the column $Mean\_size\_of\_GS$ and the average number of decision rules is recorded in the column $Exh\,rul\,no.$. Values are given in rows indexed by granulation radii values, each of which is of the form of $r = \frac{i}{|A|}$ for $i = 1, \ldots, |A|$. This format of tables holds for all data sets. Let us point to the fact that values for non-granulated data are collected under the radius of 1.000 meaning the canonical indiscernibility granulation, i.e., the canonical rough set procedure (Tables 4.10, 4.11, 4.12, 4.13 and 4.14).

In Table 4.18, we collect results for total accuracy of Algorithm 8_v1_w1. For instance, in case of *Adult* data set, for the radius of 0.5 the size of data set is reduced as 16000:182.76 and the size of decision algorithm is reduced as 110950:378; the accuracy is diminished by granulation by order of 0.04. For *Hepatitis* at the radius of 0.526, the size of data set is reduced as 124:18.56 and the size of decision algorithm is

Table 4.11 Data set: adult

r_{gran}	$kNN - acc5xCV5$	$811 - acc5xCV5$	$Mean\_size\_of\_GS$	$Exh\,rul\,no.$
0.071	0.63554	0.63561	2.16	14
0.143	0.63554	0.64018	3.64	14
0.214	0.64251	0.64211	7.96	14
0.286	0.65526	0.67502	15	13
0.357	0.67866	0.69369	34.24	36
0.429	0.69556	0.6807	78.84	133
0.500	0.6805	0.68389	182.76	378
0.571	0.67871	0.70163	428.84	1650
0.643	0.67758	0.71287	997.28	6443
0.714	0.68048	0.71386	2290.64	23609
0.786	0.6846	0.71784	5126.24	59337
0.857	0.69795	0.72191	10036.3	96687
0.929	0.71582	0.72053	14840.6	108830
1.000	0.71621	0.72026	15993.4	110950

Table 4.12 Data set: car evaluation

r_{gran}	$1NN - acc5xCV5$	$811 - acc5xCV5$	$Mean\_size\_of\_GS$	$Exh\ rul\ no.$
0.167	0.388988	0.387164	8.08	33
0.333	0.456468	0.365273	17.16	72
0.500	0.495127	0.37674	38.84	170
0.667	0.546064	0.42835	106.24	312
0.833	0.611924	0.449526	368.76	468
1.000	0.359964	0.4567	1382.4	326

Table 4.13 Data set: hepatitis—missing values filled by majority voting

r_{gran}	$kNN - acc5xCV5$	$811 - acc5xCV5$	$Mean\_size\_of\_GS$	$Exh\ rul\ no.$
0.053	0.807742	0.809032	2	18
0.105	0.807742	0.809032	2	18
0.158	0.807742	0.809032	2	18
0.211	0.807742	0.809032	2.12	18
0.263	0.809032	0.809032	2.72	18
0.316	0.811612	0.805161	3.48	21
0.368	0.812902	0.815484	5.2	23
0.421	0.832258	0.833548	7.16	28
0.474	0.847742	0.851613	11.28	43
0.526	0.815484	0.869678	18.56	85
0.579	0.812902	0.867097	29.8	266
0.632	0.832259	0.865807	46.36	550
0.684	0.83871	0.847742	69.6	1162
0.737	0.83871	0.849033	90.08	1384
0.789	0.854194	0.847743	109.68	1700
0.842	0.854194	0.842581	116.96	1769
0.895	0.854194	0.845162	121	1814
0.947	0.854194	0.843872	121.96	1823
1.000	0.854194	0.841291	124	1824

reduced as 1824:85 with the accuracy increased by 0.027. For *Mushroom* at the radius of 0.455, the size of data set is reduced as 6499.25:20.28, and the size of decision algorithm is reduced as 3825:117 with accuracy increased slightly by 0.002. For *SPECT heart* at the radius of 0.3478, the reduction in data set size is 184.76: 3.2 with reduction in decision algorithm size 16803:7 and increase in accuracy of 0.15 (Tables 4.15, 4.16 and 4.17).

In prevailing number of granulations, we observe reduction in rule number, only in case of Car Evaluation and Nursery data we have exception to this rule at passage between radii 1.0 and $\frac{|A|-1}{|A|}$. The number of rules larger at the decreased radius of

Table 4.14 Data set: congressional voting records

r_{gran}	$kNN - acc5xCV5$	$811 - acc5xCV5$	$Mean\_size\_of\_GS$	$Exh\ rul\ no.$
0.063	0.90115	0.897932	2.04	28
0.125	0.90115	0.897932	2.12	28
0.188	0.90115	0.897932	2.52	28
0.250	0.90115	0.897932	3.12	28
0.313	0.901609	0.896552	3.52	28
0.375	0.902528	0.895173	4.8	28
0.438	0.914023	0.893334	6.08	27
0.500	0.921379	0.897932	6.76	27
0.563	0.934712	0.89977	8.92	27
0.625	0.936092	0.905288	11.8	41
0.688	0.925517	0.902989	18.84	70
0.750	0.908966	0.910345	30.68	190
0.813	0.915402	0.91954	52.36	1067
0.875	0.912643	0.925517	90.92	2749
0.938	0.925976	0.92046	146.56	3437
1.000	0.935172	0.916322	214.08	3676

Table 4.15 Data set: mushroom

r_{gran}	$kNN - acc5xCV5$	$811 - acc5xCV5$	$Mean\_size\_of\_GS$	$Exh\ rul\ no.$
0.045	0.887838	0.887789	2	18
0.091	0.887838	0.887789	2	18
0.136	0.887838	0.887789	2	18
0.182	0.887838	0.887789	2.04	18
0.227	0.88801	0.887789	2.52	18
0.273	0.88806	0.887789	2.56	18
0.318	0.887814	0.886337	3.68	18
0.364	0.886041	0.886312	5.64	18
0.409	0.883358	0.88513	9.16	22
0.455	0.918243	0.892343	20.28	117
0.500	0.91551	0.890595	20.96	91
0.545	0.936856	0.895372	28.72	247
0.591	0.965288	0.90517	38.2	600
0.636	0.983482	0.903742	42.72	789
0.682	0.995298	0.905367	48.08	1698
0.727	0.995815	0.902313	46.6	1947
0.773	0.997882	0.89909	55.32	2597
0.818	0.999015	0.905957	93.16	3091
0.864	0.99968	0.899704	193.48	4281
0.909	1	0.897046	1751.88	4747
0.955	1	0.892048	1755	4794
1.000	1	0.889784	6499.2	3825

Table 4.16 Data set: nursery

r_{gran}	$kNN - acc5xCV5$	$811 - acc5xCV5$	$Mean\_size\_of\_GS$	$Exh\ rul\ no.$
0.125	0.408518	0.403395	7.56	36
0.250	0.491574	0.443364	13.76	58
0.375	0.532485	0.493826	26.24	287
0.500	0.547762	0.50872	58.92	632
0.625	0.585926	0.513504	161.24	1736
0.750	0.620479	0.514862	547	3618
0.875	0.667654	0.518765	2368.08	5492
1.000	0.463534	0.526358	10368	2076

Table 4.17 Data set: SPECT heart

r_{gran}	$kNN - acc5xCV5$	$811 - acc5xCV5$	$Mean\_size\_of\_GS$	$Exh\ rul\ no.$
0.043478	0.518407	0.548302	2	6
0.086956	0.508596	0.548302	2	6
0.130434	0.508596	0.548302	2	6
0.173913	0.511615	0.548302	2.2	6
0.217391	0.521426	0.559538	2.28	6
0.260869	0.568442	0.573612	2.76	6
0.304347	0.563159	0.575751	2.8	6
0.347826	0.610469	0.591405	3.2	7
0.391304	0.622474	0.569378	3.64	8
0.434782	0.659288	0.585087	4.32	8
0.47826	0.710748	0.573068	6.64	12
0.521739	0.711489	0.546052	6.68	17
0.565217	0.725045	0.497429	8.8	20
0.608695	0.757792	0.474843	11.4	22
0.652173	0.771488	0.484739	16.04	148
0.695652	0.78422	0.474228	32.6	372
0.73913	0.788022	0.44566	32.48	372
0.782608	0.767814	0.438882	50.08	2017
0.826086	0.741608	0.417037	71.44	6417
0.869565	0.719861	0.400657	101.28	11674
0.913043	0.704948	0.412718	162.84	16303
0.956521	0.703467	0.422474	162.88	16803
1	0.699692	0.428441	184.76	16803

granulation, seems to be caused by the peculiar characteristics of the mentioned data, viz., there are small number of attributes with a few values and relatively large number of objects. The granulation with the radius $\frac{|A|-1}{|A|}$ reduces a large number of conflicting objects and of conflicting combinations of attributes, hence the number

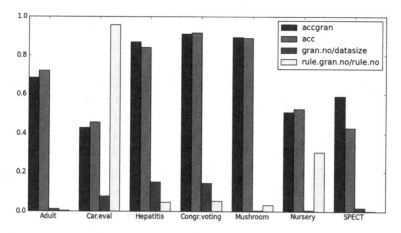

Fig. 4.15 Visualization of data from Table 4.18

of shorter rules is large and the final number of rules is greater than the number of rules obtained for $r_{gran} = 1.0$ (Fig. 4.15).

4.7.1 Layered Learning

In Artiemjew [1] the idea of *layered learning*, known in Machine Learning in some other contexts, especially in agent systems, cf., e.g., Stone [25] was applied to granular reflection by iterating the procedure until stabilization of granules. We would like to present in Table 4.19 the results of layered learning for items shown in Table 4.18. We may observe a sharp decline in size of both data set as well as decision algorithm coupled with a slight change in accuracy: in some cases decrease in other increase, e.g., in case of SPECT, 2 granules and 4 decision rules replace 183 objects and 16803 rules in non-granulated case giving an increase in accuracy of classification of 0.07.

Table 4.18 Selected best results of accuracy with comparison of data size and decision algorithm size reduction

dataset	r.gran	accgran	acc	gran.no	datasize	rule.gran.no	rule.no
Adult	0.500	0.6838	0.7202	182.76	16000	378	110950
Car	0.667	0.42835	0.4567	106.24	1382.4	312	326
Hepatitis	0.526	0.8696	0.8412	18.56	124	85	1824
Congr.voting	0.750	0.9103	0.9163	30.68	214.08	190	3676
Mushroom	0.455	0.8923	0.8897	20.28	6499.2	117	3825
Nursery	0.500	0.5087	0.5263	58.92	10368	632	20176
SPECT	0.3478	0.5914	0.4284	3.2	184.76	7	16803

Table 4.19 The effect of layered granulation for the best results from Table 4.18; l_1—first layer, l_2—second layer, l_3 the third layer

Dataset	r.gran	accgran.811	acc	gran.no	datasize	rule.gran.no	rule.no
l_1.Adult	0.500	0.6838	0.7202	182.76	16000	378	110950
l_2.Adult	0.500	0.703	0.7202	6.4	16000	18.8	110950
l_3.Adult	0.500	0.697	0.7202	2.4	16000	18.8	110950
l_1.Car	0.667	0.428	0.4567	106.24	1382.4	312	326
l_2.Car	0.667	0.389	0.4567	48	1382.4	201.8	326
l_3.Car	0.667	0.380	0.4567	30.4	1382.4	142.6	326
l_1.Hepatitis	0.526	0.870	0.8412	18.56	124	85	1824
l_2.Hepatitis	0.526	0.787	0.8412	5.4	124	38.2	1824
l_3.Hepatitis	0.526	0.736	0.8412	3.4	124	34.4	1824
l_1.Congr.voting	0.750	0.9103	0.9163	30.68	214.08	190	3676
l_2.Congr.voting	0.750	0.908	0.9163	9.2	214.08	102.6	3676
l_3.Congr.voting	0.750	0.793	0.9163	6.4	214.08	92.6	3676
l_1.Mushroom	0.455	0.8923	0.8897	20.28	6499.2	117	3825
l_2.Mushroom	0.455	0.876	0.8897	2	6499.2	19.6	3825
l_3.Mushroom	0.455	0.876	0.8897	2	6499.2	17.6	3825
l_1.Nursery	0.500	0.5087	0.5263	58.92	10368	632	20176
l_2.Nursery	0.500	0.455	0.5263	20	10368	182.2	20176
l_3.Nursery	0.500	0.428	0.5263	10.8	10368	95.8	20176
l_1.SPECT	0.3478	0.5914	0.4284	3.2	184.76	7	16803
l_2.SPECT	0.3478	0.501	0.4284	2	184.76	4	16803
l_3.SPECT	0.3478	0.501	0.4284	2	184.76	4	16803

4.8 Applications of Granular Reflections to Missing Values

A particular but important problem in data analysis is the treatment of missing values. In many data, some values of some attributes are not recorded due to many factors, like omissions, inability to take them, loss due to some events etc.

Analysis of systems with missing values requires a decision on how to treat missing values; Grzymala-Busse and Ming Hu [8] and Grzymala-Busse [9] analyze nine such methods, among them, 1. *most common attribute value*, 2. *concept restricted most common attribute value*, 3. *assigning all possible values to the missing location*, 4. *treating the unknown value as a new valid value*, etc. Their results indicate that methods 3, 4 perform very well and in a sense stand out among all nine methods.

We adopt and consider two methods, i.e., 3, 4 from the above mentioned. As usual, the question on how to use granular structures in analysis of incomplete systems, should be answered first.

The idea is to embed the missing value into a granule: by averaging the attribute value over the granule in the way already explained, it is hoped the the average value would fit in a satisfactory way into the position of the missing value.

We will use the symbol $*$, commonly used for denoting the missing value; we will use two methods 3, 4 for treating $*$, i.e., either $*$ is a *don't care* symbol meaning that any value of the respective attribute can be substituted for $*$, hence, $* = v$ for each value v of the attribute, or $*$ is a new value on its own, i.e., if $* = v$ then v can only be $*$.

Our procedure for treating missing values is based on the granular structure $(G(\mu, r), \mathcal{G}, \mathcal{S}, \{a* : a \in A\})$; the strategy $\mathcal{S}$ is the majority voting, i.e., for each attribute a, the value $a^*(g)$ is the most frequent of values in $\{a(u) : u \in g\}$. The strategy $\mathcal{G}$ consists in random selection of granules for a covering.

For an object u with the value of $*$ at an attribute a,, and a granule $g = g(v, r) \in G(\mu, r)$, the question whether u is included in g is resolved according to the adopted strategy of treating $*$: in case $* = don't\ care$, the value of $*$ is regarded as identical with any value of a hence $|IND(u, v)|$ is automatically increased by 1, which increases the granule; in case $* = *$, the granule size is decreased. Assuming that $*$ is sparse in data, majority voting on g would produce values of a^* distinct from $*$ in most cases; nevertheless the value of $*$ may appear in new objects g^*, and then in the process of classification, such value is repaired by means of the granule closest to g^* with respect to the rough inclusion μ_L, in accordance with the chosen method for treating $*$.

In plain words, objects with missing values are in a sense absorbed by close to them granules and missing values are replaced with most frequent values in objects collected in the granule; in this way the method 3 or 4 in [7] is combined with the idea of a frequent value, in a novel way.

We have thus four possible strategies:

1. *Strategy A: in building granules $* = don't\ care$, in repairing values of $*$, $* = don't\ care$;*
2. *Strategy B: in building granules $* = don't\ care$, in repairing values of $*$, $* = *$;*
3. *Strategy C: in building granules $* = *$, in repairing values of $*$, $* = don't\ care$;*
4. *Strategy D: in building granules $* = *$, in repairing values of $*$, $* = *$.*

We show how effective are these strategies by perturbing the data set Pima Indians Diabetes, from UCI Repository [26]. We perturb this data set by randomly replacing 10 % of attribute values in the data set with missing $*$ values. As algorithm for rule induction, the exhaustive algorithm has been selected. Five times 5-fold cross validation $5 \times CV5$ has been applied.

Results of granular treatment in case of Strategies A,B,C,D in terms of accuracy are reported in the Table 4.20. Tables 4.21 and 4.22 show, respectively, the bias of accuracy and the size of the granular training set. Let us again stress that results for the non-granulated case are shown in the column labeled *nil*.

These results show that accuracy of the classifier with recovery of the missing values is on par with accuracy of the classifier on the non-perturbed data for granulation radii in the range from 0.5 to 0.875 for each of strategies A, B, C, D, whereas for strategies C and D it is on par for radii from 0.125 to 0.875, i.e., practically for all non-trivial cases of granulation.

We conclude that the essential for results of classification is the strategy of treating the missing value of $*$ as $* = *$ in both strategies C and D; the repairing strategy

Table 4.20 $5 \times$ CV-5; The result of accuracy for A, B, C, D strategies of missing values handling versus complete data; **Pima Indiand Diabetes**; **Standard granulation**; **10 % of missing values**; r_{gran} = Granulation radius; nil = result for data without missing values

r_{gran}	nil	A	B	C	D
0	0.349	0.349	0.349	0.349	0.349
0.125	0.442	0.349	0.349	0.387	0.388
0.25	0.658	0.349	0.349	0.651	0.652
0.375	0.654	0.36	0.36	0.65	0.648
0.5	0.644	0.619	0.626	0.644	0.642
0.625	0.645	0.637	0.64	0.645	0.648
0.75	0.647	0.646	0.64	0.648	0.647
0.875	0.647	0.646	0.642	0.649	0.647
1	0.647	0.647	0.647	0.647	0.647

Table 4.21 $5 \times$ CV-5; Bias of Accuracy for A, B, C, D strategies of missing values handling versus complete data; **Pima Indiand Diabetes**; **Standard granulation**; **10 % of missing values**; r_{gran} = Granulation radius; nil = result for data without missing values

r_{gran}	nil	A	B	C	D
0	0	0	0	0	0
0.125	0.078	0	0	0.048	0.155
0.25	0.006	0	0	0.004	0.004
0.375	0.005	0.044	0.044	0.009	0.007
0.5	0.008	0.036	0.029	0.01	0.011
0.625	0.01	0.019	0.022	0.007	0.005
0.75	0.01	0.013	0.019	0.011	0.008
0.875	0.01	0.007	0.006	0.007	0.008
1	0.01	0.006	0.006	0.008	0.008

Table 4.22 $5 \times$ CV-5; Granular system size for A, B, C, D strategies of missing values handling versus complete data; **Pima Indiand Diabetes**; **Standard granulation**; **10 % of missing values**; r_{gran} = Granulation radius; nil = result for data without missing values

r_{gran}	nil	A	B	C	D
0	1	1	1	1	1
0.125	21	1.44	1.32	19.36	19.64
0.25	115.64	4.84	5.08	110	109.2
0.375	328.76	18.2	15.44	320.6	321.28
0.5	518.96	48.88	49.64	523.56	523.2
0.625	607.16	204.4	188.2	606.32	606.36
0.75	614.4	436.88	441.88	614.24	614.24
0.875	614.4	586.48	583.96	614.4	614.4
1	614.4	612.72	612.68	614.4	614.4

has almost no effect: C and D differ very slightly with respect to this strategy. It is worthwhile to observe a significant reduction in size of the training set when passing from the non–granulated case to granular reflections.

A more detailed account of methods of granular computing in treatment of missing values is given in Chap. 8, where four strategies introduced above are applied to all 13 data sets perturbed by randomly inserted 5 or 10 % star values.

References

1. Artiemjew, P.: Classifiers from Granulated Data Sets: Concept Dependent and Layered Granulation. In: Proceedings RSKD'07. Workshops at ECML/PKDD'07, pp. 1–9. Warsaw University Press, Warsaw (2007)
2. Artiemjew, P.: Rough mereological classifiers obtained from weak rough set inclusions. In: Proceedings of International Conference on Rough Set and Knowledge Technology RSKT'08, Chengdu China. Lecture Notes in Artificial Intelligence, vol. 5009, pp. 229–236. Springer, Berlin (2008)
3. Artiemjew, P.: On classification of data by means of rough mereological granules of objects and rules. In: Proceedings of International Conference on Rough Set and Knowledge Technology RSKT'08, Chengdu China. Lecture Notes in Artificial Intelligence, vol. 5009, pp. 221–228. Springer, Berlin (2008)
4. Artiemjew, P.: Natural versus granular computing: classifiers from granular structures. In: Proceedings of 6th International Conference on Rough Sets and Current Trends in Computing RSCTC'08, Akron OH (2008)
5. Artiemjew, P.: A review of the knowledge granulation methods: discrete vs. continuous algorithms. In: Skowron A., Suraj Z. (eds.) Rough Sets and Intelligent Systems. ISRL, vol. 43, pp. 41–59. Springer, Berlin (2013)
6. Bazan, J.G.: A comparison of dynamic and non-dynamic rough set methods for extracting laws from decision tables. In: Polkowski, L., Skowron, A. (eds.) Rough Sets in Knowledge Discovery, pp. 321–365. Physica Verlag, Heidelberg (1998)
7. Bazan, J.G., Synak, P., Nguyen, S.H., Nguyen, H.S.: Rough set algorithms in classification problems. In: Polkowski, L., Tsumoto, S., Lin, T.Y. (eds.) Rough Set Methods and Applications. New Developments in Knowledge Discovery in Information Systems, pp. 49–88. Physica Verlag, Heidelberg (2000)
8. Grzymala-Busse J.W., Hu, M.: A comparison of several approaches to missing attribute values in data mining. In: Proceedings RSCTC 2000. Lecture Notes in Artificial Intelligence, vol. 2005, pp. 378–385. Springer, Berlin (2000)
9. Grzymala-Busse, J.W.: Data with missing attribute values: generalization of rule indiscernibility relation and rule induction. Transactions on Rough Sets I. Lecture Notes in Computer Science, vol. 3100, pp. 78–95. Springer, Berlin (2004)
10. Michie, D., Spiegelhalter, D.J., Taylor, C.C. (eds.): Statlog Project. Machine Learning, Neural and Statistical Classification; http://www.is.umk.pl/projects/datasets-stat.html
11. Nguyen, S.H.: Regularity analysis and its applications in data mining. In: Polkowski, L., Tsumoto, S., Lin, T.Y. (eds.) Rough Set Methods and Applications. New Developments in Knowledge Discovery in Information Systems. pp. 289–378. Physica Verlag, Heidelberg (2000)
12. Polkowski, L.: Granulation of knowledge in decision systems: the approach based on rough inclusions. The method and its applications. In: Proceedings RSEISP'07. Lecture Notes in Artificial Intelligence, vol. 4585, p. 69. Springer, Berlin (2004)
13. Polkowski, L.: The paradigm of granular rough computing. In: Proceedings ICCI'07, Lake Tahoe NV. pp. 145–163. IEEE Computer Society, Los Alamitos (2007)

14. Polkowski, L.: A unified approach to granulation of knowledge and granular computing based on rough mereology. In: Pedrycz, W., Skowron, A., Kreinovich, V. (eds.) Handbook of Granular Computing. Wiley, New York, Chapter 16 (2008)
15. Polkowski, L.: Granulation of knowledge: similarity based approach in information and decision systems. In: Meyers, R.A. (ed.) Encyclopedia of Complexity and System Sciences. Springer, Berlin, article 00788 (2009)
16. Polkowski, L., Artiemjew, P.: Granular computing: granular classifiers and missing values. In: Proceedings ICCI'07, Lake Tahoe NV. pp. 186–194. IEEE Computer Society, Los Alamitos (2007)
17. Polkowski, L., Artiemjew, P.: On granular rough computing with missing values. In: Proceedings RSEISP'07. Lecture Notes in Artificial Intelligence, vol. 4585, pp. 271–279. Springer, Berlin (2007)
18. Polkowski, L., Artiemjew, P.: On granular rough computing: factoring classifiers through granular structures. In: Proceedings RSEISP 2007. Lecture Notes in Artificial Intelligence, vol. 4585, pp. 280–290. Springer, Berlin (2007)
19. Polkowski, L., Artiemjew, P.: Towards granular computing: classifiers induced from granular structures. In: Proceedings RSKD'07. The Workshops at ECML/PKDD'07, pp. 43–53. Warsaw University Press, Warsaw (2007)
20. Polkowski, L., Artiemjew, P.: Classifiers based on granular structures from rough inclusions. In: Proceedings of 12th International Conference on Information Processing and Management of Uncertainty in Knowledge-Based Systems IPMU'08, pp. 1786–1794. Torremolinos (Malaga), Spain (2008)
21. Polkowski, L., Artiemjew P.: Rough sets in data analysis: foundations and applications. In: Smoliński, T.G., Milanova, M., Hassanien, A.-E. (eds.) Applications of Computational Intelligence in Biology: Current Trends and open Problems, SCI, vol. 122, pp. 33–54. Springer, Berlin (2008)
22. Polkowski, L., Artiemjew, P.: Rough mereology in classification of data: Voting by means of residual rough inclusions. In: Proceedings of 6th International Conference on Rough Sets and Current Trends in Computing RSCTC'08, Akron OH, USA. Lecture Notes in Artificial Intelligence, vol. 5306, pp. 113–120. Springer, Berlin (2008)
23. Polkowski, L., Artiemjew, P.: A study in granular computing: on classifiers induced from granular reflections of data. Transactions on Rough Sets IX. Lecture Notes in Computer Science, vol. 5390, pp. 230–263. Springer, Berlin (2008)
24. Polkowski, L., Artiemjew, P.: On classifying mappings induced by granular structures. Transactions on Rough Sets IX. Lecture Notes in Computer Science, vol. 5390, pp. 264–286. Springer, Berlin (2008)
25. Stone, P.: Layered Learning in Multiagent Systems. A Winning Approach to Robotic Soccer. MIT Press, Cambridge (2000)
26. UCI Repository. http://www.archive.ics.uci.edu/ml/. Accessed 11 Nov 2014
27. Wróblewski, J.: Adaptive aspects of combining approximation spaces. In: Pal, S.K., Polkowski, L., Skowron, A. (eds.) Rough Neural Computing. Techniques for Computing with Words, pp. 139–156. Springer, Berlin (2004)

Chapter 5
Covering Strategies

They will never agree, because they are arguing from different premises

[Samuel Johnson]

In this chapter, we review strategies for the choice of a granular covering for concept dependent granulation and we study the impact of a covering strategy, see Sect. 4.20, on the accuracy of the respective classifier. As observed in Chaps. 3 and 4, we are at liberty in selecting the strategy $\mathcal{G}$ which defines the method of choice of a covering of the universe of the data set with granules. We focus our attention on three main classes of coverings, viz.,

1. Class of coverings obtained by selecting granules in a specific order, without additional features, which does encompass subclasses Cov1, Cov2.
2. Class of coverings based on choice of granules by their size, which consists of subclasses Cov3, Cov4, Cov5.
3. Class of coverings obtained by accounting for the number of new objects transferred by a chosen granule to the constructed covering, which does contain subclasses Cov6, Cov7, Cov8.

In addition to the three main classes, we consider coverings obtained as in Classes 1, 2, 3, and modified by taking into consideration the size of the decision class corresponding to the granule. This yields subclasses Cov9–Cov16.

We evaluate the impact on classification of those coverings with data sets from UCI Repository. We apply the kNN classifier, cf., a discussion in Sect. 3.11.2 with the parameter k optimized, and we validate the results with multiple CV5 cross-validation, cf., Sect. 3.18.

5.1 Description of the Chosen Classifier

We apply kNN classifier, in our study of the impact of covering strategy on the quality of the classifier. The procedure applied is as follows.

© Springer International Publishing Switzerland 2015

L. Polkowski and P. Artiemjew, *Granular Computing in Decision Approximation*,
Intelligent Systems Reference Library 77, DOI 10.1007/978-3-319-12880-1_5

Step 1:   The training granular decision system $(G^{trn}_{r_{gran}}, A, d)$ and the test decision
             system (U_{tst}, A, d) have been input, where A is a set of conditional
             attributes, d the decision attribute, and, r_{gran} a granulation radius.
Step 2:   Classification of test objects by means of granules of training objects is
             performed as follows.

For all conditional attributes $a \in A$, training objects $v \in G^{trn}$, and test objects
$u \in U_{tst}$, we compute weights $w(u, v)$ based on the Hamming metric.

In the voting procedure of the kNN classifier, we use optimal k estimated by CV5,
details of the procedure are highlighted in Sect. 5.2.

If the cardinality of the smallest training decision class is less than k, we apply
the value for $k = |the\ smallest\ training\ decision\ class|$.

The test object u is classified by means of weights computed for all training objects
v, see the description of algorithms in Sect. 3.2. Weights are sorted in increasing
order as,

$$w_1^{C1}(u, v_1^{C1}) \leq w_2^{C1}(u, v_2^{C1}) \leq \cdots \leq w_{|C_1|}^{C1}(u, v_{|C_1|}^{C1});$$

$$w_1^{C2}(u, v_1^{C2}) \leq w_2^{C2}(u, v_2^{C2}) \leq \cdots \leq w_{|C_2|}^{C2}(u, v_{|C_2|}^{C2});$$

$$\cdots$$

$$w_1^{Cm}(u, v_1^{Cm}) \leq w_2^{Cm}(u, v_2^{Cm}) \leq \cdots \leq w_{|C_m|}^{Cm}(u, v_{|C_m|}^{Cm}),$$

where $C_1, C_2, \ldots, C_m$ are all decision classes in the training set.

Based on computed and sorted weights, training decision classes vote by means
of the following parameter, where c runs over decision classes in the training set,

$$Concept\_weight_c(u) = \sum_{i=1}^{k} w_i^c(u, v_i^c). \qquad (5.1)$$

Finally, the test object u is classified into the class c with a minimal value of *Concept_weight_c(u)*.

After all test objects u are classified, the quality parameter of *accuracy, acc* is
computed, according to the formula

$$acc = \frac{number\ of\ correctly\ classified\ objects}{number\ of\ classified\ objects}.$$

The results of classification for tested covering methods with real data sets—
selected from UCI Repository—see Table A.1, are reported starting from Sect. 5.3.
In Table A.1, we have the detailed information on data sets chosen. Successive
columns in Table A.1 give the name of the data *name*, the type of attributes *attr type*,

the number of attributes *attr no.*, the number of objects *obj no.*, the number of decision classes *class no.*. The method of granulation chosen for experiments require data of categorical attribute type, hence, integer and real attributes are treated as categorical.

5.2 Parameter Estimation in kNN Classifier

In our experiments, we use the classical version of kNN classifier based on the Hamming metric, cf., Chap. 3. In the first step, we estimate the optimal k based on $5 \times$ CV5 cross-validation on the part of data set. In the next step, we use the estimated value of k in order to find k nearest objects for each decision class and then we vote for decision. If the value of k is larger than the smallest training decision class cardinality than k is mapped on the cardinality of this class (Table 5.1).

In Table 5.2 we can see the estimated values of k for all examined data sets. These values were chosen as optimal on basis of experiments with various values of k and results estimated by means of multiple 5CV.

5.3 Granular Covering Methods

In this section, we address in detail the mentioned above in this chapter strategies Cov1–Cov16 for covering synthesis. We illustrate methods Cov1–Cov16 with the exemplary simple decision system in Table 5.3.

Table 5.1 Data sets description

Name	Attr type	Attr no.	Obj no.	Class no.
Adult	*Categorical, integer*	15	48,842	2
Australian-credit	*Categorical, integer, real*	15	690	2
Car Evaluation	*Categorical*	7	1,728	4
Diabetes	*Categorical, integer*	9	768	2
Fertility_Diagnosis	*Real*	10	100	2
German-credit	*Categorical, integer*	21	1,000	2
Heartdisease	*Categorical, real*	14	270	2
Hepatitis	*Categorical, integer, real*	20	155	2
Congressional Voting Records	*Categorical*	17	435	2
Mushroom	*Categorical*	23	8,124	2
Nursery	*Categorical*	9	12,960	5
Soybean-large	*Categorical*	36	307	19
SPECT Heart	*Categorical*	23	267	2
SPECTF Heart	*Integer*	45	267	2

Table 5.2 Estimated parameters for kNN based on $5 \times$ CV5 cross-validation

Name	Optimal k
Adult	79
Australian-credit	5
Car Evaluation	8
Diabetes	3
Fertility_Diagnosis	5
German-credit	18
Heartdisease	19
Hepatitis	3
Congressional Voting Records	3
Mushroom	1
Nursery	4
Soybean-large	3
SPECT Heart	19
SPECTF Heart	14

Table 5.3 Exemplary decision system

Object	a_1	a_2	a_3	a_4	d
u_1	2	1	1	1	1
u_2	2	1	1	1	1
u_3	2	1	1	3	2
u_4	1	2	2	3	2
u_5	3	2	2	2	2
u_6	4	3	4	6	2

Granulation is carried out for the granulation radius $r_{gran} = 0.25$ and granules are defined by the standard rough inclusion based on the Łukasiewicz t-norm, i.e.,

$$g_{.25}(u) = \{v \in U : \frac{|IND(u, v)|}{|A|} \geq 0.25\}. \qquad (5.2)$$

We recall that $IND(u, v) = \{a \in A : a(u) = a(v)\}$. Granules computed in this setting are: $g_{0.25}(u_1) = \{u_1, u_2, u_3\}$, $g_{0.25}(u_2) = \{u_1, u_2, u_3\}$, $g_{0.25}(u_3) = \{u_1, u_2, u_3, u_4\}$, $g_{0.25}(u_4) = \{u_3, u_4, u_5\}$, $g_{0.25}(u_5) = \{u_4, u_5\}$, $g_{0.25}(u_6) = \{u_6\}$.

5.3.1 Order-Preserving Coverings: Cov1

The first method of covering finding is based on the choice of granules one by one, with respect to a given ordering on the set of granules, until a covering of the universe

U of objects is obtained. The covering is constructed in a deterministic way based on the granules of the radius of 0.25 found from Table 5.3 in the following steps.

$U_{cover} \leftarrow \emptyset$

Step1: $\quad gr_{gran}(u_1) \rightarrow U_{cover}, \ U_{cover} = \{u_1, u_2, u_3\}.$
Step2: $\quad gr_{gran}(u_2) \nrightarrow U_{cover},$ no change.
Step3: $\quad gr_{gran}(u_3) \rightarrow U_{cover}, \ U_{cover} = \{u_1, u_2, u_3, u_4\}.$
Step4: $\quad gr_{gran}(u_4) \rightarrow U_{cover}, \ U_{cover} = \{u_1, u_2, u_3, u_4, u_5\}.$
Step5: $\quad gr_{gran}(u_5) \nrightarrow U_{cover},$ no change.
Step6: $\quad gr_{gran}(u_6) \rightarrow U_{cover}, \ U_{cover} = U,$ the original decision system is covered.

Finally, $U_{cover} = \{gr_{gran}(u_1), gr_{gran}(u_3), gr_{gran}(u_4), gr_{gran}(u_6)\}.$

5.3.2 Random Coverings: Cov2

Assuming that probability of the random choice is the same for all granules, exemplary random covering process may look as follows.

Draw1 $\quad \{1, 2, 3, 4, 5, 6\} = 2, \ gr_{gran}(u_2) \rightarrow U_{cover}, \ U_{cover} = \{u_1, u_2, u_3\}.$
Draw2 $\quad \{1, 3, 4, 5, 6\} = 4, \ gr_{gran}(u_4) \rightarrow U_{cover}, \ U_{cover} = \{u_1, u_2, u_3, u_4, u_5\}.$
Draw3 $\quad \{1, 3, 5, 6\} = 1, \ gr_{gran}(u_1) \nrightarrow U_{cover},$ no change.
Draw4 $\quad \{3, 5, 6\} = 6, \ gr_{gran}(u_6) \rightarrow U_{cover}, \ U_{cover} = U,$ the original decision system is covered.

Finally, $U_{cover} = \{gr_{gran}(u_2), gr_{gran}(u_4), gr_{gran}(u_6)\}.$

In the next group, we discuss methods based on cardinalities of granules. The first one is based on the selection in order of increasing cardinalities (the shortest granules first).

5.3.3 Coverings by Granules of a Minimal Size: Cov3

The third method of covering is based on the choice of granules in order of increasing cardinality. Ties are resolved according to the ordering of granules: in case of a tie, we choose granules from the smallest index to the highest. For our exemplary data set, with indiscernibility matrix from Table 5.3, the covering is constructed in a deterministic way.

Step1: $\quad gr_{gran}(u_6) \rightarrow U_{cover}, \ U_{cover} = \{u_6\}.$
Step2: $\quad gr_{gran}(u_5) \rightarrow U_{cover}, \ U_{cover} = \{u_5, u_6\}.$
Step3: $\quad gr_{gran}(u_1) \rightarrow U_{cover}, \ U_{cover} = \{u_1, u_2, u_3, u_5, u_6\}.$
Step4: $\quad gr_{gran}(u_2) \nrightarrow U_{cover},$ no change.
Step5: $\quad gr_{gran}(u_4) \rightarrow U_{cover}, \ U_{cover} = U,$ the original decision system has been covered.

Finally $U_{cover} = \{gr_{gran}(u_6), gr_{gran}(u_5), gr_{gran}(u_1), gr_{gran}(u_4)\}.$

5.3.4 Coverings by Granules of Average Size: Cov4

To choose the granules with an average size (average number of objects), we have to define the average size. By average size we understand the arithmetic mean of size of all granules rounded to the nearest integer value. If the fraction part is equal $\frac{1}{2}$ the value is rounded up to the nearest integer value. For the average value equal to φ, the granule possible size is $1, 2, \ldots, \varphi - 1, \varphi, \varphi + 1, \ldots, card\{U\}$. An exemplary policy in search of a covering is the following.

If $\varphi - 1 < card\{U\} - \varphi$, then granules are chosen in the following order of their size:

$$\varphi, \varphi - 1, \varphi + 1, \varphi - 2, \varphi + 2, \ldots, 1, 2\varphi - 1, 2\varphi, \ldots, card\{U\}.$$

In case $\varphi - 1 > card\{U\} - \varphi$, the order of choice is the following:

$$\varphi, \varphi - 1, \varphi + 1, \varphi - 2, \varphi + 2, \ldots, 2\varphi - card\{U\}, card\{U\}, 2\varphi - card\{U\} - 1, \ldots, 1,$$

whereas in case of equality $\varphi - 1 = card\{U\} - \varphi$, selection is as follows:

$$\varphi, \varphi - 1, \varphi + 1, \varphi - 2, \varphi + 2, \ldots, 1, card\{U\}.$$

If we have more than one granule of the same size, such tie is resolved according to the ordering of granules. To clarify our approach we consider the following example, referring to the indiscernibility matrix Table 5.3.
In this case, sizes of granules are:

$card\{g_{r_{gran}}(u_1)\} = 3;$
$card\{g_{r_{gran}}(u_2)\} = 3;$
$card\{g_{r_{gran}}(u_3)\} = 4;$
$card\{g_{r_{gran}}(u_4)\} = 3;$
$card\{g_{r_{gran}}(u_5)\} = 2;$
$card\{g_{r_{gran}}(u_6)\} = 1,$

hence, the arithmetic mean is

$$\frac{3 + 3 + 4 + 3 + 2 + 1}{6} = \frac{16}{6} = 2\frac{2}{3}.$$

After rounding up we get $\varphi = 3$. As $card\{U\} = 6$, our example fulfils the property $\varphi - 1 < card\{U\} - \varphi$, hence, during granule selection we have the following order:

$$\varphi, \varphi - 1, \varphi + 1, \varphi - 2, \varphi + 2, \ldots, 1, 2\varphi - 1, 2\varphi, \ldots, card\{U\}.$$

We choose granules in order determined by their size, for our policy it is

$$3, 2, 4, 1, 5, 6.$$

The exemplary covering based on the described method is constructed as follows.

Step1: $gr_{gran}(u_1) \to U_{cover}, \ U_{cover} = \{u_1, u_2, u_3\}$.
Step2: $gr_{gran}(u_2) \nrightarrow U_{cover}$, no change.
Step3: $gr_{gran}(u_4) \to U_{cover}, \ U_{cover} = \{u_1, u_2, u_3, u_4, u_5\}$.
Step4: $gr_{gran}(u_5) \nrightarrow U_{cover}$, no change.
Step5: $gr_{gran}(u_3) \nrightarrow U_{cover}$, no change.
Step6: $gr_{gran}(u_6) \to U_{cover}, \ U_{cover} = U$, the original decision system is covered.

We have obtained the following covering $U_{cover} = \{gr_{gran}(u_1), gr_{gran}(u_4), gr_{gran}(u_6)\}$. Naturally, the policy of granule selection can be modified; one of those versions can look as follows.
In case $\varphi - 1 < card\{U\} - \varphi$, granules can be selected in the following way:

$$\varphi, \varphi + 1, \varphi - 1, \varphi + 2, \varphi - 2, \ldots, 2\varphi - 1, 1, 2\varphi, \ldots, card\{U\}.$$

In case $\varphi - 1 > card\{U\} - \varphi$, granules of the following sizes have the priority:

$$\varphi, \varphi+1, \varphi-1, \varphi+2, \varphi-2, \ldots, card\{U\}, 2\varphi-card\{U\}, 2\varphi-card\{U\}-1, \ldots, 1.$$

In case of equality $\varphi-1 = card\{U\}-\varphi$, we choose into the covering granules of sizes

$$\varphi, \varphi + 1, \varphi - 1, \varphi + 2, \varphi - 2, \ldots, card\{U\}, 1.$$

This seemingly subtle change in granule selection order can significantly change the form of covering of the universe of objects.

5.3.5 Coverings by Granules of Maximal Size: Cov5

The reverse method to minimal size selection works in the following way:

Step1: $gr_{gran}(u_3) \to U_{cover}, \ U_{cover} = \{u_1, u_2, u_3, u_4\}$.
Step2: $gr_{gran}(u_1) \nrightarrow U_{cover}$, no change.
Step3: $gr_{gran}(u_2) \nrightarrow U_{cover}$, no change.
Step4: $gr_{gran}(u_4) \to U_{cover}, \ U_{cover} = \{u_1, u_2, u_3, u_4, u_5\}$.
Step5: $gr_{gran}(u_5) \nrightarrow U_{cover}$, no change.
Step6: $gr_{gran}(u_6) \to U_{cover}, \ U_{cover} = U$, the original decision system is covered.

The covering is $U_{cover} = \{gr_{gran}(u_3), gr_{gran}(u_4), gr_{gran}(u_6)\}$.

5.3.6 Coverings by Granules Which Transfer
the Smallest Number of New Objects: Cov6

This method of covering finding is based on the choice of granules according to the increasing number of new objects transferred to the covering. As with the previous methods, ties are resolved hierarchically. We select granules from the smallest index up to the highest. For our exemplary granule data set, the covering is built in a deterministic way.

Step1: $g_{r_{gran}}(u_6) \rightarrow U_{cover}$, $U_{cover} = \{u_6\}$.
Step2: $g_{r_{gran}}(u_5) \rightarrow U_{cover}$, $U_{cover} = \{u_4, u_5, u_6\}$.
Step3: $g_{r_{gran}}(u_4) \rightarrow U_{cover}$, $U_{cover} = \{u_3, u_4, u_5, u_6\}$.
Step4: $g_{r_{gran}}(u_1) \rightarrow U_{cover}$, $U_{cover} = U$, the original decision system is covered.

The covering obtained is $U_{cover} = \{g_{r_{gran}}(u_6), g_{r_{gran}}(u_5), g_{r_{gran}}(u_4), g_{r_{gran}}(u_1)\}$.

5.3.7 Coverings by Granules Which Transfer
an Average Number of New Objects: Cov7

The method is similar to the previous one, in this case we select granules which transfer the average number of new objects. This method works dynamically, we compute the average number of possible new objects in each step. In practice it looks as follows.

Step1: $\frac{3+3+4+2+1}{6} = 2\frac{2}{3}$, our rounded average number of samples, which could be transferred to the covering is $\lambda = 3$, thus we choose to the covering the granule with three new objects, or the granule with the most similar number of new objects.
$g_{r_{gran}}(u_1) \rightarrow U_{cover}$, $U_{cover} = \{u_1, u_2, u_3\}$.
In the next step, we compute new average value and choose a next granule.
Step2: $\frac{1+2+2+1}{4} = 1\frac{1}{2}$, thus $\lambda = 2$.
We choose a granule with two new objects.
$g_{r_{gran}}(u_5) \rightarrow U_{cover}$, $U_{cover} = \{u_1, u_2, u_3, u_4, u_5\}$.
Step3:   In the next step the average λ is equal 1, it is only one granule, which could transfer to one new object.
$g_{r_{gran}}(u_6) \rightarrow U_{cover}$, $U_{cover} = U$, the original decision system is covered,

Finally, $U_{cover} = \{g_{r_{gran}}(u_1), g_{r_{gran}}(u_5), g_{r_{gran}}(u_6)\}$.

5.3.8 Coverings by Granules Which Transfer Maximal Number of New Objects: Cov8

In the first step of this new strategy, we select granules which transfer to the covering a maximal number of new objects. The ties are resolved similarly as with the previous strategies. Steps of this method are as follows.

Step1: $g_{r_{gran}}(u_3) \rightarrow U_{cover}$, $U_{cover} = \{u_1, u_2, u_3, u_4\}$.
Step2: $g_{r_{gran}}(u_4) \rightarrow U_{cover}$, $U_{cover} = \{u_1, u_2, u_3, u_4, u_5\}$.
Step3: $g_{r_{gran}}(u_6) \rightarrow U_{cover}$, $U_{cover} = U$, the original decision system is covered.

The covering is as follows, $U_{cover} = \{g_{r_{gran}}(u_3), g_{r_{gran}}(u_4), g_{r_{gran}}(u_6)\}$.

5.3.9 Order-Preserving Coverings Proportional to the Size of Decision Classes: Cov9

For decision classes $X = \{u_1, u_2\}$ and $Y = \{u_3, u_4, u_5, u_6\}$, disproportion is as 1 to 2, which leads us to the strategy of choice that for each granule from the class X chooses two granules from the class Y. The granules are chosen one by one proportionally to cardinality of decision classes.

The method works in deterministic way, and its steps are as follows.
We select a granule from the class X.

Step1: $g_{r_{gran}}(u_1) \rightarrow U_{cover}$, $U_{cover} = \{u_1, u_2, u_3\}$.
   Next, we select two consecutive granules in the class Y.
Step2: $g_{r_{gran}}(u_3) \rightarrow U_{cover}$, $U_{cover} = \{u_1, u_2, u_3, u_4\}$.
Step3: $g_{r_{gran}}(u_4) \rightarrow U_{cover}$, $U_{cover} = \{u_1, u_2, u_3, u_4, u_5\}$.
   The next granule comes from the class X.
Step4: $g_{r_{gran}}(u_2) \nrightarrow U_{cover}$, no change.
   Now, two consecutive granules are selected from the class Y.
Step5: $g_{r_{gran}}(u_5) \nrightarrow U_{cover}$, no change.
Step6: $g_{r_{gran}}(u_6) \rightarrow U_{cover}$, $U_{cover} = U$, the original decision system is covered.

Finally, $U_{cover} = \{g_{r_{gran}}(u_1), g_{r_{gran}}(u_3), g_{r_{gran}}(u_4), g_{r_{gran}}(u_6)\}$.

5.3.10 Random Coverings Proportional to the Size of Decision Classes: Cov10

The next method is similar to the previous one, but the proportional choice is now random, and the results of granulation could be distinct. In our example, we have only two decision classes $X = \{u_1, u_2\}$ and $Y = \{u_3, u_4, u_5, u_6\}$, so the disproportion in size is as 1 to 2 in favor of Y.

We make the first random choice in the class X.

Draw1   from $\{1, 2\} = 2$, $g_{r_{gran}}(u_2) \to U_{cover}$, $U_{cover} = \{u_1, u_2, u_3\}$.
   Now, we have two consecutive random choices from the class Y.
Draw2   from $\{3, 4, 5, 6\} = 4$, $g_{r_{gran}}(u_4) \to U_{cover}$, $U_{cover} = \{u_1, u_2, u_3, u_4, u_5\}$.
Draw3   from $\{3, 5, 6\} = 5$, $g_{r_{gran}}(u_5) \not\to U_{cover}$, no change.
   The next random choice is from class X.
Draw4   from $\{1\} = 1$, $g_{r_{gran}}(u_1) \not\to U_{cover}$, no change.
   Now, two consecutive random choices from the class Y follow.
Draw5   from $\{3, 6\} = 6$, $g_{r_{gran}}(u_6) \to U_{cover}$, $U_{cover} = U$, the original decision
   system is covered.

The second random choice is redundant.
The covering of the universe of objects is as follows:

$$U_{cover} = \{g_{r_{gran}}(u_2), g_{r_{gran}}(u_4), g_{r_{gran}}(u_6)\}.$$

The next group of methods extend Cov3–Cov5 to the proportional choice of granules with respect to the size of decision classes in the granular reflection.

5.3.11 Coverings Proportional to the Size of Decision Classes by Granules of a Minimal Size: Cov11

In this method we select granules of a current minimal cardinality, in proportion to the size of decision classes.
   We select a granule in the class X.

Step1: $g_{r_{gran}}(u_1) \to U_{cover}$, $U_{cover} = \{u_1, u_2, u_3\}$.
   Next, we select two consecutive granules in the class Y.
Step2: $g_{r_{gran}}(u_6) \to U_{cover}$, $U_{cover} = \{u_1, u_2, u_3, u_6\}$.
Step3: $g_{r_{gran}}(u_5) \to U_{cover}$, $U_{cover} = U$, the original decision system is covered.

Finally, we obtain $U_{cover} = \{g_{r_{gran}}(u_1), g_{r_{gran}}(u_6), g_{r_{gran}}(u_5)\}$.

5.3.12 Coverings Proportional to the Size of Decision Classes by Granules of the Average Size: Cov12

This method is similar to the previous one, where we first select granules with the average cardinality. For exemplary decision class $K = \{v_1, v_2, \ldots, v_k\}$, after computing the average value of granules and rounding to the nearest integer value, we get φ_K. Granules may have sizes $1, 2, \ldots, \varphi_K - 1, \varphi_K, \varphi_K + 1, \ldots, |U|$. In order to include a granule into the covering, we have to determine the policy of choice.

In case $\varphi_K - 1 < |U| - \varphi_K$, granules are chosen in the following order:

$$\varphi_K, \varphi_K - 1, \varphi_K + 1, \varphi_K - 2, \varphi_K + 2, \ldots, 1, 2\varphi_K - 1, 2\varphi_K, \ldots, |U|.$$

In case $\varphi_K - 1 > |U| - \varphi_K$, the order of choice is different:

$$\varphi_K, \varphi_K - 1, \varphi_K + 1, \varphi_K - 2, \varphi_K + 2, \ldots, 2\varphi_K - |U|, |U|, 2\varphi_K - |U| - 1, \ldots, 1.$$

In case of equality $\varphi_K - 1 = |U| - \varphi_K$, the order is:

$$\varphi_K, \varphi_K - 1, \varphi_K + 1, \varphi_K - 2, \varphi_K + 2, \ldots, 1, |U|.$$

As with the previous methods, ties are resolved in order–preserving style.

In our example, the covering could be find in the following way. For decision classes $X = \{u_1, u_2\}$ and $Y = \{u_3, u_4, u_5, u_6\}$, the average size of granules in the class X is equal to $\frac{3+3}{2} = 3$, hence, $\varphi_X = 3$ and in the class Y it is equal to $\frac{4+3+2+1}{4} = 2\frac{1}{2}$, hence, $\varphi_Y = 3$. The politics of granule choice is identical for concepts X and Y, the granules are chosen in the order 3, 2, 4, 1, 5, 6. If there are no granules of considered size, we select granules of the next smaller size. Considering sizes of decision classes, for each granule in the class X, two granules in the class Y are selected.

We choose in the class X the first granule of size 3.

Step1: $g_{r_{gran}}(u_1) \rightarrow U_{cover}$, $U_{cover} = \{u_1, u_2, u_3\}$.

Next, we choose two consecutive granules in the class Y, we begin with granules of size 3.

Step2: $g_{r_{gran}}(u_4) \rightarrow U_{cover}$, $U_{cover} = \{u_1, u_2, u_3, u_4, u_5\}$.

In the decision class Y we have now no granule of size 3, so we choose from granules of size 2.

Step3: $g_{r_{gran}}(u_5) \nrightarrow U_{cover}$, no change.

The next granule of size 3 comes from the class X.

Step4: $g_{r_{gran}}(u_2) \nrightarrow U_{cover}$, no change.

We now choose a granule with size 1 in the class Y.

Step5: $g_{r_{gran}}(u_6) \rightarrow U_{cover}$, $U_{cover} = U$, the original decision system is covered.

The covering is $U_{cover} = \{g_{r_{gran}}(u_1), g_{r_{gran}}(u_4), g_{r_{gran}}(u_6)\}$.

5.3.13 Coverings Proportional to the Size of Decision Classes by Granules of a Maximal Size: Cov13

For this strategy, derived from the method for Cov5, the covering proportional to cardinalities od decision classes is chosen as follows.

First, we choose a granule in the class X.

Step1: $gr_{gran}(u_1) \rightarrow U_{cover}$, $U_{cover} = \{u_1, u_2, u_3\}$.
    Next, we select two consecutive granules in the class Y.
Step2: $gr_{gran}(u_3) \rightarrow U_{cover}$, $U_{cover} = \{u_1, u_2, u_3, u_4\}$.
Step3: $gr_{gran}(u_4) \rightarrow U_{cover}$, $U_{cover} = \{u_1, u_2, u_3, u_4, u_5\}$.
    Now, we choose a granule in the class X.
Step4: $gr_{gran}(u_2) \nrightarrow U_{cover}$, no change.
    Finally, we select two granules in the class Y.
Step5: $gr_{gran}(u_5) \nrightarrow U_{cover}$, no change.
Step6: $gr_{gran}(u_6) \rightarrow U_{cover}$, $U_{cover} = U$, the original decision system is covered.

The granular covering is $U_{cover} = \{gr_{gran}(u_1), gr_{gran}(u_3), gr_{gran}(u_4), gr_{gran}(u_6)\}$.

5.3.14 Coverings Proportional to the Size of Decision Classes, by Granules Which Transfer the Smallest Number of New Objects: Cov14

From the class X, we select a granule minimal with respect to the number of transferred objects.

Step1: $gr_{gran}(u_1) \rightarrow U_{cover}$, $U_{cover} = \{u_1, u_2, u_3\}$.
    From the class Y, we select two consecutive granules with the minimal transfer property.
Step2: $gr_{gran}(u_3) \rightarrow U_{cover}$, $U_{cover} = \{u_1, u_2, u_3, u_4\}$.
Step3: $gr_{gran}(u_4) \rightarrow U_{cover}$, $U_{cover} = \{u_1, u_2, u_3, u_4, u_5\}$.
    Next, we select a granule from the class X.
Step4: $gr_{gran}(u_2) \nrightarrow U_{cover}$, no change.
    Finally, a granule from the class Y.
Step5: $gr_{gran}(u_6) \rightarrow U_{cover}$, $U_{cover} = U$, the original decision system is covered.

The covering is $U_{cover} = \{gr_{gran}(u_1), gr_{gran}(u_3), gr_{gran}(u_4), gr_{gran}(u_6)\}$.

5.3.15 Coverings Proportional to the Size of Decision Classes, by Granules Which Transfer the Average Number of New Objects: Cov15

This method is a variation on the method Cov7, and the choice is proportional to the size of decision classes. The average number of new objects, which could be transferred from granules into the covering is computed dynamically in each step. The proportion of decision classes is 1 to 2.

First, we select one granule in the class X, the average number of new possible objects in covering equals $\frac{3+3}{2}$, hence $\varphi = 3$.

Based on the average value, we select in X the following granule.

Step1: $gr_{gran}(u_1) \to U_{cover}$, $U_{cover} = \{u_1, u_2, u_3\}$.
   Next, in the class Y, the average is equal to $\frac{1+2+2+1}{4} = \frac{6}{4}$, thus $\varphi = 2$.
   In the class Y we choose
Step2: $gr_{gran}(u_4) \to U_{cover}$, $U_{cover} = \{u_1, u_2, u_3, u_4, u_5\}$, and then as still $\varphi = 3$,
Step3: $gr_{gran}(u_6) \to U_{cover}$, $U_{cover} = U$, the original decision system is covered.

Finally, $U_{cover} = \{gr_{gran}(u_1), gr_{gran}(u_4), gr_{gran}(u_6)\}$.

5.3.16 Coverings Proportional to the Size of Decision Classes, by Granules Which Transfer a Maximal Number of New Objects: Cov16

The last method is modeled on that of Cov8, with cardinality of decision classes accounted for.
   The first granule comes from the class X.

Step1: $gr_{gran}(u_1) \to U_{cover}$, $U_{cover} = \{u_1, u_2, u_3\}$.
   Two consecutive granules are taken in the class Y.
Step2: $gr_{gran}(u_4) \to U_{cover}$, $U_{cover} = \{u_1, u_2, u_3, u_4, u_5\}$.
Step3: $gr_{gran}(u_6) \to U_{cover}$, $U_{cover} = U$, the original decision system is covered.

The covering is $U_{cover} = \{gr_{gran}(u_1), gr_{gran}(u_4), gr_{gran}(u_6)\}$.

5.4 Experimental Session with Real World Data Sets

We carry out experiments with data from UCI Repository. The missing values have been filled in with the most common values in decision classes. A short description is as follows: the original data were split into five sub-data sets each time and used in the five time repeated CV5 method. For an exemplary fold, the chosen training set consisted of four parts of original data and the test set consisted of the fifth part. The training data set have been granulated and used for classification of test data by means of the kNN classifier. Granular reflections have been formed from covering granules with use of the majority voting strategy. The distance between objects in kNN classifier is computed by the Hamming distance function.
   The value of k for each data set is estimated based on the samples of data, cf., Sect. 5.2.

5.5 Summary of Results for Discrete Data Sets
from UCI Repository

Before showing the results of classifications on granular reflections of the chosen data sets, we give a summary of interpretations of these results.

1. Comparison between Cov1 and Cov2 shows that results are almost identical due to multiple Cross Validation method applied. The result is an average value from 25 tests (5 × CV5), it means that hierarchical covering was performed on folds formed in random way, thus the result is equal to random choice. Random covering is statistically better than hierarchical method, because the complexity of granulation is independent of the order of objects in training decision system. In case of the hierarchical method the order of objects plays significant role and has influence on the speed of granulation and on the quality of granular decision system in the sense of accuracy of classification. In the random method any preprocessing of training data does not matter. In the hierarchical method, preprocessing could improve the quality od granular data. Considering the equality of Cov1 and Covr2 methods for multiple Cross Validation the result for Cov1 was not included, consequently the result for Cov9 was not included, too. Taking into account the comparison between Cov2 and Cov10 methods, we have skipped the result for Cov10.
2. In case of Cov3 versus Cov11, Cov3 method seems better. The possible explanation is that a choice proportional to decision classes size forces the choice of longer granules after using the shortest one from the particular class. It seems better to choose the shortest granule globally, but the quality of result depends significantly on data set content.
3. In Case of Cov3 versus Cov4 versus Cov5, Cov3 and Cov4 win in most cases. Cov5 method wins from time to time for larger radii, because for large radii, we meet almost exclusively singleton granules, and only small number of larger granules. Cov5 method takes into account such larger granules, but still small considering the size of training decision system. That is the reason for good classification for larger radii. In case of Cov3 and Cov4 there is no effect of granulation for radii close to 1, hence, Cov5 could win with significant reduction of object number in the training data set.
4. In case of Cov11 versus Cov12 versus Cov13, the result is fully comparable to the comparison in pt. 3.
5. In case of Cov4 versus Cov12, the result is highly dependent on the data set content.
6. In case of Cov5 versus Cov13, results are fully comparable, in some cases Cov13 wins slightly.
7. In case of Cov6 versus Cov7 versus Cov8, Cov7 and Cov8 seem to perform better.

8. In case of Cov6 versus Cov14, results are fully comparable and seem to depend on the data set content.
9. In case of Cov7 versus Cov15, it is Cov15 which seems to be slightly better.
10. In case of Cov8 versus Cov16, in most cases Cov16 wins slightly, but the result seems to be highly dependent on the data set content.

Detailed results for real data sets described in Table 5.1 are shown in Tables 5.4, 5.5, 5.6, 5.7, 5.8, 5.9, 5.10, 5.11, 5.12, 5.13, 5.14, 5.15 and 5.16. These tables include additional visualizations of results, and, comparisons of random coverings (Cov2) with two groups of methods:

(a) dependent on the granule size (Cov3, Cov4 and Covr5);
(b) dependent on the number of new granules transferred to the covering (Cov6, Cov7 and Cov8).

There are additional results for respective mentioned methods, when the decision system is covered in proportion to decision classes cardinality (Cov11–Cov16).

Table 5.4 5 *times* CV-5; Average global accuracy of classification (Acc) and size of granular decision systems (GranSize); Result of experiments for Covering finding methods of concept-dependent granulation with use of kNN classifier; data set: **australian**; r_{gran} = Granulation radius, Cov_i = Average accuracy for method Cov_i

r_{gran}	Cov2	Cov3	Cov4	Cov5	Cov6	Cov7	Cov8	Cov11	Cov12	Cov13	Cov14	Cov15	Cov16
Acc													
0	0.774	0.777	0.773	0.772	0.774	0.773	0.775	0.772	0.772	0.773	0.771	0.775	0.772
0.071	0.773	0.791	0.774	0.772	0.797	0.794	0.775	0.794	0.773	0.773	0.795	0.775	0.772
0.143	0.779	0.806	0.776	0.772	0.801	0.797	0.775	0.804	0.773	0.773	0.806	0.776	0.772
0.214	0.786	0.818	0.779	0.772	0.802	0.799	0.778	0.817	0.778	0.772	0.8	0.789	0.775
0.286	0.804	0.853	0.785	0.785	0.822	0.815	0.783	0.849	0.787	0.783	0.822	0.801	0.784
0.357	0.813	0.819	0.794	0.793	0.814	0.813	0.821	0.814	0.798	0.789	0.808	0.813	0.817
0.429	0.828	0.82	0.819	0.809	0.818	0.836	0.836	0.825	0.818	0.807	0.822	0.842	0.845
0.5	0.845	0.772	0.842	0.841	0.842	0.847	0.846	0.797	0.842	0.837	0.845	0.841	0.844
0.571	0.838	0.829	0.844	0.838	0.846	0.85	0.832	0.83	0.842	0.845	0.842	0.833	0.838
0.643	0.845	0.849	0.848	0.848	0.848	0.847	0.849	0.848	0.849	0.852	0.853	0.845	0.846
0.714	0.854	0.85	0.851	0.849	0.858	0.846	0.853	0.852	0.852	0.849	0.856	0.856	0.856
0.786	0.858	0.856	0.855	0.855	0.859	0.854	0.856	0.857	0.851	0.855	0.857	0.856	0.861
0.857	0.861	0.859	0.859	0.857	0.86	0.858	0.857	0.859	0.853	0.857	0.86	0.857	0.864
0.929	0.863	0.861	0.862	0.859	0.863	0.861	0.859	0.861	0.856	0.86	0.861	0.859	0.865
1	0.861	0.86	0.861	0.857	0.86	0.861	0.858	0.86	0.854	0.858	0.859	0.859	0.865
GranSize													
0	2	2	2	2	2	2	2	2	2	2	2	2	2
0.071	2.64	4.16	3	2	4.32	4.08	2.44	4.16	3	2	4.36	2.8	2

(continued)

Table 5.4 (continued)

r_{gran}	Cov2	Cov3	Cov4	Cov5	Cov6	Cov7	Cov8	Cov11	Cov12	Cov13	Cov14	Cov15	Cov16
0.143	3.92	9.08	5	2	11.48	7.8	3.68	9.16	4.6	2	11.72	4.76	2
0.214	5.36	15.4	4.36	2.72	23.56	8.96	5.48	15.16	4.36	2.72	23.76	6.6	2.68
0.286	9.12	29.2	8.4	11.08	48.8	10.6	8.48	29.36	8.68	10.84	49.84	10.52	4
0.357	16.12	51.68	18.24	25.16	93.4	15.76	16.16	51.2	18.72	24.76	94	17.88	7.32
0.429	32.44	77.72	39.32	46.8	155.72	29.44	32.32	76.88	37.96	46.92	154.8	34.28	15.28
0.5	71.64	133.88	80.72	88.68	250.12	66.8	73.12	135.52	80.36	88.48	251.48	70.84	37.8
0.571	157.96	228.68	163.96	167.04	348.84	152.52	157.08	226.88	165.36	168.88	346.36	154.44	105.8
0.643	318.96	356.4	328.32	310.72	423.68	332.52	319.04	356.24	327.92	311.44	425.04	333.52	275.8
0.714	468.16	477.76	477.76	455.8	492.64	481.84	467.08	476.8	477.24	455.72	492.6	481.4	452.08
0.786	535.84	536.56	536.76	535.04	536.36	536.56	536	536.44	536.52	534.92	536.64	536.6	535.36
0.857	547.2	547.32	547.24	547.28	547.2	547.2	547.12	547.08	547.24	547.2	547.28	547.28	547.24
0.929	548.8	548.96	548.88	548.96	548.84	548.8	548.8	548.76	548.8	548.76	548.8	548.8	548.8
1	552	552	552	552	552	552	552	552	552	552	552	552	552

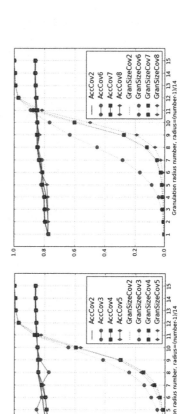

Table 5.5 5 *times* CV-5; Average global accuracy of classification (Acc) and size of granular decision systems (GranSize); Result of experiments for Covering finding methods of concept-dependent granulation with use of kNN classifier; data set: **car evaluation**; r_{gran} = Granulation radius, Cov_i = Average accuracy for method Cov_i

r_{gran}	$Cov2$	$Cov3$	$Cov4$	$Cov5$	$Cov6$	$Cov7$	$Cov8$	$Cov11$	$Cov12$	$Cov13$	$Cov14$	$Cov15$	$Cov16$
Acc													
0	0.327	0.319	0.323	0.329	0.328	0.336	0.32	0.333	0.319	0.33	0.327	0.325	0.324
0.167	0.396	0.392	0.436	0.331	0.418	0.376	0.334	0.398	0.428	0.332	0.437	0.401	0.337
0.333	0.539	0.567	0.533	0.448	0.599	0.441	0.452	0.561	0.541	0.438	0.614	0.526	0.449
0.5	0.681	0.605	0.625	0.576	0.7	0.573	0.62	0.601	0.627	0.573	0.693	0.678	0.617
0.667	0.804	0.754	0.751	0.701	0.828	0.789	0.805	0.749	0.738	0.704	0.828	0.812	0.814
0.833	0.864	0.924	0.802	0.813	0.919	0.864	0.87	0.925	0.803	0.82	0.921	0.863	0.871
1	0.944	0.944	0.943	0.944	0.943	0.942	0.944	0.946	0.943	0.947	0.944	0.946	0.946

(continued)

Table 5.5 (continued)

r_{gran}	Cov2	Cov3	Cov4	Cov5	Cov6	Cov7	Cov8	Cov11	Cov12	Cov13	Cov14	Cov15	Cov16
GranSize													
0	4	4	4	4	4	4	4	4	4	4	4	4	4
0.167	8.32	11.56	10.28	6.68	25.64	6.8	6	11.32	9.68	6.84	25.72	9.88	5.96
0.333	16.96	30.72	26.96	21	92.04	14.68	10.36	27.12	27	20.44	93.76	18.68	10.28
0.5	38.2	85.88	81.6	60.68	223.12	36.44	20.12	87.16	81.8	60.12	222.56	40.56	20.12
0.667	107.04	229.24	191.88	173.64	482.6	98.04	54.96	230.48	193.08	174.84	485.28	106.4	55.84
0.833	371.64	500.52	427.32	394.92	959.96	343.48	212.76	497	427.08	394.16	954.36	361.72	212.48
1	1382.4	1382.4	1382.4	1382.4	1382.4	1382.4	1382.4	1382.4	1382.4	1382.4	1382.4	1382.4	1382.4

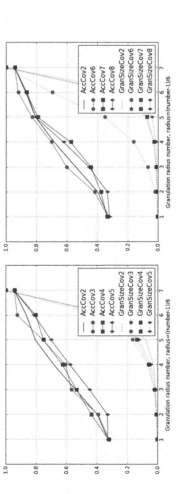

Table 5.6 5 *times* CV-5; Average global accuracy of classification (Acc) and size of granular decision systems (GranSize); Result of experiments for Covering finding methods of concept-dependent granulation with use of kNN classifier; data set: **diabetes**; r_{gran} = Granulation radius, Cov_i = Average accuracy for method Cov_i

r_{gran}	$Cov2$	$Cov3$	$Cov4$	$Cov5$	$Cov6$	$Cov7$	$Cov8$	$Cov11$	$Cov12$	$Cov13$	$Cov14$	$Cov15$	$Cov16$
Acc													
0	0.599	0.597	0.606	0.595	0.6	0.602	0.603	0.603	0.597	0.6	0.605	0.598	0.602
0.125	0.606	0.611	0.639	0.585	0.599	0.613	0.582	0.608	0.631	0.59	0.592	0.594	0.605
0.25	0.62	0.643	0.622	0.619	0.641	0.608	0.615	0.64	0.624	0.615	0.642	0.618	0.624
0.375	0.641	0.631	0.635	0.647	0.647	0.648	0.628	0.646	0.647	0.648	0.644	0.645	0.638
0.5	0.655	0.641	0.636	0.651	0.648	0.639	0.654	0.647	0.646	0.649	0.651	0.653	0.656
0.625	0.657	0.648	0.641	0.65	0.648	0.647	0.646	0.648	0.648	0.652	0.652	0.656	0.656
0.75	0.656	0.648	0.64	0.65	0.647	0.647	0.646	0.648	0.649	0.652	0.652	0.654	0.655
0.875	0.656	0.648	0.64	0.65	0.647	0.647	0.646	0.648	0.649	0.652	0.652	0.654	0.655
1	0.656	0.648	0.64	0.65	0.647	0.647	0.646	0.648	0.649	0.652	0.652	0.654	0.655

(continued)

Table 5.6 (continued)

r_{gran}	Cov2	Cov3	Cov4	Cov5	Cov6	Cov7	Cov8	Cov11	Cov12	Cov13	Cov14	Cov15	Cov16
GranSize													
0	2	2	2	2	2	2	2	2	2	2	2	2	2
0.125	34.28	69.72	55.96	43.72	114	24.24	16.04	70.92	56.44	42.48	114.28	24.08	16.04
0.25	155.64	243.88	159.84	176.88	342.16	124.96	97.4	244.52	159.52	177.68	343.36	136.68	97.32
0.375	365.32	416.48	374.36	372.32	474.76	370.68	323.36	417.92	374.44	372.64	474.28	377.88	324.88
0.5	539.76	548.44	548.36	534.64	562.84	554.6	526.16	548.2	549.16	533.88	562.88	554.56	526.52
0.625	609.96	609.88	609.64	609.64	610.28	610.44	609.64	609.84	609.92	609.68	610.32	610.36	609.8
0.75	614.4	614.4	614.4	614.4	614.4	614.4	614.4	614.4	614.4	614.4	614.4	614.4	614.4
0.875	614.4	614.4	614.4	614.4	614.4	614.4	614.4	614.4	614.4	614.4	614.4	614.4	614.4
1	614.4	614.4	614.4	614.4	614.4	614.4	614.4	614.4	614.4	614.4	614.4	614.4	614.4

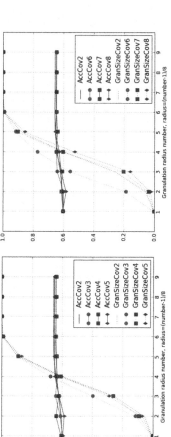

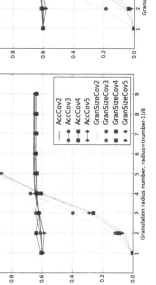

Table 5.7 5 *times* CV-5; Average global accuracy of classification (Acc) and size of granular decision systems (GranSize); Result of experiments for Covering finding methods of concept-dependent granulation with use of kNN classifier; data set: **fertility diagnosis**; r_{gran} = Granulation radius, Cov_i = Average accuracy for method Cov_i

r_{gran}	Cov2	Cov3	Cov4	Cov5	Cov6	Cov7	Cov8	Cov11	Cov12	Cov13	Cov14	Cov15	Cov16
Acc													
0	0.406	0.41	0.428	0.422	0.4	0.4	0.408	0.444	0.394	0.422	0.424	0.362	0.376
0.111	0.434	0.474	0.48	0.422	0.48	0.484	0.408	0.516	0.454	0.422	0.464	0.364	0.376
0.222	0.478	0.486	0.6	0.432	0.608	0.562	0.434	0.504	0.554	0.418	0.568	0.464	0.414
0.333	0.568	0.684	0.66	0.528	0.744	0.628	0.534	0.72	0.646	0.484	0.732	0.632	0.56
0.444	0.716	0.844	0.722	0.692	0.808	0.74	0.568	0.834	0.716	0.684	0.82	0.748	0.618
0.556	0.862	0.854	0.866	0.86	0.862	0.834	0.82	0.854	0.844	0.842	0.852	0.836	0.786
0.667	0.836	0.866	0.864	0.848	0.866	0.856	0.83	0.864	0.86	0.834	0.862	0.83	0.828
0.778	0.868	0.866	0.874	0.868	0.872	0.862	0.874	0.87	0.866	0.85	0.872	0.868	0.864
0.889	0.87	0.874	0.872	0.878	0.882	0.866	0.88	0.874	0.872	0.86	0.87	0.874	0.872
1	0.872	0.874	0.87	0.878	0.878	0.868	0.88	0.874	0.874	0.862	0.876	0.87	0.868

(continued)

Table 5.7 (continued)

r_{gran}	Cov2	Cov3	Cov4	Cov5	Cov6	Cov7	Cov8	Cov11	Cov12	Cov13	Cov14	Cov15	Cov16
GranSize													
0	2	2	2	2	2	2	2	2	2	2	2	2	2
0.111	2.4	3.56	3.56	2	3.56	3	2	3.6	3.56	2	3.6	2.4	2
0.222	3.84	4.64	3.56	2.2	8.96	4.64	2.2	4.44	3.84	2.24	8.52	4.08	2.2
0.333	6	8.16	5.36	3.76	15.16	6.28	3.16	8.16	5.08	3.52	15.48	6.36	3.12
0.444	10.48	15.2	9.4	8.6	26.04	11.4	6.16	15.08	9.68	8.6	26.4	10.72	6.08
0.556	18.56	25.48	19.04	19.72	42	18.4	11.6	25.68	18.08	19.44	42.08	19.08	11.68
0.667	35.84	44.32	37.72	36.8	59.12	35.2	26.44	44.32	38.4	35.32	59.88	35.52	26.64
0.778	57.76	60.72	60.4	56.4	64.6	61.44	54.4	60.28	60.36	56.08	64.56	61.16	53.88
0.889	74.12	74.16	73.96	73.56	74.56	74.6	73.56	74.28	74	73.32	74.52	74.64	73.52
1	80	80	80	80	80	80	80	80	80	80	80	80	80

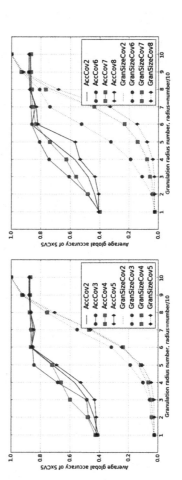

Table 5.8 5 *times* CV-5; Average global accuracy of classification (Acc) and size of granular decision systems (GranSize); Result of experiments for Covering finding methods of concept-dependent granulation with use of kNN classifier; data set: **german credit**; r_{gran} = Granulation radius, Cov_i = Average accuracy for method Cov_i

r_{gran} Acc	Cov2	Cov3	Cov4	Cov5	Cov6	Cov7	Cov8	Cov11	Cov12	Cov13	Cov14	Cov15	Cov16
0	0.561	0.561	0.563	0.566	0.566	0.559	0.56	0.561	0.556	0.562	0.564	0.558	0.57
0.05	0.561	0.565	0.563	0.566	0.566	0.559	0.56	0.563	0.557	0.562	0.568	0.558	0.57
0.1	0.561	0.568	0.568	0.566	0.57	0.568	0.56	0.568	0.56	0.562	0.574	0.558	0.57
0.15	0.561	0.589	0.599	0.566	0.592	0.578	0.56	0.595	0.59	0.562	0.586	0.558	0.57
0.2	0.567	0.621	0.627	0.566	0.619	0.582	0.56	0.622	0.625	0.562	0.624	0.562	0.57
0.25	0.589	0.651	0.631	0.566	0.656	0.595	0.56	0.649	0.628	0.562	0.655	0.576	0.57
0.3	0.606	0.652	0.655	0.575	0.667	0.578	0.584	0.661	0.653	0.571	0.671	0.605	0.59
0.35	0.642	0.654	0.641	0.591	0.674	0.545	0.608	0.652	0.644	0.586	0.674	0.643	0.609
0.4	0.656	0.701	0.642	0.645	0.675	0.552	0.652	0.703	0.633	0.642	0.668	0.68	0.647
0.45	0.678	0.701	0.636	0.675	0.677	0.583	0.676	0.7	0.643	0.68	0.666	0.684	0.671
0.5	0.695	0.708	0.655	0.681	0.696	0.617	0.697	0.699	0.661	0.682	0.705	0.691	0.697
0.55	0.708	0.714	0.695	0.694	0.711	0.67	0.71	0.704	0.693	0.696	0.702	0.711	0.703
0.6	0.718	0.716	0.708	0.715	0.713	0.702	0.719	0.727	0.724	0.716	0.714	0.721	0.72
0.65	0.718	0.717	0.722	0.71	0.725	0.718	0.718	0.72	0.723	0.712	0.722	0.717	0.713
0.7	0.732	0.725	0.722	0.722	0.723	0.721	0.735	0.726	0.727	0.722	0.72	0.722	0.727
0.75	0.73	0.733	0.734	0.723	0.732	0.732	0.73	0.733	0.733	0.726	0.724	0.724	0.729
0.8	0.733	0.731	0.734	0.731	0.734	0.733	0.734	0.73	0.733	0.738	0.73	0.728	0.733

(continued)

Table 5.8 (continued)

r_{gran}	Cov2	Cov3	Cov4	Cov5	Cov6	Cov7	Cov8	Cov11	Cov12	Cov13	Cov14	Cov15	Cov16
0.85	0.728	0.73	0.73	0.731	0.732	0.732	0.729	0.726	0.731	0.731	0.725	0.731	0.727
0.9	0.728	0.731	0.73	0.732	0.73	0.731	0.73	0.725	0.731	0.731	0.726	0.731	0.727
0.95	0.728	0.731	0.73	0.732	0.73	0.731	0.73	0.725	0.731	0.731	0.726	0.73	0.727
1	0.728	0.731	0.73	0.731	0.73	0.731	0.73	0.725	0.73	0.73	0.726	0.73	0.727
GranSize													
0	2	2	2	2	2	2	2	2	2	2	2	2	2
0.05	2	2.6	2.6	2	2.64	2.64	2	2.6	2.64	2	2.72	2	2
0.1	2.08	4.12	3.16	2	4.28	3	2	4.2	3.24	2	4.4	2	2
0.15	2.44	4.24	3.24	2	6.92	3	2	4.28	3.32	2	6.84	3	2
0.2	3.16	6.24	4.6	2	12.68	4	2	6.24	4.8	2	12.8	4	2
0.25	4.96	9.24	4.84	2	19.64	5.72	2	9.36	5.4	2	19.92	5.8	2
0.3	6.92	16.04	7.72	4.4	32.8	9.36	3.84	16.16	7.36	4.32	32.44	8.6	3.76
0.35	11.44	25.92	13.36	11.16	57.2	14.64	5.08	25.72	13.16	10.72	56.88	12.96	5.16
0.4	18.96	42.4	19.08	20.88	94.68	20.84	8.32	42.8	19	21.36	93.56	20.44	8.16
0.45	32.6	70.72	36.08	40.32	154	33.88	14.36	70.96	36.32	39.84	154.2	34.24	14.2
0.5	58.2	112.24	62.72	70.16	219.04	56.88	26.08	111	63.32	71.4	220.4	59.48	26
0.55	107.12	181.72	112	121.04	332.76	103.68	53.16	179.8	110.28	122	334.48	104.44	52.48
0.6	188.4	282.2	190.36	199.56	443	190.56	112.6	282.36	187.6	200.28	445.08	193.32	112.24

(continued)

Table 5.8 (continued)

r_{gran}	Cov2	Cov3	Cov4	Cov5	Cov6	Cov7	Cov8	Cov11	Cov12	Cov13	Cov14	Cov15	Cov16
0.65	320.44	414.08	324.24	310.88	576.04	339.4	231.72	412.2	326.24	311.52	576.04	341.72	231.36
0.7	485.8	539.48	509.04	464.36	664.76	520.32	417.28	539.08	511.2	466.4	662.2	522.4	418.24
0.75	649.68	666.24	666.32	627.68	708.36	681.4	615.24	666.6	666.24	627.56	708.72	680.8	614.28
0.8	750.8	753.88	753.76	746.36	761.56	760.56	744.6	753.72	754	746.36	761.04	760.76	744.4
0.85	789.6	789.72	789.52	789.64	789.68	789.52	789.52	789.32	789.52	789.72	789.68	789.6	789.52
0.9	796.48	796.6	796.48	796.04	796.76	796.72	796	796.56	796.48	796.28	796.72	796.6	796.08
0.95	798.8	798.72	798.56	798.64	798.76	798.76	798.72	798.6	798.64	798.8	798.8	798.68	798.8
1	800	800	800	800	800	800	800	800	800	800	800	800	800

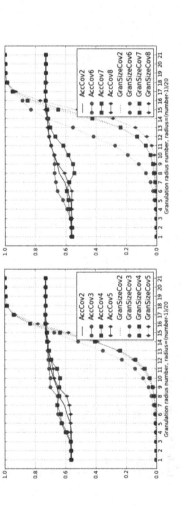

Table 5.9 5 *times* CV-5; Average global accuracy of classification (Acc) and size of granular decision systems (GranSize); Result of experiments for Covering finding methods of concept-dependent granulation with use of kNN classifier; data set: **heart disease**; r_{gran} = Granulation radius, Cov_i = Average accuracy for method Cov_i

r_{gran}	Cov2	Cov3	Cov4	Cov5	Cov6	Cov7	Cov8	Cov11	Cov12	Cov13	Cov14	Cov15	Cov16
Acc													
0	0.804	0.801	0.8	0.799	0.804	0.804	0.799	0.796	0.796	0.798	0.803	0.802	0.791
0.077	0.804	0.792	0.8	0.799	0.802	0.803	0.799	0.787	0.796	0.798	0.801	0.802	0.791
0.154	0.798	0.79	0.798	0.799	0.8	0.796	0.799	0.806	0.793	0.798	0.804	0.804	0.791
0.231	0.799	0.811	0.794	0.799	0.808	0.805	0.799	0.818	0.793	0.798	0.807	0.797	0.791
0.308	0.803	0.811	0.804	0.801	0.804	0.809	0.804	0.816	0.806	0.794	0.815	0.815	0.796
0.385	0.819	0.83	0.819	0.802	0.811	0.825	0.813	0.825	0.813	0.801	0.806	0.812	0.806
0.462	0.819	0.822	0.826	0.822	0.833	0.823	0.816	0.817	0.823	0.826	0.815	0.821	0.823
0.538	0.824	0.827	0.823	0.837	0.813	0.827	0.824	0.813	0.829	0.837	0.816	0.832	0.821
0.615	0.817	0.828	0.824	0.816	0.821	0.813	0.813	0.816	0.82	0.815	0.815	0.812	0.81
0.692	0.827	0.829	0.821	0.819	0.823	0.819	0.821	0.822	0.827	0.825	0.826	0.818	0.827
0.769	0.822	0.824	0.825	0.824	0.825	0.824	0.828	0.817	0.824	0.825	0.828	0.824	0.82
0.846	0.826	0.825	0.823	0.824	0.823	0.824	0.828	0.818	0.824	0.827	0.829	0.826	0.825
0.923	0.826	0.825	0.823	0.824	0.823	0.824	0.828	0.818	0.824	0.827	0.829	0.826	0.825
1	0.826	0.825	0.823	0.824	0.823	0.824	0.828	0.818	0.824	0.827	0.829	0.826	0.825
GranSize													
0	2	2	2	2	2	2	2	2	2	2	2	2	2
0.077	2.2	4	3	2	4	3.12	2	4	3	2	4	2	2

(continued)

Table 5.9 (continued)

r_{gran}	Cov2	Cov3	Cov4	Cov5	Cov6	Cov7	Cov8	Cov11	Cov12	Cov13	Cov14	Cov15	Cov16
0.154	2.96	4.68	3.16	2	7.72	5.64	2	4.76	3.2	2	7.76	4	2
0.231	5.12	8.28	4.24	2.04	13.36	7.76	2.04	8.08	4.24	2.04	14.44	6.08	2.04
0.308	8.84	15.04	9.72	7.36	28.28	10.48	4.6	15.28	9.76	7.28	28.24	9.92	4.76
0.385	16.76	32	17.36	17.4	53.56	17.04	8.8	32.08	17.44	17.68	53.64	17.8	8.68
0.462	34.08	57.2	36.84	39.92	89.52	34.52	20.16	57.2	37.6	39.8	89.2	36.2	19.8
0.538	71.68	95.28	68.68	74.4	133.4	69.64	47.64	95.04	67.24	74.24	134.36	73.76	47.24
0.615	126.56	140.4	130.04	122.96	166.88	131.8	110.84	140.32	129.8	123.36	168.76	134.96	111.04
0.692	180.92	184.68	184.4	179.76	191.8	188.56	176.12	184.52	184.8	178.6	192.12	187.2	176.2
0.769	210	210.72	210.8	209.48	210.6	210.8	209.6	210.8	210.72	209.72	210.64	210.76	209.32
0.846	216	216	216	216	216	216	216	216	216	216	216	216	216
0.923	216	216	216	216	216	216	216	216	216	216	216	216	216
1	216	216	216	216	216	216	216	216	216	216	216	216	216

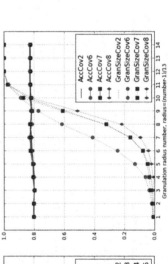

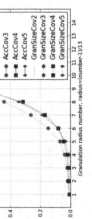

Table 5.10 5 *times* CV-5; Average global accuracy of classification (Acc) and size of granular decision systems (GranSize); Result of experiments for Covering finding methods of concept-dependent granulation with use of kNN classifier; data set: **hepatitis-filled**; r_{gran} = Granulation radius, Cov_i = Average accuracy for method Cov_i

r_{gran}	$Cov2$	$Cov3$	$Cov4$	$Cov5$	$Cov6$	$Cov7$	$Cov8$	$Cov11$	$Cov12$	$Cov13$	$Cov14$	$Cov15$	$Cov16$
Acc													
0	0.803	0.818	0.815	0.803	0.81	0.808	0.809	0.814	0.806	0.814	0.81	0.806	0.806
0.053	0.803	0.818	0.815	0.803	0.81	0.808	0.809	0.814	0.806	0.814	0.81	0.806	0.806
0.105	0.803	0.818	0.815	0.803	0.81	0.808	0.809	0.814	0.806	0.814	0.81	0.806	0.806
0.158	0.803	0.818	0.814	0.803	0.81	0.808	0.809	0.814	0.806	0.814	0.81	0.806	0.806
0.211	0.804	0.818	0.817	0.803	0.812	0.808	0.809	0.812	0.808	0.814	0.812	0.806	0.806
0.263	0.806	0.846	0.828	0.803	0.831	0.828	0.809	0.823	0.831	0.814	0.831	0.806	0.806
0.316	0.814	0.861	0.84	0.803	0.85	0.826	0.809	0.859	0.837	0.814	0.858	0.812	0.806
0.368	0.83	0.893	0.84	0.806	0.889	0.822	0.809	0.907	0.831	0.814	0.888	0.821	0.812
0.421	0.854	0.883	0.837	0.808	0.893	0.83	0.839	0.886	0.839	0.821	0.888	0.84	0.831
0.474	0.875	0.881	0.845	0.835	0.893	0.854	0.865	0.881	0.858	0.843	0.885	0.872	0.872
0.526	0.876	0.886	0.861	0.867	0.877	0.868	0.881	0.883	0.88	0.874	0.877	0.872	0.893
0.579	0.881	0.898	0.888	0.866	0.88	0.872	0.861	0.893	0.884	0.881	0.875	0.871	0.868
0.632	0.893	0.893	0.879	0.881	0.89	0.87	0.884	0.899	0.886	0.893	0.893	0.886	0.883
0.684	0.877	0.879	0.884	0.881	0.897	0.886	0.889	0.884	0.895	0.892	0.895	0.88	0.894
0.737	0.888	0.897	0.884	0.892	0.899	0.886	0.88	0.888	0.902	0.886	0.884	0.881	0.89
0.789	0.892	0.899	0.894	0.905	0.899	0.898	0.884	0.892	0.901	0.897	0.89	0.895	0.897
0.842	0.892	0.899	0.894	0.903	0.898	0.898	0.885	0.893	0.902	0.898	0.89	0.893	0.895
0.895	0.895	0.901	0.893	0.903	0.899	0.901	0.89	0.894	0.903	0.903	0.892	0.892	0.899
0.947	0.895	0.901	0.893	0.903	0.899	0.901	0.89	0.894	0.903	0.903	0.892	0.892	0.899
1	0.895	0.901	0.893	0.903	0.899	0.901	0.89	0.894	0.903	0.903	0.892	0.892	0.899

(continued)

Table 5.10 (continued)

r_{gran}	Cov2	Cov3	Cov4	Cov5	Cov6	Cov7	Cov8	Cov11	Cov12	Cov13	Cov14	Cov15	Cov16
GranSize													
0	2	2	2	2	2	2	2	2	2	2	2	2	2
0.053	2	2	2	2	2	2	2	2	2	2	2	2	2
0.105	2	2	2	2	2	2	2	2	2	2	2	2	2
0.158	2.04	2.84	2.88	2	2.92	2.8	2	2.84	2.84	2	2.96	2	2
0.211	2.28	3.24	3.08	2	4.04	3	2	3.2	3.12	2	4.08	2.24	2
0.263	2.68	5.52	4.68	2	7.4	3	2	5.56	4.76	2	7.36	3	2
0.316	3.68	7.92	3.56	2	9.24	3.16	2	8.08	3.32	2	9.12	4.08	2
0.368	5.24	12.16	5.36	2.88	17.36	5	2.76	11.96	5.12	2.92	17.28	6	2.68
0.421	7.56	15.92	6.8	5.76	25.12	7.52	3.96	16.16	6.68	5.68	25	8.8	4.04
0.474	11.6	24.48	10.72	11.64	34.2	11.36	5.64	24.76	10.56	11.64	33.92	12.96	5.68
0.526	18.88	35.04	16.52	20.24	49.2	18.64	9.76	34.44	17.2	20.24	49.16	20.24	9.56
0.579	31.08	48.08	31.92	31.2	62.92	30.32	19.56	47.32	32	30.76	63.24	32.6	20.2
0.632	46.4	63.24	46.96	45.64	77.84	47.24	34.8	62.36	46.96	45.64	78.28	49.32	34.72

(continued)

Table 5.10 (continued)

r_{gran}	Cov2	Cov3	Cov4	Cov5	Cov6	Cov7	Cov8	Cov11	Cov12	Cov13	Cov14	Cov15	Cov16
0.684	69.64	80	69.88	66.96	88.88	73	60.84	79.56	70.16	66.6	88.6	75	60.88
0.737	89.64	95.72	92.8	87.64	101.64	93.64	84.64	96.04	92.6	87.48	101.64	93.84	84.96
0.789	109.8	112.12	112.08	108.36	114.2	113.12	107.2	112.36	112.2	108.56	114.4	113.2	107.24
0.842	116.8	117.8	117.84	116.8	118.6	118.56	116.28	117.92	117.76	116.96	118.56	118.44	116.32
0.895	121	121	121	121	121	121	121	121	121	121	121	121	121
0.947	122	121.96	122	122	122	121.88	121.96	122	122	121.96	121.96	121.96	121.96
1	124	124	124	124	124	124	124	124	124	124	124	124	124

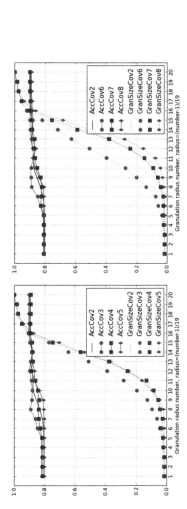

Table 5.11 5 *times* CV-5; Average global accuracy of classification (Acc) and size of granular decision systems (GranSize); Result of experiments for Covering finding methods of concept-dependent granulation with use of kNN classifier; data set: **congressional voting records**; r_{gran} = Granulation radius, Cov_i = Average accuracy for method Cov_i

r_{gran}	$Cov2$	$Cov3$	$Cov4$	$Cov5$	$Cov6$	$Cov7$	$Cov8$	$Cov11$	$Cov12$	$Cov13$	$Cov14$	$Cov15$	$Cov16$
Acc													
0	0.901	0.901	0.9	0.9	0.901	0.901	0.898	0.9	0.901	0.901	0.899	0.898	0.9
0.063	0.901	0.901	0.901	0.9	0.901	0.901	0.898	0.901	0.901	0.901	0.899	0.898	0.9
0.125	0.901	0.902	0.901	0.9	0.901	0.901	0.898	0.902	0.9	0.901	0.9	0.898	0.9
0.188	0.901	0.901	0.901	0.9	0.902	0.901	0.898	0.902	0.899	0.901	0.901	0.898	0.9
0.25	0.901	0.911	0.905	0.9	0.911	0.902	0.898	0.912	0.906	0.901	0.911	0.898	0.9
0.313	0.901	0.914	0.904	0.9	0.916	0.897	0.898	0.917	0.905	0.901	0.916	0.898	0.9
0.375	0.902	0.931	0.907	0.9	0.93	0.893	0.898	0.93	0.908	0.901	0.926	0.897	0.9
0.438	0.902	0.946	0.904	0.9	0.945	0.892	0.9	0.943	0.904	0.901	0.946	0.899	0.902
0.5	0.907	0.944	0.899	0.901	0.954	0.903	0.903	0.947	0.902	0.901	0.948	0.905	0.904
0.563	0.913	0.942	0.901	0.899	0.948	0.908	0.913	0.943	0.903	0.898	0.944	0.915	0.915
0.625	0.928	0.951	0.911	0.918	0.95	0.926	0.918	0.941	0.909	0.914	0.946	0.935	0.909
0.688	0.942	0.951	0.942	0.943	0.95	0.95	0.939	0.948	0.936	0.941	0.95	0.949	0.943
0.75	0.944	0.951	0.956	0.954	0.946	0.945	0.948	0.949	0.953	0.957	0.942	0.954	0.946
0.813	0.95	0.955	0.95	0.95	0.947	0.948	0.948	0.952	0.95	0.954	0.947	0.955	0.949
0.875	0.942	0.945	0.947	0.947	0.942	0.95	0.948	0.947	0.951	0.947	0.944	0.948	0.944
0.938	0.949	0.945	0.947	0.943	0.942	0.948	0.946	0.943	0.942	0.944	0.941	0.943	0.947
1	0.947	0.946	0.947	0.948	0.947	0.95	0.949	0.946	0.948	0.948	0.944	0.949	0.949
GranSize													
0	2	2	2	2	2	2	2	2	2	2	2	2	2
0.063	2	2.8	2.8	2	2.8	2.8	2	2.8	2.8	2	2.8	2	2
0.125	2.12	3.6	3.6	2	3.6	3	2	3.68	3.64	2	3.64	2	2
0.188	2.48	5.16	3.36	2	6.32	3	2	5.04	3.56	2	6.72	2.96	2

(continued)

Table 5.11 (continued)

r_{gran}	Cov2	Cov3	Cov4	Cov5	Cov6	Cov7	Cov8	Cov11	Cov12	Cov13	Cov14	Cov15	Cov16
0.25	2.8	7.96	4.12	2	8.24	3	2	8	4.28	2	8.44	3.68	2
0.313	3.44	12.84	4.2	2	13.92	5.32	2	12.88	4.44	2	13.72	5.48	2
0.375	4.8	17.96	5.28	2	19.64	5	2	17.64	5.2	2	19.8	6.88	2
0.438	6.12	24.72	4.76	2.92	27.8	4.96	2.88	24.8	5	2.8	28	7.76	2.84
0.5	7.16	31.6	5.28	5.04	38.2	7.08	3.56	31.84	5.32	5.12	37.84	8.68	3.44
0.563	9.36	36.36	10.24	11.84	48.12	9	4.2	36.4	10.04	12.28	48.28	10.8	4.12
0.625	13.56	43.4	17.64	25.16	59.24	12.32	5.8	42.88	17.12	25.08	58.96	14.88	5.64
0.688	19.92	46.6	29.56	33.56	73.72	19.04	9.04	47.2	30.04	34.16	73.84	20.12	8.92
0.75	31.08	64.84	38.64	45.24	92	27.84	16.84	65.52	38.72	44.76	91.72	32.04	16.88
0.813	53.16	87.52	61.12	66.72	114.12	46.44	34.08	87.36	60.12	67.04	114.64	56.72	34.64
0.875	90.44	112.8	91.92	94.52	135.32	80.84	71.2	113.6	91.64	94.84	134.64	87.44	70.76
0.938	145.08	161.8	144.32	140.92	177.2	135.2	127.28	161.92	145.56	140.96	177.92	137.32	127.12
1	214.28	213.88	214	213.92	214.32	214.08	214	213.8	214.08	214.24	213.88	214.16	213.8

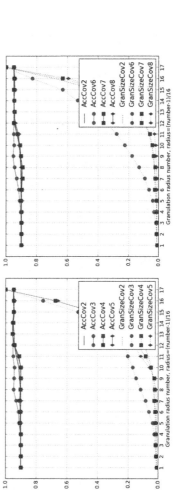

Table 5.12 5 *times* CV-5; Average global accuracy of classification (Acc) and size of granular decision systems (GranSize); Result of experiments for Covering finding methods of concept-dependent granulation with use of kNN classifier; data set: **mushroom-modified-filled**; r_{gran} = Granulation radius, Cov_i = Average accuracy for method Cov_i

r_{gran}	$Cov2$	$Cov3$	$Cov4$	$Cov5$	$Cov6$	$Cov7$	$Cov8$	$Cov11$	$Cov12$	$Cov13$	$Cov14$	$Cov15$	$Cov16$
Acc													
0	0.888	0.887	0.888	0.887	0.888	0.887	0.887	0.887	0.888	0.888	0.888	0.888	0.887
0.045	0.888	0.887	0.888	0.887	0.888	0.887	0.887	0.887	0.888	0.888	0.888	0.888	0.887
0.091	0.888	0.887	0.888	0.887	0.888	0.887	0.887	0.887	0.888	0.888	0.888	0.888	0.887
0.136	0.888	0.887	0.888	0.887	0.888	0.887	0.887	0.887	0.888	0.888	0.888	0.888	0.887
0.182	0.888	0.88	0.888	0.887	0.88	0.887	0.887	0.88	0.888	0.888	0.88	0.888	0.887
0.227	0.887	0.877	0.888	0.887	0.879	0.887	0.887	0.878	0.888	0.888	0.88	0.888	0.887
0.273	0.888	0.882	0.889	0.887	0.879	0.884	0.887	0.884	0.891	0.888	0.879	0.888	0.887
0.318	0.888	0.879	0.893	0.887	0.877	0.882	0.887	0.881	0.893	0.888	0.875	0.885	0.887
0.364	0.886	0.902	0.891	0.887	0.896	0.885	0.887	0.901	0.893	0.888	0.893	0.887	0.887
0.409	0.886	0.896	0.888	0.887	0.896	0.89	0.881	0.896	0.888	0.888	0.895	0.888	0.881
0.455	0.894	0.931	0.895	0.883	0.896	0.892	0.886	0.929	0.896	0.882	0.896	0.887	0.882
0.5	0.912	0.946	0.895	0.881	0.94	0.913	0.9	0.948	0.897	0.886	0.942	0.905	0.895
0.545	0.943	0.959	0.904	0.895	0.971	0.944	0.934	0.958	0.905	0.895	0.971	0.93	0.939
0.591	0.968	0.988	0.93	0.931	0.99	0.957	0.944	0.988	0.928	0.929	0.99	0.958	0.937
0.636	0.985	0.99	0.975	0.974	0.992	0.979	0.978	0.989	0.974	0.975	0.991	0.978	0.974
0.682	0.99	0.996	0.99	0.991	0.998	0.99	0.992	0.996	0.99	0.991	0.998	0.99	0.991
0.727	0.995	0.994	0.994	0.995	0.998	0.992	0.993	0.995	0.994	0.995	0.998	0.993	0.994
0.773	0.998	0.999	0.999	0.998	0.999	0.996	0.997	0.999	0.999	0.999	0.999	0.997	0.998
0.818	0.999	0.999	0.999	0.999	0.999	0.999	0.999	0.999	0.999	0.998	0.999	0.999	0.999

(continued)

Table 5.12 (continued)

r_{gran}	Cov2	Cov3	Cov4	Cov5	Cov6	Cov7	Cov8	Cov11	Cov12	Cov13	Cov14	Cov15	Cov16
0.864	1	1	0.999	0.999	1	0.999	1	1	0.999	0.999	1	1	1
0.909	1	1	1	1	1	1	1	1	1	1	1	1	1
0.955	1	1	1	1	1	1	1	1	1	1	1	1	1
1	1	1	1	1	1	1	1	1	1	1	1	1	1
GranSize													
0	2	2	2	2	2	2	2	2	2	2	2	2	2
0.045	2	2	2	2	2	2	2	2	2	2	2	2	2
0.091	2	2	2	2	2	2	2	2	2	2	2	2	2
0.136	2	2	2	2	2	2	2	2	2	2	2	2	2
0.182	2.16	5	2	2	4.96	2	2	4.96	2	2	4.96	2	2
0.227	2.24	6.84	2	2	7.6	3	2	6.76	2	2	7.52	3	2
0.273	2.32	17.48	5.84	2	20.28	6	2	17.48	5.8	2	20.28	4.8	2
0.318	3.8	9.92	4.4	2	19.2	9.72	2	9.76	3.96	2	19	5.16	2
0.364	5.64	20.4	7.52	2	59.96	11.16	2	20.2	6.08	2	60.48	8.08	2
0.409	9.52	37.6	10.56	5.6	106.04	13.68	3	37.96	11.76	5.48	106.24	12.04	3
0.455	15.16	81.2	23.8	18.16	265.04	17.72	5.24	81.4	22.64	18	263.44	18.44	5.24
0.5	21.44	183	29.64	35.24	369.12	23	8.04	183.68	29.04	34.84	371.32	25	8.04
0.545	29.6	283.72	60.2	68.8	552.28	26	11.24	284.92	59.96	67.84	554.96	27.04	11.2
0.591	37.56	286.36	94.16	92.28	654.56	25.84	13.96	287.88	95.48	91.52	650.2	26.28	13.88
0.636	46.16	375	170.56	116.28	529.92	29.08	19.04	376.52	171.84	116.4	530.88	30.36	19.08

(continued)

Table 5.12 (continued)

r_{gran}	Cov2	Cov3	Cov4	Cov5	Cov6	Cov7	Cov8	Cov11	Cov12	Cov13	Cov14	Cov15	Cov16
0.682	46.08	362.68	160.2	123.44	511.68	35.56	25.96	363	161.36	123.92	511.8	44.76	25.8
0.727	48.32	346.52	181.16	128.12	466.72	41.12	29.96	349.44	182.68	125.72	466.36	47.64	30.08
0.773	57.08	289.52	169.32	152.68	456.16	54.56	35.32	290.4	171.44	153.12	454.12	59.44	35.24
0.818	90.48	297.12	224.96	221.72	601	67.96	47.44	293.16	231.92	219.92	600.76	83.52	47.4
0.864	193.88	445.12	420.68	397.24	1036.48	145.8	101.44	448.52	414.6	398.8	1034.88	173.76	100.92
0.909	512.84	858.76	801.6	801.04	2004.76	414.24	266.28	852.6	798.56	797.76	2003.48	463.64	266.48
0.955	1756.68	2133.12	1832.84	1744.44	3974.36	1503.36	1000.96	2124.6	1828.52	1734.96	3972	1541.88	1001.24
1	6499.2	6499.2	6499.2	6499.2	6499.2	6499.2	6499.2	6499.2	6499.2	6499.2	6499.2	6499.2	6499.2

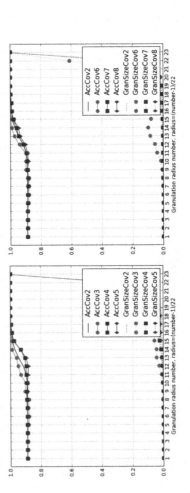

Table 5.13 5 *times* CV-5; Average global accuracy of classification (Acc) and size of granular decision systems (GranSize); Result of experiments for Covering finding methods of concept-dependent granulation with use of kNN classifier; data set: **nursery**; r_{gran} = Granulation radius, Cov_i = Average accuracy for method Cov_i

r_{gran}	$Cov2$	$Cov3$	$Cov4$	$Cov5$	$Cov6$	$Cov7$	$Cov8$	$Cov11$	$Cov12$	$Cov13$	$Cov14$	$Cov15$	$Cov16$
Acc													
0	0.399	0.413	0.39	0.391	0.373	0.398	0.39	0.372	0.362	0.393	0.4	0.392	0.371
0.125	0.418	0.441	0.419	0.41	0.422	0.427	0.408	0.395	0.379	0.414	0.464	0.435	0.395
0.25	0.5	0.512	0.508	0.522	0.542	0.512	0.5	0.474	0.457	0.514	0.586	0.527	0.474
0.375	0.529	0.53	0.544	0.548	0.524	0.54	0.519	0.494	0.498	0.55	0.565	0.548	0.49
0.5	0.55	0.494	0.575	0.591	0.532	0.565	0.538	0.459	0.532	0.587	0.578	0.557	0.515
0.625	0.585	0.501	0.62	0.624	0.546	0.606	0.598	0.462	0.57	0.628	0.593	0.564	0.576
0.75	0.622	0.585	0.668	0.648	0.554	0.641	0.632	0.538	0.615	0.65	0.606	0.624	0.608
0.875	0.669	0.641	0.664	0.601	0.494	0.663	0.639	0.59	0.609	0.6	0.534	0.665	0.613
1	0.569	0.583	0.569	0.569	0.556	0.57	0.57	0.555	0.547	0.569	0.58	0.569	0.557

(continued)

Table 5.13 (continued)

r_{gran}	Cov2	Cov3	Cov4	Cov5	Cov6	Cov7	Cov8	Cov11	Cov12	Cov13	Cov14	Cov15	Cov16
GranSize													
0	4.96	5	4.96	4.96	4.92	4.96	4.96	4.92	4.88	4.96	5	4.96	4.92
0.125	7.56	9.72	8.6	8.36	25.92	7.68	6.96	8.28	7.96	8.88	25.88	8.96	6.92
0.25	13.92	19.48	15.08	23.4	122.6	12.68	7.96	20.08	13.88	23.24	121.64	16.4	7.92
0.375	26.44	62.12	37.24	60.48	356.36	23.64	12.96	65.76	35.44	61.08	356.8	30.64	12.92
0.5	59.4	208.52	101.16	164.4	850.8	48.32	27.8	211.68	94.48	167.08	856.08	64.28	27.48
0.625	160.92	555.68	264.36	432.16	1813.92	134.4	76.28	561.88	257.44	440.52	1817.32	168.2	75.72
0.75	545.6	1226.68	739.8	1054.52	3589.16	480.92	266.76	1231.96	738.8	1042.88	3591.24	521.08	266.28
0.875	2370.48	3137.76	2533.56	2561.16	6886.56	2027.52	1296.92	3131.64	2528.04	2553	6902.44	2196.6	1293.04
1	10368	10368	10368	10368	10368	10368	10368	10368	10368	10368	10368	10368	10368

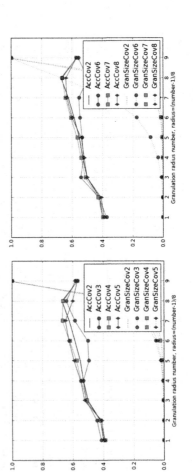

Table 5.14 5 *times* CV-5; Average global accuracy of classification (Acc) and size of granular decision systems (GranSize); Result of experiments for Covering finding methods of concept-dependent granulation with use of kNN classifier; data set: **soybean-large-all-filled**; r_{gran} = Granulation radius, Cov_i = Average accuracy for method Cov_i

r_{gran}	$Cov2$	$Cov3$	$Cov4$	$Cov5$	$Cov6$	$Cov7$	$Cov8$	$Cov11$	$Cov12$	$Cov13$	$Cov14$	$Cov15$	$Cov16$
Acc													
0	0.681	0.682	0.683	0.686	0.683	0.684	0.683	0.683	0.671	0.682	0.69	0.686	0.677
...	...	...	...	...	...	...	...	...	...	...	...	...	...
0.4	0.681	0.682	0.683	0.686	0.683	0.684	0.683	0.683	0.671	0.682	0.69	0.686	0.677
0.429	0.681	0.681	0.683	0.686	0.682	0.683	0.683	0.683	0.671	0.682	0.69	0.686	0.677
0.457	0.681	0.685	0.684	0.686	0.683	0.683	0.683	0.684	0.671	0.682	0.691	0.686	0.677
0.486	0.681	0.692	0.683	0.686	0.691	0.683	0.683	0.69	0.671	0.682	0.697	0.687	0.677
0.514	0.682	0.693	0.683	0.686	0.691	0.683	0.683	0.693	0.671	0.682	0.7	0.688	0.677
0.543	0.687	0.702	0.681	0.686	0.698	0.683	0.683	0.707	0.669	0.682	0.709	0.691	0.677
0.571	0.69	0.719	0.692	0.687	0.712	0.695	0.684	0.713	0.672	0.682	0.72	0.699	0.678
0.6	0.699	0.739	0.696	0.687	0.726	0.702	0.685	0.731	0.683	0.683	0.738	0.699	0.679
0.629	0.709	0.776	0.719	0.7	0.769	0.734	0.703	0.776	0.709	0.692	0.772	0.72	0.688
0.657	0.721	0.758	0.748	0.706	0.748	0.734	0.705	0.754	0.732	0.699	0.763	0.726	0.699
0.686	0.748	0.755	0.735	0.712	0.766	0.741	0.716	0.756	0.717	0.707	0.772	0.737	0.706
0.714	0.761	0.773	0.74	0.729	0.766	0.769	0.733	0.761	0.731	0.734	0.769	0.755	0.735
0.743	0.787	0.793	0.75	0.737	0.799	0.787	0.772	0.797	0.745	0.739	0.795	0.783	0.751
0.771	0.824	0.824	0.8	0.786	0.824	0.818	0.82	0.82	0.798	0.78	0.83	0.834	0.817
0.8	0.842	0.851	0.832	0.823	0.854	0.842	0.847	0.842	0.832	0.825	0.858	0.839	0.834
0.829	0.859	0.876	0.863	0.859	0.874	0.846	0.847	0.862	0.864	0.858	0.874	0.865	0.849
0.857	0.865	0.874	0.867	0.871	0.871	0.878	0.861	0.875	0.87	0.861	0.875	0.859	0.857

(continued)

Table 5.14 (continued)

r_{gran}	Cov2	Cov3	Cov4	Cov5	Cov6	Cov7	Cov8	Cov11	Cov12	Cov13	Cov14	Cov15	Cov16
0.886	0.875	0.877	0.893	0.876	0.888	0.891	0.866	0.873	0.885	0.877	0.886	0.886	0.873
0.914	0.89	0.886	0.9	0.893	0.893	0.888	0.877	0.887	0.894	0.898	0.89	0.885	0.881
0.943	0.882	0.883	0.892	0.891	0.894	0.889	0.884	0.885	0.886	0.894	0.891	0.888	0.886
0.971	0.918	0.916	0.919	0.916	0.919	0.918	0.911	0.913	0.917	0.918	0.916	0.913	0.911
1	0.917	0.913	0.919	0.918	0.917	0.915	0.914	0.914	0.915	0.916	0.916	0.912	0.911
GranSize													
0	19	19	19	19	19	19	19	19	19	19	19	19	19
...	...	...	...	...	...	...	...	...	...	...	...	...	...
0.4	19	19	19	19	19	19	19	19	19	19	19	19	19
0.429	19.04	19.68	19.6	19	19.64	19.76	19	19.6	19.68	19	19.6	19	19
0.457	19.2	20.68	19.8	19	20.64	19.8	19	20.6	19.8	19	20.6	19	19
0.486	19.32	20.76	19.8	19	20.8	19.8	19	20.8	19.8	19	20.8	19.08	19
0.514	19.52	21.32	20	19	22.08	19.84	19	21.36	19.88	19	22.2	19.8	19
0.543	20.56	23.84	20.84	19	25.36	20.76	19	23.84	20.92	19	25.24	20.16	19
0.571	22	26.24	22.08	19.16	30.48	21.68	19.16	26.36	22.24	19.12	30.76	23.56	19.16
0.6	24	31.6	24.04	19.16	37.24	22.6	19.2	31.48	24.32	19.24	37.2	25.08	19.2
0.629	27.4	39.04	29.2	22.08	47.92	25.2	21.36	39	29	22.08	48.32	29.24	21.4
0.657	31.04	50.32	35.48	25.36	63.8	28	22.64	50.76	35.48	25.68	63.76	33.92	22.68
0.686	35.48	67.2	42.12	30.6	84.12	30.76	25.24	66.68	41.44	30.52	84.8	38.76	25.24
0.714	41.92	81.36	50.84	40.24	104.88	35.12	28.56	81.8	51.32	40.48	105.36	45.28	28.52
0.743	48.16	91.64	58.88	50.92	122.08	40.72	33.08	92.84	59.08	50.96	120.48	53.36	33.28
0.771	56.84	96.84	63.72	60.2	139.96	48.8	40.12	98.24	64.44	60.56	140.32	63.24	40.48
0.8	71.6	114.2	77.56	75.72	156.48	61.76	50.32	113.32	78.04	76	157.32	75.8	50.48

(continued)

Table 5.14 (continued)

r_{gran}	Cov2	Cov3	Cov4	Cov5	Cov6	Cov7	Cov8	Cov11	Cov12	Cov13	Cov14	Cov15	Cov16
0.829	91.32	130	93.88	91.96	175.04	75.36	64.36	129.08	92.68	91.76	174.84	94.84	63.6
0.857	119.16	160.36	119.96	116.4	209.76	103.12	86.56	159.48	120.2	116.8	209.2	124.88	86.08
0.886	160.96	198.68	168.76	159.88	260.28	145.2	128.96	198.68	167.56	159.08	259.8	170.48	128.04
0.914	222.92	256.76	227.84	222.88	314.88	207.44	185.64	257.48	227.08	221.76	314	231.84	185.92
0.943	300.28	322.16	306.56	291.76	369.64	295.52	270.08	321.88	306.72	291.56	370.92	310.56	270.76
0.971	405.6	416.04	415.56	393.84	432.76	419.08	385.68	416.2	416.52	392.92	431.92	415.08	385.52
1	509.64	509.6	509.76	509.96	509.84	509.48	509.44	509.64	510.36	509.8	509.68	509.88	509.64

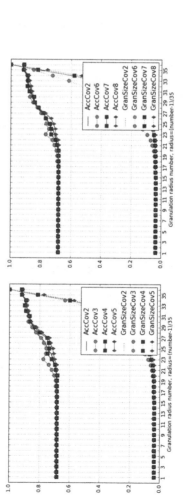

Table 5.15 5 *times* CV-5; Average global accuracy of classification (Acc) and size of granular decision systems (GranSize); Result of experiments for Covering finding methods of concept-dependent granulation with use of kNN classifier; data set: **SPECT-all-modif**; r_{gran} = Granulation radius, Cov_i = Average accuracy for method Cov_i

r_{gran}	$Cov2$	$Cov3$	$Cov4$	$Cov5$	$Cov6$	$Cov7$	$Cov8$	$Cov11$	$Cov12$	$Cov13$	$Cov14$	$Cov15$	$Cov16$
Acc													
0	0.529	0.557	0.505	0.551	0.568	0.522	0.497	0.542	0.569	0.544	0.557	0.531	0.528
0.045	0.529	0.562	0.522	0.551	0.568	0.526	0.497	0.564	0.586	0.544	0.564	0.531	0.528
0.091	0.529	0.606	0.573	0.551	0.587	0.558	0.497	0.601	0.605	0.544	0.576	0.531	0.528
0.136	0.53	0.638	0.604	0.551	0.599	0.585	0.497	0.63	0.638	0.544	0.608	0.531	0.528
0.182	0.538	0.644	0.619	0.551	0.617	0.584	0.497	0.628	0.636	0.544	0.633	0.538	0.528
0.227	0.536	0.651	0.648	0.551	0.631	0.582	0.497	0.638	0.639	0.544	0.639	0.549	0.528
0.273	0.567	0.642	0.64	0.551	0.64	0.599	0.497	0.638	0.626	0.544	0.642	0.599	0.528
0.318	0.603	0.634	0.664	0.551	0.643	0.623	0.497	0.616	0.677	0.544	0.645	0.611	0.528
0.364	0.645	0.703	0.679	0.554	0.677	0.67	0.579	0.7	0.67	0.59	0.676	0.656	0.567
0.409	0.67	0.737	0.703	0.588	0.739	0.699	0.632	0.741	0.695	0.613	0.724	0.676	0.611
0.455	0.68	0.747	0.706	0.636	0.756	0.654	0.657	0.746	0.706	0.641	0.745	0.706	0.652
0.5	0.707	0.761	0.694	0.652	0.747	0.721	0.684	0.754	0.694	0.66	0.733	0.745	0.685
0.545	0.715	0.771	0.746	0.725	0.753	0.684	0.682	0.77	0.736	0.721	0.741	0.756	0.661
0.591	0.753	0.77	0.763	0.775	0.786	0.703	0.721	0.762	0.751	0.777	0.778	0.767	0.739
0.636	0.777	0.773	0.756	0.79	0.772	0.724	0.761	0.78	0.77	0.792	0.786	0.792	0.774
0.682	0.782	0.789	0.762	0.795	0.784	0.719	0.735	0.792	0.774	0.805	0.792	0.79	0.748
0.727	0.789	0.795	0.776	0.796	0.804	0.747	0.75	0.776	0.777	0.792	0.801	0.788	0.77
0.773	0.794	0.795	0.798	0.792	0.813	0.78	0.785	0.796	0.787	0.793	0.813	0.801	0.803
0.818	0.792	0.794	0.794	0.794	0.792	0.8	0.794	0.797	0.796	0.792	0.799	0.794	0.796

(continued)

Table 5.15 (continued)

r_{gran}	Cov2	Cov3	Cov4	Cov5	Cov6	Cov7	Cov8	Cov11	Cov12	Cov13	Cov14	Cov15	Cov16
0.864	0.794	0.794	0.794	0.794	0.798	0.795	0.794	0.794	0.794	0.794	0.794	0.794	0.794
0.909	0.794	0.794	0.794	0.794	0.794	0.794	0.794	0.794	0.794	0.794	0.794	0.794	0.794
0.955	0.793	0.794	0.794	0.794	0.798	0.795	0.794	0.794	0.794	0.794	0.797	0.794	0.795
1	0.803	0.799	0.796	0.795	0.792	0.792	0.796	0.799	0.793	0.797	0.792	0.797	0.796
GranSize													
0	2	2	2	2	2	2	2	2	2	2	2	2	2
0.045	2	2.64	2.64	2	2.64	2.64	2	2.68	2.68	2	2.6	2	2
0.091	2.12	2.96	3	2	3	2.96	2	3	3	2	3	2	2
0.136	2.24	3.96	3.92	2	4	3	2	3.88	3.92	2	3.96	2	2
0.182	2.4	5.2	5.2	2	5.64	3	2	5.28	5.16	2	5.76	2.96	2
0.227	2.64	5.88	5.68	2	6.84	3	2	5.8	5.68	2	6.72	3	2
0.273	2.92	6.36	5.4	2	8.44	3	2	6.2	5.16	2	8.28	3.32	2
0.318	3.32	8.08	4.2	2	11.08	3	2	8	4	2	10.96	3.64	2
0.364	3.92	9.8	4.36	2.92	14.48	3.68	2.84	9.68	4.16	2.72	14.48	4.16	2.8
0.409	4.64	12.28	4.68	3.8	19.68	4	3	12.12	4.4	3.8	19.76	4.96	3
0.455	5.76	15.52	5.6	5.96	24.44	4.92	3.04	15.8	5.28	6.2	24.44	6.6	3.04
0.5	6.96	20.32	6.28	8.68	31.4	6.28	3.4	20.88	6.04	9	31.08	8.24	3.48
0.545	8.52	26.04	8.12	16.36	41.76	7.28	5.44	26.16	7.76	16.32	41.96	10.16	5.36
0.591	11.08	31.76	10.2	25.96	52.8	9.92	6.52	31.56	9.96	25.88	51.8	12.8	6.6
0.636	15.36	39.44	17.28	33.44	65.68	14.44	8.6	39.72	17.16	32.84	65.68	18.12	8.6
0.682	23.48	48.12	27.32	43.52	78.4	19.68	12.04	47.48	27.56	43.2	78	24.44	11.88
0.727	35.12	62.4	38.72	54.32	91.28	28.68	19.12	62.52	39.8	54.4	91.56	34.56	19.12

(continued)

Table 5.15 (continued)

r_{gran}	Cov2	Cov3	Cov4	Cov5	Cov6	Cov7	Cov8	Cov11	Cov12	Cov13	Cov14	Cov15	Cov16
0.773	50.36	78.44	56.48	59.48	108	43.24	33.52	79.88	55.76	60.04	108.44	52.16	33.08
0.818	72.92	96.2	76.88	74.24	124.56	62.88	54.04	96.8	77	74.16	125.8	73.36	54.04
0.864	100.8	118.04	101.4	95.24	137.52	91.2	83.88	117.08	100.88	95.88	137.4	100.4	83.76
0.909	132.16	144.16	134.4	127.44	152.6	128.44	121.24	143.92	135.2	127.84	152.24	135.4	121.04
0.955	162.68	165.08	164.52	160.12	170.2	164.32	159.56	165.6	165	160.04	170.36	167.52	159.4
1	184.8	184.84	184.72	184.72	184.92	184.88	184.84	184.84	184.76	184.72	184.8	184.8	184.68

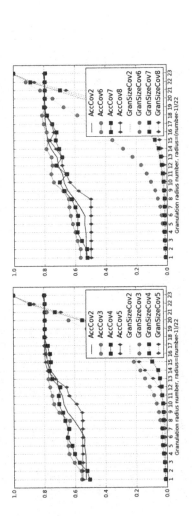

Table 5.16 5 *times* CV-5; Average global accuracy of classification (Acc) and size of granular decision systems (GranSize); Result of experiments for Covering finding methods of concept-dependent granulation with use of kNN classifier; data set: **SPECTF-all-modif**; r_{gran} = Granulation radius, Cov_i = Average accuracy for method Cov_i

r_{gran}	$Cov2$	$Cov3$	$Cov4$	$Cov5$	$Cov6$	$Cov7$	$Cov8$	$Cov11$	$Cov12$	$Cov13$	$Cov14$	$Cov15$	$Cov16$
Acc													
0	0.434	0.447	0.426	0.452	0.439	0.413	0.432	0.445	0.445	0.435	0.429	0.416	0.44
0.023	0.581	0.656	0.617	0.542	0.632	0.522	0.487	0.682	0.608	0.526	0.646	0.589	0.503
0.045	0.671	0.732	0.705	0.638	0.739	0.581	0.62	0.756	0.693	0.662	0.724	0.657	0.617
0.068	0.744	0.774	0.754	0.73	0.76	0.662	0.727	0.78	0.759	0.735	0.76	0.75	0.723
0.091	0.783	0.789	0.787	0.766	0.775	0.736	0.767	0.791	0.787	0.775	0.77	0.774	0.761
0.114	0.79	0.779	0.784	0.784	0.757	0.757	0.788	0.778	0.791	0.782	0.773	0.787	0.788
0.136	0.783	0.781	0.788	0.77	0.774	0.779	0.799	0.792	0.786	0.79	0.772	0.789	0.786
0.159	0.788	0.784	0.789	0.784	0.775	0.777	0.797	0.786	0.79	0.793	0.781	0.786	0.792
0.182	0.794	0.787	0.79	0.787	0.795	0.79	0.796	0.796	0.794	0.79	0.792	0.793	0.79
0.205	0.793	0.793	0.784	0.781	0.782	0.79	0.793	0.795	0.793	0.787	0.79	0.791	0.786
0.227	0.793	0.791	0.784	0.784	0.782	0.787	0.794	0.796	0.791	0.787	0.787	0.789	0.791
0.25	0.795	0.792	0.784	0.786	0.783	0.789	0.795	0.797	0.793	0.787	0.787	0.791	0.792
...	...	...	...	...	...	...	...	...	...	...	...	...	...
1	0.795	0.792	0.784	0.786	0.783	0.789	0.795	0.797	0.793	0.787	0.787	0.791	0.792
GranSize													
0	2	2	2	2	2	2	2	2	2	2	2	2	2
0.023	6.6	11.28	6.36	4.8	14.8	5.24	3.4	11.16	6.24	4.64	15.16	7.04	3.36

(continued)

Table 5.16 (continued)

r_{gran}	Cov2	Cov3	Cov4	Cov5	Cov6	Cov7	Cov8	Cov11	Cov12	Cov13	Cov14	Cov15	Cov16
0.045	14.08	23.72	13.88	11.84	33.48	11.24	7	23.6	13.44	12.44	33.08	14.92	6.96
0.068	28.36	44.24	29.92	26.2	59.76	25.44	14.72	44.8	29.56	26.6	59.2	30.24	14.44
0.091	50.6	70.52	52.96	44.36	95.56	49.44	30.36	69.52	52.92	45.2	94.76	53.96	30.52
0.114	84.2	103.76	88.12	72.28	142.72	87.16	57.8	103.56	88.04	73.28	142.16	91.96	58.44
0.136	125.72	137.24	133.32	115	181.36	135.16	102.76	138	134.6	113.84	181.08	137.76	103.12
0.159	166.24	171.6	171.84	158.68	185.56	175.4	155.52	171.6	171	158.68	185.88	175.68	155.12
0.182	195.56	196.68	196.28	193.12	197.2	197.4	193.16	196.4	196.12	193.28	197.56	197.4	193.04
0.205	209.4	209.6	209.52	208.92	210.04	210.12	208.96	209.68	209.68	208.8	210.04	210.08	208.88
0.227	212.96	213	212.96	213	213	212.96	212.96	212.96	212.96	212.88	212.96	213	212.92
0.25	213.6	213.6	213.6	213.6	213.6	213.6	213.6	213.6	213.6	213.6	213.6	213.6	213.6
...	...	...	...	...	...	...	...	...	...	...	...	...	...
1	213.6	213.6	213.6	213.6	213.6	213.6	213.6	213.6	213.6	213.6	213.6	213.6	213.6

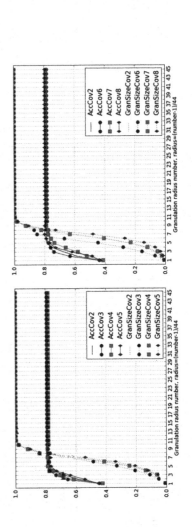

5.6 Validation of Results: Combined Average Accuracy with Percentage of Reduction of Object Number, and, 5 × CV5 Accuracy Bias

In this section, we exhibit, for results in Tables 5.4, 5.5, 5.6, 5.7, 5.8, 5.9, 5.10, 5.11, 5.12, 5.13, 5.14, 5.15 and 5.16, the respective combined parameters, which show overall effectiveness of our methods. Each parameter combines accuracy of classification (acc) and percentage reduction of training data set size after granulation. The best result is chosen only from bolded entries, where accuracy error with comparison with original accuracy is less than 5 % (0.05 of accuracy). Original accuracy is the value for classification without granulation (for the radius equal 1). Results are shown in Tables 5.17, 5.18, 5.19, 5.20, 5.21, 5.22, 5.23, 5.24, 5.25, 5.26, 5.27 and 5.28.

The combined parameter is defined as follows.

$$CombAGS = \frac{acc + (1 - \frac{GranSize}{GranSize_{r=1}})}{2}.$$

In Tables 5.17, 5.18, 5.19, 5.20, 5.21, 5.22, 5.23, 5.24, 5.25, 5.26, 5.27 and 5.28, we report respective values of bias of accuracy (AccBias) for five times Cross Validation five method defined as follows, cf., Chap. 3.

$$AccBias = \frac{\sum_{i=1}^{5}(max(acc_1^{CV5}, acc_2^{CV5}, ..., acc_5^{CV5}) - acc_i^{CV5})}{5},$$

where,

$$Acc = \frac{\sum_{i=1}^{5} acc_i^{CV5}}{5}.$$

Considering results evaluated by the parameter *CombAGS* chosen for *acc* values fulfilling inequality $(acc_{r=1} - acc) \leq 0.05$, where *AccBias* helps in tie resolution in cases of the same value of *CombAGS*, the most impressive results for examined data sets in Table 5.1 are as follows.

1. For **Australian Credit** data set, $Cov3$ wins over other methods; for radius $r_{gran} = 0.286$, we have $ComAGS = 0.9$ with $AccBias = 0.001$, error in accuracy $acc_{r=1} - acc = 0.007$ we have reduction of training objects in range of 94.7 %. See the result in Tables 5.4 and 5.17.
2. For **Car Evaluation** data set, $Cov3$ wins over other methods; for radius $r_{gran} = 0.833$ we have $ComAGS = 0.781$ with $AccBias = 0.005$, error in accuracy $acc_{r=1} - acc = 0.02$, we have reduction of training objects in range of 63.79 %. See the result in Tables 5.5 and 5.18.
3. For **Pima Indians Diabetes** data set, $Cov4$ wins over other methods; for radius $r_{gran} = 0.00$ we have $ComAGS = 0.801$ with $AccBias = 0.009$, error in

accuracy $acc_{r=1} - acc = 0.036$, we have reduction of training objects in range of 99.67 %. See the result in Tables 5.6 and 5.19.

4. For **Fertility Diagnosis** data set, $Cov4$ wins over other methods; for radius $r_{gran} = 0.556$, we have $ComAGS = 0.814$ with $AccBias = 0.014$, error in accuracy $acc_{r=1} - acc = 0.004$, we have reduction of the number of training objects in range of 75.75 %. See the result in Tables 5.7 and 5.20.

5. For **German Credit** data set, $Cov3$ wins over other methods; for radius $r_{gran} = 0.4$ we have $ComAGS = 0.824$ with $AccBias = 0.014$, error in accuracy $acc_{r=1} - acc = 0.03$, we have reduction of training objects in range of 94.7 %. See the result in Tables 5.8 and 5.21.

6. For **Heart Disease** data set, $Cov4$ wins over other methods; for radius $r_{gran} = 0$ we have $ComAGS = 0.895$ with $AccBias = 0.011$, error in accuracy $acc_{r=1} - acc = 0.023$, we have reduction of training objects in range of 99.07 %. See the result in Tables 9.3 and 5.22.

7. For **Hepatitis** data set, $Cov8$ wins over other methods; for radius $r_{gran} = 0.474$, we have $ComAGS = 0.91$ with $AccBias = 0.013$, error in accuracy $acc_{r=1} - acc = 0.025$, we have reduction of training objects in range of 95.6 %. See the result in Tables 5.10 and 5.23.

8. For **Congressional Voting Records** data set, $Cov8$ wins over other methods; for radius $r_{gran} = 0.438$ we have $ComAGS = 0.943$ with $AccBias = 0.003$, error in accuracy $acc_{r=1} - acc = 0.049$, we have reduction of training objects in range of 98.65 %. See the result in Tables 5.11 and 5.24.

9. For **Mushroom** data set, $Cov8$ wins over other methods; for radius $r_{gran} = 0.773$ we have $ComAGS = 0.996$ with $AccBias = 0.001$, error in accuracy $acc_{r=1} - acc = 0.003$, we have reduction of training objects in range of 99.25 %. See the result in Tables 5.12 and 5.25.

10. For **Nursery** data set, $Cov8$ wins over other methods; for radius $r_{gran} = 0.75$, we have $ComAGS = 0.803$ with $AccBias = 0.03$, gain in accuracy $acc - acc_{r=1} = 0.062$, we have reduction of training objects in range of 97.43 %. See the result in Tables 5.13 and 5.26.

11. For **Soybean Large** data set, $Cov7$ wins over other methods; for radius $r_{gran} = 0.857$ we have $ComAGS = 0.838$ with $AccBias = 0.006$, error in accuracy $acc_{r=1} - acc = 0.037$, we have reduction of training objects in range of 79.76 %. See the result in Tables 5.14 and 5.27.

12. For **SPECT** data set, $Cov4$ wins over other methods; for radius $r_{gran} = 0.591$, we have $ComAGS = 0.854$ with $AccBias = 0.005$, error in accuracy $acc_{r=1} - acc = 0.033$, we have reduction of training objects in range of 94.48 %. See the result in Tables 5.15 and 5.28.

13. For **SPECTF** data set, $Cov4$ wins over other methods; for radius $r_{gran} = 0.068$, we have $ComAGS = 0.807$ with $AccBias = 0.017$, error in accuracy $acc_{r=1} - acc = 0.03$, we have reduction of training objects in range of 85.99 %. See the result in Tables 5.16 and 5.29.

Table 5.17 5 *times* CV-5; Combination of accuracy and percentage reduction of objects in training data set (CombAGS) and Bias of accuracy form 5 × CV5 (AccBias); Result of experiments for Covering finding methods of concept-dependent granulation with use of kNN classifier; data set: **australian**; r_{gran} = Granulation radius, Cov_i = Average accuracy for method Cov_i

r_{gran}	$Cov2$	$Cov3$	$Cov4$	$Cov5$	$Cov6$	$Cov7$	$Cov8$	$Cov11$	$Cov12$	$Cov13$	$Cov14$	$Cov15$	$Cov16$
CombAGS													
0	0.885	0.887	0.885	0.884	0.885	0.885	0.886	0.884	0.884	0.885	0.884	0.886	0.884
0.071	0.884	0.892	0.884	0.884	0.895	0.893	0.886	0.893	0.884	0.885	0.894	0.885	0.884
0.143	0.886	0.895	0.883	0.884	0.89	0.891	0.886	0.894	0.882	0.885	0.892	0.884	0.884
0.214	0.888	0.895	0.886	0.884	0.88	0.891	0.887	0.895	0.885	0.884	0.878	0.889	0.885
0.286	0.894	0.9	0.885	0.882	0.867	0.898	0.888	0.898	0.886	0.882	0.866	0.891	0.888
0.357	0.892	0.863	0.88	0.874	0.822	0.892	0.904	0.861	0.882	0.872	0.819	0.89	0.902
0.429	0.885	0.84	0.874	0.862	0.768	0.891	0.904	0.843	0.875	0.861	0.771	0.89	0.909
0.5	0.858	0.765	0.848	0.84	0.694	0.863	0.889	0.776	0.848	0.838	0.695	0.856	0.888
0.571	0.776	0.707	0.773	0.768	0.607	0.787	0.82	0.709	0.771	0.77	0.607	0.777	0.823
0.643	0.634	0.602	0.627	0.643	0.54	0.622	0.674	0.601	0.627	0.644	0.542	0.62	0.673
0.714	0.503	0.492	0.493	0.512	0.483	0.487	0.518	0.494	0.494	0.512	0.482	0.492	0.519
0.786	0.444	0.442	0.441	0.443	0.444	0.441	0.443	0.443	0.44	0.443	0.442	0.442	0.446
0.857	0.435	0.434	0.434	0.433	0.434	0.433	0.433	0.434	0.431	0.433	0.434	0.433	0.436
0.929	0.434	0.433	0.434	0.432	0.434	0.433	0.432	0.433	0.431	0.433	0.433	0.432	0.435
1	0.431	0.43	0.431	0.429	0.43	0.431	0.429	0.43	0.427	0.429	0.43	0.43	0.433
AccBias													
0	0.003	0.009	0.002	0.004	0.008	0.004	0.002	0.002	0.002	0.004	0.003	0.002	0.006
0.071	0.001	0.003	0.002	0.004	0.006	0.007	0.002	0.001	0.004	0.004	0.003	0.002	0.006

(continued)

Table 5.17 (continued)

r_{gran}	Cov2	Cov3	Cov4	Cov5	Cov6	Cov7	Cov8	Cov11	Cov12	Cov13	Cov14	Cov15	Cov16
0.143	0.011	0.004	0.005	0.004	0.007	0.011	0.002	0.003	0.002	0.004	0.013	0.004	0.006
0.214	0.003	0.004	0.005	0.005	0.01	0.016	0.005	0.015	0.002	0.004	0.008	0.008	0.003
0.286	0.008	0.001	0.008	0.012	0.008	0.005	0.019	0.008	0.01	0.014	0.012	0.008	0.019
0.357	0.011	0.016	0.006	0.014	0.008	0.025	0.014	0.003	0.006	0.005	0.011	0.017	0.011
0.429	0.013	0.012	0.009	0.003	0.008	0.017	0.009	0.007	0.009	0.013	0.007	0.022	0.017
0.5	0.02	0.016	0.01	0.006	0.009	0.016	0.017	0.018	0.006	0.008	0.014	0.008	0.005
0.571	0.008	0.019	0.005	0.011	0.006	0.006	0.011	0.012	0.009	0.008	0.01	0.006	0.006
0.643	0.01	0.007	0.008	0.007	0.015	0.006	0.006	0.01	0.005	0.009	0.01	0.007	0.005
0.714	0.004	0.017	0.004	0.007	0.003	0.009	0.007	0.009	0.005	0.011	0.011	0.009	0.006
0.786	0.009	0.009	0.001	0.003	0.005	0.006	0.011	0.007	0.001	0.01	0.002	0.01	0.002
0.857	0.008	0.015	0.005	0.007	0.009	0.004	0.011	0.005	0.002	0.01	0.004	0.011	0.003
0.929	0.008	0.013	0.003	0.006	0.008	0.004	0.012	0.006	0.002	0.013	0.004	0.012	0.003
1	0.007	0.013	0.003	0.008	0.009	0.004	0.012	0.005	0.002	0.008	0.003	0.014	0.003

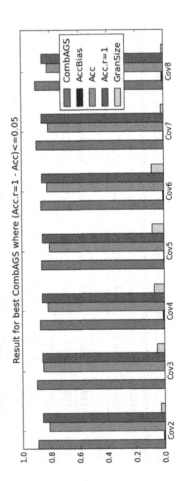

Result for best CombAGS where (Acc.r=1 - Acc)<=0.05

Table 5.18 5 *times* CV-5; Combination of accuracy and percentage reduction of objects in training data set (CombAGS) and Bias of accuracy form $5 \times$ CV5 (AccBias); Result of experiments for Covering finding methods of concept-dependent granulation with use of kNN classifier; data set: **car evaluation**; $r_{gran} =$ Granulation radius, $Cov_i =$ Average accuracy for method Cov_i

r_{gran}	$Cov2$	$Cov3$	$Cov4$	$Cov5$	$Cov6$	$Cov7$	$Cov8$	$Cov11$	$Cov12$	$Cov13$	$Cov14$	$Cov15$	$Cov16$
CombAGS													
0	0.662	0.658	0.66	0.663	0.663	0.667	0.659	0.665	0.658	0.664	0.662	0.661	0.661
0.167	0.695	0.692	0.714	0.663	0.7	0.686	0.665	0.695	0.71	0.664	0.709	0.697	0.666
0.333	0.763	0.772	0.757	0.716	0.766	0.715	0.722	0.771	0.761	0.712	0.773	0.756	0.721
0.5	0.827	0.771	0.783	0.766	0.769	0.773	0.803	0.769	0.784	0.765	0.766	0.824	0.801
0.667	0.863	0.794	0.806	0.788	0.739	0.859	0.883	0.791	0.799	0.789	0.738	0.868	0.887
0.833	0.798	**0.781**	0.746	0.764	**0.612**	0.808	0.858	**0.783**	0.747	0.767	**0.615**	0.801	0.859
1	**0.472**	**0.472**	**0.472**	**0.472**	**0.472**	**0.471**	**0.472**	**0.473**	**0.472**	**0.474**	**0.472**	**0.473**	**0.473**

(continued)

Table 5.18 (continued)

r_{gran}	Cov2	Cov3	Cov4	Cov5	Cov6	Cov7	Cov8	Cov11	Cov12	Cov13	Cov14	Cov15	Cov16
AccBias													
0	0.004	0.024	0.017	0.008	0.009	0.01	0.01	0.013	0.016	0.009	0.012	0.012	0.005
0.167	0.012	0.012	0.032	0.018	0.021	0.007	0.045	0.029	0.025	0.024	0.012	0.024	0.021
0.333	0.008	0.021	0.01	0.014	0.023	0.025	0.016	0.054	0.007	0.014	0.009	0.005	0.017
0.5	0.013	0.01	0.009	0.021	0.019	0.047	0.017	0.008	0.011	0.01	0.009	0.014	0.026
0.667	0.024	0.014	0.007	0.022	0.015	0.011	0.015	0.006	0.009	0.007	0.012	0.007	0.019
0.833	0.014	**0.005**	0.003	0.002	**0.006**	0.016	0.006	**0.002**	0.01	0.007	**0.004**	0.01	0.003
1	**0.005**	**0.003**	**0.003**	**0.003**	**0.003**	**0.003**	**0.002**	**0.003**	**0.006**	**0.006**	**0.005**	**0.004**	**0.009**

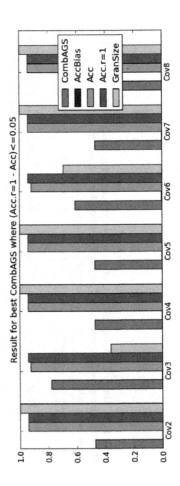

Result for best CombAGS where (Acc.r=1 - Acc)<=0.05

Table 5.19 5 *times* CV-5; Combination of accuracy and percentage reduction of objects in training data set (CombAGS) and Bias of accuracy form 5 × CV5 (AccBias); Result of experiments for Covering finding methods of concept-dependent granulation with use of kNN classifier; data set: **diabetes**; r_{gran} = Granulation radius, Cov_i = Average accuracy for method Cov_i

r_{gran}	Cov2	Cov3	Cov4	Cov5	Cov6	Cov7	Cov8	Cov11	Cov12	Cov13	Cov14	Cov15	Cov16
CombAGS													
0	0.798	0.797	**0.801**	0.796	**0.798**	**0.799**	**0.8**	**0.8**	0.797	0.798	**0.801**	0.797	0.799
0.125	0.775	0.749	0.774	0.757	0.707	0.787	0.778	0.746	**0.77**	0.76	0.703	0.777	0.789
0.25	**0.683**	0.623	0.681	0.666	0.542	0.702	0.728	0.621	0.682	0.663	0.542	0.698	0.733
0.375	0.523	0.477	0.513	0.521	0.437	0.522	0.551	0.483	0.519	0.521	0.436	0.515	0.555
0.5	0.388	0.374	0.372	0.39	0.366	0.368	0.399	0.377	0.376	0.39	0.367	0.375	0.4
0.625	0.332	0.328	0.324	0.329	0.327	0.327	0.327	0.328	0.328	0.33	0.329	0.331	0.332
0.75	0.328	0.324	0.32	0.325	0.324	0.324	0.323	0.324	0.325	0.326	0.326	0.327	0.328
0.875	0.328	0.324	0.32	0.325	0.324	0.324	0.323	0.324	0.325	0.326	0.326	0.327	0.328
1	**0.328**	0.324	0.32	0.325	0.324	0.324	0.323	0.324	0.325	0.326	0.326	0.327	0.328

(continued)

Table 5.19 (continued)

r_{gran} AccBias	Cov2	Cov3	Cov4	Cov5	Cov6	Cov7	Cov8	Cov11	Cov12	Cov13	Cov14	Cov15	Cov16
0	0.007	0.012	0.009	0.007	0.016	0.002	0.008	0.006	0.009	0.012	0.008	0.014	0.007
0.125	0.015	0.018	0.004	0.016	0.009	0.019	0.019	0.016	0.011	0.024	0.025	0.007	0.025
0.25	0.018	0.006	0.013	0.01	0.01	0.012	0.021	0.018	0.016	0.012	0.012	0.017	0.013
0.375	0.015	0.02	0.019	0.012	0.007	0.015	0.011	0.012	0.015	0.013	0.01	0.015	0.021
0.5	0.011	0.017	0.015	0.004	0.012	0.011	0.009	0.004	0.008	0.005	0.017	0.003	0.01
0.625	0.006	0.004	0.007	0.008	0.004	0.008	0.009	0.008	0.016	0.014	0.004	0.008	0.003
0.75	0.008	0.005	0.006	0.008	0.006	0.01	0.011	0.008	0.014	0.014	0.006	0.007	0.003
0.875	0.008	0.005	0.006	0.008	0.006	0.01	0.011	0.008	0.014	0.014	0.006	0.007	0.003
1	0.008	0.005	0.006	0.008	0.006	0.01	0.011	0.008	0.014	0.014	0.006	0.007	0.003

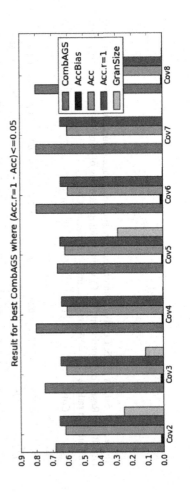

Result for best CombAGS where (Acc.r=1 - Acc)<=0.05

Table 5.20 5 *times* CV-5; Combination of accuracy and percentage reduction of objects in training data set (CombAGS) and Bias of accuracy form 5 × CV5 (AccBias); Result of experiments for Covering finding methods of concept-dependent granulation with use of kNN classifier; data set: **fertility diagnosis**; r_{gran} = Granulation radius, Cov_i = Average accuracy for method Cov_i

r_{gran}	Cov2	Cov3	Cov4	Cov5	Cov6	Cov7	Cov8	Cov11	Cov12	Cov13	Cov14	Cov15	Cov16
CombAGS													
0	0.691	0.693	0.702	0.699	0.688	0.688	0.692	0.71	0.685	0.699	0.7	0.669	0.676
0.111	0.702	0.715	0.718	0.699	0.718	0.723	0.692	0.736	0.705	0.699	0.71	0.667	0.676
0.222	0.715	0.714	0.778	0.702	0.748	0.752	0.703	0.724	0.753	0.695	0.731	0.707	0.693
0.333	0.747	0.791	0.797	0.741	0.777	0.775	0.747	0.809	0.791	0.72	0.769	0.776	0.761
0.444	0.793	0.827	0.802	0.792	0.741	0.799	0.746	0.823	0.798	0.788	0.745	0.807	0.771
0.556	0.815	0.768	0.814	0.807	0.669	0.802	0.838	0.767	0.809	0.8	0.663	0.799	0.82
0.667	0.694	0.656	0.696	0.694	0.564	0.708	0.75	0.655	0.69	0.696	0.557	0.693	0.748
0.778	0.573	0.554	0.56	0.582	0.532	0.547	0.597	0.558	0.556	0.575	0.533	0.552	0.595
0.889	0.472	0.474	0.474	0.479	0.475	0.467	0.48	0.473	0.474	0.472	0.469	0.471	0.477
1	0.436	0.437	0.435	0.439	0.439	0.434	0.44	0.437	0.437	0.431	0.438	0.435	0.434

(continued)

Table 5.20 (continued)

r_{gran}	Cov2	Cov3	Cov4	Cov5	Cov6	Cov7	Cov8	Cov11	Cov12	Cov13	Cov14	Cov15	Cov16
AccBias													
0	0.044	0.02	0.032	0.028	0.04	0.07	0.022	0.046	0.046	0.028	0.026	0.068	0.074
0.111	0.046	0.016	0.02	0.028	0.06	0.066	0.022	0.054	0.036	0.028	0.026	0.066	0.074
0.222	0.082	0.054	0.05	0.058	0.062	0.018	0.026	0.056	0.056	0.032	0.042	0.056	0.056
0.333	0.072	0.056	0.08	0.062	0.036	0.062	0.046	0.05	0.064	0.076	0.028	0.058	0.06
0.444	0.024	**0.016**	0.078	0.038	0.032	0.04	0.072	**0.026**	0.074	0.066	0.05	0.092	0.012
0.556	**0.038**	**0.026**	**0.014**	**0.04**	**0.018**	**0.066**	0.03	**0.016**	**0.026**	**0.018**	**0.028**	**0.024**	**0.032**
0.667	**0.014**	**0.034**	**0.016**	**0.012**	**0.024**	**0.024**	0.01	**0.016**	**0.02**	**0.016**	**0.008**	**0.04**	**0.016**
0.778	**0.012**	**0.014**	**0.006**	**0.022**	**0.008**	**0.008**	**0.016**	**0.02**	**0.014**	**0.01**	**0.018**	**0.012**	**0.016**
0.889	**0.01**	**0.016**	**0.008**	**0.012**	**0.008**	**0.004**	**0.01**	**0.016**	**0.008**	**0.02**	**0.01**	**0.016**	**0.018**
1	**0.008**	**0.016**	**0.01**	**0.012**	**0.002**	**0.012**	**0.01**	**0.016**	**0.006**	**0.018**	**0.014**	**0.01**	**0.012**

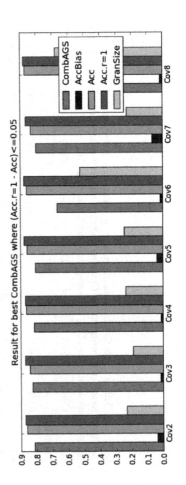

Result for best CombAGS where (Acc.r=1 − Acc)<=0.05

Table 5.21 5 *times* CV-5; Combination of accuracy and percentage reduction of objects in training data set (CombAGS) and Bias of accuracy form 5 *times* CV5 (AccBias); Result of experiments for Covering finding methods of concept-dependent granulation with use of kNN classifier; data set: **german credit**; $r_{gran} =$ Granulation radius, Cov_i = Average accuracy for method Cov_i

r_{gran}	$Cov2$	$Cov3$	$Cov4$	$Cov5$	$Cov6$	$Cov7$	$Cov8$	$Cov11$	$Cov12$	$Cov13$	$Cov14$	$Cov15$	$Cov16$
CombAGS													
0	0.779	0.779	0.78	0.782	0.782	0.778	0.779	0.779	0.777	0.78	0.781	0.778	0.784
0.05	0.779	0.781	0.78	0.782	0.781	0.778	0.779	0.78	0.777	0.78	0.782	0.778	0.784
0.1	0.779	0.781	0.782	0.782	0.782	0.782	0.779	0.781	0.778	0.78	0.784	0.778	0.784
0.15	0.779	0.792	0.797	0.782	0.792	0.787	0.779	0.795	0.793	0.78	0.789	0.777	0.784
0.2	0.782	0.807	0.811	0.782	0.802	0.789	0.779	0.807	0.81	0.78	0.804	0.779	0.784
0.25	0.791	0.82	0.812	0.782	0.816	0.794	0.779	0.819	0.811	0.78	0.815	0.784	0.784
0.3	0.799	0.816	0.823	0.785	0.813	0.783	0.79	0.82	0.822	0.783	0.815	0.797	0.793
0.35	0.814	0.811	0.812	0.789	0.801	0.763	0.801	0.81	0.814	0.786	0.801	0.813	0.801
0.4	0.816	0.824	0.809	0.809	0.778	0.763	0.821	0.825	0.805	0.808	0.776	0.827	0.818
0.45	0.819	0.806	0.795	0.812	0.742	0.77	0.829	0.806	0.799	0.815	0.737	0.821	0.827
0.5	0.811	0.784	0.788	0.797	0.711	0.773	0.832	0.78	0.791	0.796	0.715	0.808	0.832
0.55	0.787	0.743	0.778	0.771	0.648	0.77	0.822	0.74	0.778	0.772	0.642	0.79	0.819
0.6	0.741	0.682	0.735	0.733	0.58	0.732	0.789	0.687	0.745	0.733	0.579	0.74	0.79
0.65	0.659	0.6	0.658	0.661	0.502	0.647	0.714	0.602	0.658	0.661	0.501	0.645	0.712
0.7	0.562	0.525	0.543	0.571	0.446	0.535	0.607	0.526	0.544	0.57	0.446	0.535	0.602
0.75	0.459	0.45	0.451	0.469	0.423	0.44	0.48	0.45	0.45	0.471	0.419	0.437	0.481
0.8	0.397	0.394	0.396	0.399	0.391	0.391	0.402	0.394	0.395	0.403	0.389	0.389	0.401

(continued)

Table 5.21 (continued)

r_{gran}	Cov2	Cov3	Cov4	Cov5	Cov6	Cov7	Cov8	Cov11	Cov12	Cov13	Cov14	Cov15	Cov16
0.85	0.371	0.371	0.372	0.372	0.372	0.373	0.371	0.37	0.372	0.372	0.369	0.372	0.37
0.9	0.366	0.368	0.367	0.368	0.367	0.368	0.368	0.365	0.368	0.368	0.365	0.368	0.366
0.95	0.365	0.366	0.366	0.367	0.366	0.366	0.366	0.363	0.366	0.366	0.364	0.366	0.364
1	0.364	0.366	0.365	0.366	0.365	0.366	0.365	0.363	0.365	0.365	0.363	0.365	0.364
AccBias													
0	0.018	0.008	0.015	0.01	0.012	0.009	0.005	0.011	0.018	0.015	0.011	0.002	0.007
0.05	0.018	0.015	0.015	0.01	0.012	0.009	0.005	0.021	0.017	0.015	0.011	0.002	0.007
0.1	0.018	0.018	0.014	0.01	0.017	0.018	0.005	0.021	0.023	0.015	0.025	0.002	0.007
0.15	0.018	0.01	0.009	0.01	0.01	0.018	0.005	0.01	0.023	0.015	0.01	0.002	0.007
0.2	0.02	0.019	0.011	0.01	0.015	0.007	0.005	0.007	0.018	0.015	0.011	0.007	0.007
0.25	0.023	0.011	0.014	0.01	0.015	0.014	0.005	0.01	0.013	0.015	0.013	0.015	0.007
0.3	0.007	0.012	0.003	0.011	0.014	0.02	0.014	0.016	0.022	0.017	0.009	0.009	0.014
0.35	0.019	0.011	0.025	0.017	0.008	0.019	0.012	0.01	0.018	0.017	0.008	0.009	0.019
0.4	0.023	0.016	0.015	0.011	0.014	0.033	0.009	0.014	0.031	0.015	0.016	0.023	0.006
0.45	0.023	0.018	0.009	0.009	0.009	0.039	0.015	0.011	0.006	0.008	0.027	0.009	0.041
0.5	0.016	0.009	0.008	0.014	0.012	0.028	0.028	0.018	0.014	0.004	0.005	0.008	0.005
0.55	0.012	0.01	0.013	0.011	0.004	0.041	0.014	0.016	0.003	0.004	0.013	0.008	0.013
0.6	0.005	0.015	0.007	0.01	0.013	0.014	0.007	0.003	0.011	0.01	0.009	0.015	0.015
0.65	0.011	0.004	0.01	0.016	0.007	0.01	0.02	0.011	0.014	0.011	0.014	0.008	0.011

(continued)

Table 5.21 (continued)

r_{gran}	Cov2	Cov3	Cov4	Cov5	Cov6	Cov7	Cov8	Cov11	Cov12	Cov13	Cov14	Cov15	Cov16
0.7	0.009	0.007	0.016	0.009	0.013	0.013	0.016	0.009	0.025	0.011	0.004	0.002	0.01
0.75	0.003	0.005	0.01	0.003	0.008	0.01	0.005	0.007	0.009	0.003	0.008	0.01	0.009
0.8	0.004	0.007	0.013	0.009	0.006	0.007	0.008	0.008	0.004	0.01	0.005	0.01	0.003
0.85	0.009	0.009	0.011	0.005	0.006	0.007	0.01	0.007	0.007	0.007	0.006	0.004	0.004
0.9	0.009	0.007	0.008	0.003	0.004	0.008	0.012	0.006	0.007	0.006	0.006	0.009	0.005
0.95	0.009	0.007	0.007	0.003	0.004	0.007	0.012	0.006	0.007	0.006	0.006	0.009	0.005
1	0.009	0.007	0.007	0.004	0.003	0.007	0.012	0.006	0.007	0.007	0.006	0.008	0.004

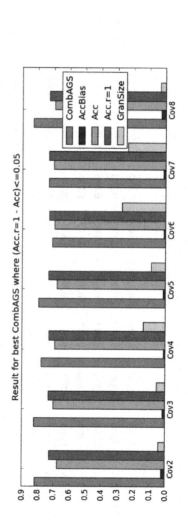

Result for best CombAGS where (Acc.r=1 - Acc)<=0.05

Table 5.22 5 *times* CV-5; Combination of accuracy and percentage reduction of objects in training data set (CombAGS) and Bias of accuracy from 5 × CV5 (AccBias); Result of experiments for Covering finding methods of concept-dependent granulation with use of kNN classifier; data set: **heart disease**; r_{gran} = Granulation radius, Cov_i = Average accuracy for method Cov_i

r_{gran}	Cov2	Cov3	Cov4	Cov5	Cov6	Cov7	Cov8	Cov11	Cov12	Cov13	Cov14	Cov15	Cov16
CombAGS													
0	0.897	0.896	0.895	0.895	0.897	0.897	0.895	0.893	0.893	0.894	0.897	0.896	0.891
0.077	0.897	0.887	0.893	0.895	0.892	0.894	0.895	0.884	0.891	0.894	0.891	0.896	0.891
0.154	0.892	0.884	0.892	0.895	0.882	0.885	0.895	0.892	0.889	0.894	0.884	0.893	0.891
0.231	0.888	0.886	0.887	0.895	0.873	0.885	0.895	0.89	0.887	0.894	0.87	0.884	0.891
0.308	0.881	0.871	0.88	0.883	0.837	0.88	0.891	0.873	0.88	0.88	0.842	0.885	0.887
0.385	0.871	0.841	0.869	0.861	0.782	0.873	0.886	0.838	0.866	0.86	0.779	0.865	0.883
0.462	0.831	0.779	0.828	0.819	0.709	0.832	0.861	0.776	0.824	0.821	0.701	0.827	0.866
0.538	0.746	0.693	0.753	0.746	0.598	0.752	0.802	0.687	0.759	0.747	0.597	0.745	0.801
0.615	0.616	0.589	0.611	0.623	0.524	0.601	0.65	0.583	0.61	0.622	0.517	0.594	0.648
0.692	0.495	0.487	0.484	0.493	0.468	0.473	0.503	0.484	0.486	0.499	0.468	0.476	0.506
0.769	0.425	0.424	0.425	0.427	0.425	0.424	0.429	0.421	0.424	0.427	0.426	0.424	0.425
0.846	0.413	0.413	0.412	0.412	0.412	0.412	0.414	0.409	0.412	0.414	0.415	0.413	0.413
0.923	0.413	0.413	0.412	0.412	0.412	0.412	0.414	0.409	0.412	0.414	0.415	0.413	0.413
1	0.413	0.413	0.412	0.412	0.412	0.412	0.414	0.409	0.412	0.414	0.415	0.413	0.413
AccBias													
0	0.011	0.025	0.011	0.016	0.021	0.022	0.009	0.011	0.016	0.01	0.012	0.009	0.013
0.077	0.011	0.016	0.011	0.016	0.016	0.027	0.009	0.02	0.015	0.01	0.014	0.009	0.013
0.154	0.017	0.017	0.013	0.016	0.011	0.022	0.009	0.013	0.019	0.01	0.004	0.007	0.013

(continued)

Table 5.22 (continued)

r_{gran}	Cov2	Cov3	Cov4	Cov5	Cov6	Cov7	Cov8	Cov11	Cov12	Cov13	Cov14	Cov15	Cov16
0.231	0.009	0.019	0.024	0.016	0.007	0.013	0.009	0.008	0.014	0.01	0.008	0.01	0.013
0.308	0.023	0.011	0.015	0.013	0.01	0.036	0.019	0.006	0.02	0.006	0.011	0.022	0.011
0.385	0.007	0.01	0.014	0.009	0.015	0.012	0.013	0.016	0.013	0.014	0.009	0.018	0.013
0.462	0.025	0.015	0.007	0.015	0.019	0.01	0.021	0.009	0.007	0.011	0.011	0.009	0.01
0.538	0.005	0.006	0.007	0.011	0.005	0.018	0.013	0.005	0.004	0.011	0.017	0.009	0.019
0.615	0.02	0.013	0.013	0.007	0.009	0.013	0.013	0.01	0.006	0.007	0.007	0.01	0.012
0.692	0.01	0.008	0.004	0.003	0.01	0.007	0.013	0.007	0.007	0.004	0.004	0.012	0.007
0.769	0.004	0.006	0.008	0.009	0.008	0.006	0.005	0.009	0.005	0.001	0.005	0.005	0.006
0.846	0.004	0.004	0.007	0.01	0.007	0.005	0.009	0.008	0.009	0.006	0.004	0.004	0.004
0.923	0.004	0.004	0.007	0.01	0.007	0.005	0.009	0.008	0.009	0.006	0.004	0.004	0.004
1	0.004	0.004	0.007	0.01	0.007	0.005	0.009	0.008	0.009	0.006	0.004	0.004	0.004

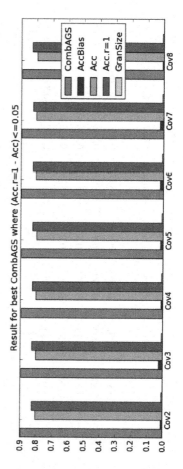

Result for best CombAGS where (Acc.r=1 - Acc)<=0.05

Table 5.23 5 *times* CV-5; Combination of accuracy and percentage reduction of objects in training data set (CombAGS) and Bias of accuracy form 5 × CV5 (AccBias); Result of experiments for Covering finding methods of concept-dependent granulation with use of kNN classifier; data set: **hepatitis-filled**; r_{gran} = Granulation radius, Cov_i = Average accuracy for method Cov_i

r_{gran}	$Cov2$	$Cov3$	$Cov4$	$Cov5$	$Cov6$	$Cov7$	$Cov8$	$Cov11$	$Cov12$	$Cov13$	$Cov14$	$Cov15$	$Cov16$
CombAGS													
0	0.893	0.901	0.899	0.893	0.897	0.896	0.896	0.899	0.895	0.899	0.897	0.895	0.895
0.053	0.893	0.901	0.899	0.893	0.897	0.896	0.896	0.899	0.895	0.899	0.897	0.895	0.895
0.105	0.893	0.901	0.899	0.893	0.897	0.896	0.896	0.899	0.895	0.899	0.897	0.895	0.895
0.158	0.893	0.898	0.895	0.893	0.893	0.893	0.896	0.896	0.892	0.899	0.893	0.895	0.895
0.211	0.893	0.896	0.896	0.893	0.89	0.892	0.896	0.893	0.891	0.899	0.89	0.894	0.895
0.263	0.892	0.901	0.895	0.893	0.886	0.902	0.896	0.889	0.896	0.899	0.886	0.891	0.895
0.316	0.892	0.899	0.906	0.893	0.888	0.9	0.896	0.897	0.905	0.899	0.892	0.89	0.895
0.368	0.894	0.897	0.898	0.891	0.875	0.891	0.893	0.905	0.895	0.895	0.874	0.886	0.895
0.421	0.897	0.877	0.891	0.881	0.845	0.885	0.904	0.878	0.893	0.888	0.843	0.885	0.899
0.474	0.891	0.842	0.879	0.871	0.809	0.881	0.91	0.841	0.886	0.875	0.806	0.884	0.913
0.526	0.862	0.802	0.864	0.852	0.74	0.859	0.901	0.803	0.871	0.855	0.74	0.854	0.908
0.579	0.815	0.755	0.815	0.807	0.686	0.814	0.852	0.756	0.813	0.816	0.683	0.804	0.853
0.632	0.759	0.692	0.75	0.756	0.631	0.745	0.802	0.698	0.754	0.762	0.631	0.744	0.802
0.684	0.658	0.617	0.66	0.671	0.59	0.649	0.699	0.621	0.665	0.677	0.59	0.638	0.702
0.737	0.583	0.563	0.568	0.593	0.54	0.565	0.599	0.557	0.578	0.59	0.532	0.562	0.602
0.789	0.503	0.497	0.495	0.516	0.489	0.493	0.51	0.493	0.498	0.511	0.484	0.491	0.516
0.842	0.475	0.475	0.472	0.481	0.471	0.471	0.474	0.471	0.476	0.477	0.467	0.469	0.478
0.895	0.46	0.463	0.459	0.464	0.462	0.463	0.457	0.459	0.464	0.464	0.458	0.458	0.462
0.947	0.456	0.459	0.455	0.46	0.458	0.459	0.453	0.455	0.46	0.46	0.454	0.454	0.458
1	0.448	0.451	0.447	0.452	0.45	0.451	0.445	0.447	0.452	0.452	0.446	0.446	0.45

(continued)

Table 5.23 (continued)

r_{gran}	Cov2	Cov3	Cov4	Cov5	Cov6	Cov7	Cov8	Cov11	Cov12	Cov13	Cov14	Cov15	Cov16
AccBias													
0	0.01	0.021	0.03	0.01	0.015	0.012	0.01	0.012	0.006	0.012	0.015	0.013	0.006
0.053	0.01	0.021	0.03	0.01	0.015	0.012	0.01	0.012	0.006	0.012	0.015	0.013	0.006
0.105	0.01	0.021	0.03	0.01	0.015	0.012	0.01	0.012	0.006	0.012	0.015	0.013	0.006
0.158	0.01	0.021	0.031	0.01	0.015	0.012	0.01	0.012	0.006	0.012	0.015	0.013	0.006
0.211	0.009	0.021	0.028	0.01	0.014	0.012	0.01	0.014	0.005	0.012	0.014	0.013	0.006
0.263	0.019	0.012	0.023	0.01	0.008	0.01	0.01	0.009	0.014	0.012	0.034	0.013	0.006
0.316	0.025	**0.004**	0.018	0.01	**0.014**	0.019	0.01	**0.005**	0.021	0.012	**0.019**	0.008	0.006
0.368	0.015	**0.017**	0.018	0.006	**0.021**	0.01	0.01	**0.022**	0.001	0.012	**0.009**	0.005	0.014

(continued)

Table 5.23 (continued)

r_{gran}	Cov2	Cov3	Cov4	Cov5	Cov6	Cov7	Cov8	Cov11	Cov12	Cov13	Cov14	Cov15	Cov16
0.421	0.036	0.021	0.014	0.018	0.01	0.009	0.026	0.01	0.013	0.018	0.022	0.012	0.034
0.474	0.022	0.009	0.013	0.017	0.023	0.023	0.013	0.009	0.026	0.009	0.012	0.018	0.031
0.526	0.034	0.017	0.023	0.036	0.026	0.048	0.015	0.034	0.023	0.004	0.019	0.031	0.023
0.579	0.015	0.018	0.015	0.018	0.01	0.012	0.023	0.01	0.026	0.015	0.015	0.026	0.009
0.632	0.01	0.01	0.018	0.028	0.019	0.014	0.006	0.017	0.01	0.017	0.01	0.01	0.027
0.684	0.026	0.025	0.019	0.015	0.006	0.01	0.014	0.019	0.021	0.018	0.008	0.01	0.009
0.737	0.022	0.019	0.026	0.012	0.01	0.01	0.01	0.015	0.014	0.036	0.006	0.009	0.026
0.789	0.018	0.023	0.009	0.018	0.01	0.012	0.019	0.018	0.015	0.013	0.006	0.008	0.013
0.842	0.018	0.023	0.009	0.013	0.012	0.012	0.012	0.017	0.014	0.012	0.006	0.01	0.014
0.895	0.014	0.022	0.01	0.013	0.01	0.015	0.013	0.015	0.013	0.013	0.005	0.012	0.01
0.947	0.014	0.022	0.01	0.013	0.01	0.015	0.013	0.015	0.013	0.013	0.005	0.012	0.01
1	0.014	0.022	0.01	0.013	0.01	0.015	0.013	0.015	0.013	0.013	0.005	0.012	0.01

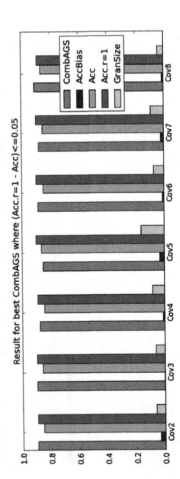

Result for best CombAGS where (Acc.r=1 - Acc)<=0.05

Table 5.24 5 *times* CV-5; Combination of accuracy and percentage reduction of objects in training data set (CombAGS) and Bias of accuracy form 5 × CV5 (AccBias); Result of experiments for Covering finding methods of concept-dependent granulation with use of kNN classifier; data set: congresional voting records; r_{gran} = Granulation radius, Cov_i = Average accuracy for method Cov_i

r_{gran}	Cov2	Cov3	Cov4	Cov5	Cov6	Cov7	Cov8	Cov11	Cov12	Cov13	Cov14	Cov15	Cov16
CombAGS													
0	0.946	0.946	0.945	0.945	0.946	0.946	0.944	0.945	0.946	0.946	0.945	0.944	0.945
0.063	0.946	0.944	0.944	0.945	0.944	0.944	0.944	0.944	0.944	0.946	0.943	0.944	0.945
0.125	0.946	0.943	0.942	0.945	0.942	0.943	0.944	0.942	0.941	0.946	0.941	0.944	0.945
0.188	0.945	0.938	0.943	0.945	0.936	0.943	0.944	0.939	0.941	0.946	0.935	0.942	0.945
0.25	0.944	0.937	0.943	0.945	0.936	0.944	0.944	0.937	0.943	0.946	0.936	0.94	0.945
0.313	0.942	0.927	0.942	0.945	0.926	0.936	0.944	0.928	0.942	0.946	0.926	0.936	0.945
0.375	0.94	0.924	0.941	0.945	0.919	0.935	0.944	0.924	0.942	0.946	0.917	0.932	0.945
0.438	0.937	0.915	0.941	0.943	0.908	0.934	0.943	0.914	0.94	0.944	0.908	0.931	0.944
0.5	0.937	0.898	0.937	0.939	0.888	0.935	0.943	0.899	0.939	0.939	0.886	0.932	0.944
0.563	0.935	0.886	0.927	0.922	0.862	0.933	0.947	0.886	0.928	0.92	0.859	0.932	0.948
0.625	0.932	0.874	0.914	0.9	0.837	0.934	0.945	0.87	0.915	0.898	0.835	0.933	0.941
0.688	0.925	0.867	0.902	0.893	0.803	0.931	0.948	0.864	0.898	0.891	0.802	0.928	0.951
0.75	0.899	0.824	0.888	0.871	0.758	0.907	0.935	0.821	0.886	0.874	0.757	0.902	0.934
0.813	0.851	0.773	0.832	0.819	0.707	0.866	0.894	0.772	0.835	0.821	0.705	0.845	0.893
0.875	0.76	0.709	0.759	0.753	0.655	0.786	0.808	0.708	0.761	0.752	0.657	0.77	0.807
0.938	0.636	0.594	0.636	0.642	0.558	0.658	0.676	0.593	0.631	0.643	0.555	0.651	0.676
1	0.474	0.473	0.474	0.474	0.474	0.475	0.475	0.473	0.474	0.474	0.472	0.475	0.475

(continued)

Table 5.24 (continued)

r_{gran}	Cov2	Cov3	Cov4	Cov5	Cov6	Cov7	Cov8	Cov11	Cov12	Cov13	Cov14	Cov15	Cov16
AccBias													
0	0	0	0.001	0.001	0	0	0.003	0.001	0	0	0.002	0.003	0.001
0.063	0	0	0	0.001	0	0	0.003	0	0	0	0.002	0.003	0.001
0.125	0	0.001	0.002	0.001	0	0	0.003	0.002	0.001	0	0.003	0.003	0.001
0.188	0	0.005	0.002	0.001	0.004	0	0.003	0.004	0.002	0	0.003	0.003	0.001
0.25	0	0.004	0.003	0.001	0.004	0.004	0.003	0.003	0.002	0	0.001	0.003	0.001
0.313	0	0.008	0.004	0.001	0.006	0.002	0.003	0.003	0.003	0	0.003	0.003	0.001
0.375	0.001	0.005	0.003	0.001	0.006	0.008	0.003	0.008	0.003	0	0.007	0.002	0.001
0.438	0.006	0.003	0.001	0.001	0.004	0.006	0.003	0.002	0.006	0	0.004	0.005	0.004
0.5	0.008	0.007	0.006	0	0.005	0.005	0.01	0.007	0.004	0	0.006	0	0.017
0.563	0.009	0.012	0.003	0.004	0.006	0.012	0.002	0.005	0.005	0.003	0.01	0.004	0.011
0.625	0.006	0.008	0.004	0.004	0.006	0.005	0.02	0.001	0.003	0.004	0.008	0.003	0.034
0.688	0.012	0.003	0.008	0.011	0.011	0.013	0.006	0.006	0.006	0.006	0.004	0.005	0.005
0.75	0.014	0.003	0.002	0.003	0.008	0.006	0.013	0.006	0.003	0.006	0.003	0.002	0.012
0.813	0.004	0.008	0.002	0.004	0.007	0.006	0.008	0.009	0.009	0.005	0.011	0.006	0.003

(continued)

Table 5.24 (continued)

r_{gran}	Cov2	Cov3	Cov4	Cov5	Cov6	Cov7	Cov8	Cov11	Cov12	Cov13	Cov14	Cov15	Cov16
0.875	0.003	0.006	0.007	0.005	0.007	0.009	0.006	0.005	0.006	0.002	0.005	0.011	0.005
0.938	0.003	0.006	0.007	0.006	0.01	0.006	0.008	0.007	0.007	0.006	0.011	0.006	0.005
1	0.009	0.008	0.007	0.004	0.005	0.004	0.008	0.001	0.004	0.004	0.005	0.006	0.003

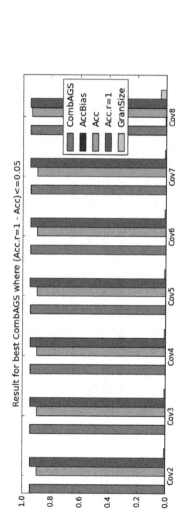

Result for best CombAGS where (Acc.r=1 - Acc) <= 0.05

Table 5.25 5 *times* CV-5; Combination of accuracy and percentage reduction of objects in training data set (CombAGS) and Bias of accuracy form 5 × CV5 (AccBias); Result of experiments for Covering finding methods of concept-dependent granulation with use of kNN classifier; data set: mushroom – modified – filled; r_{gran} = Granulation radius, Cov_i = Average accuracy for method Cov_i

r_{gran}	Cov2	Cov3	Cov4	Cov5	Cov6	Cov7	Cov8	Cov11	Cov12	Cov13	Cov14	Cov15	Cov16
CombAGS													
0	0.944	0.943	0.944	0.943	0.944	0.943	0.943	0.943	0.944	0.944	0.944	0.944	0.943
0.045	0.944	0.943	0.944	0.943	0.944	0.943	0.943	0.943	0.944	0.944	0.944	0.944	0.943
0.091	0.944	0.943	0.944	0.943	0.944	0.943	0.943	0.943	0.944	0.944	0.944	0.944	0.943
0.136	0.944	0.943	0.944	0.943	0.944	0.943	0.943	0.943	0.944	0.944	0.944	0.944	0.943
0.182	0.944	0.94	0.944	0.943	0.94	0.943	0.943	0.94	0.944	0.944	0.94	0.944	0.943
0.227	0.943	0.938	0.944	0.943	0.939	0.943	0.943	0.938	0.944	0.944	0.939	0.944	0.943
0.273	0.944	0.94	0.944	0.943	0.938	0.942	0.943	0.941	0.945	0.944	0.938	0.944	0.943
0.318	0.944	0.939	0.946	0.943	0.937	0.94	0.943	0.94	0.946	0.944	0.936	0.942	0.943
0.364	0.943	0.949	0.945	0.943	0.943	0.942	0.943	0.949	0.946	0.944	0.942	0.943	0.943
0.409	0.942	0.945	0.943	0.943	0.94	0.944	0.94	0.945	0.943	0.944	0.939	0.943	0.94
0.455	0.946	0.959	0.946	0.94	0.928	0.945	0.943	0.958	0.946	0.94	0.928	0.942	0.941
0.5	0.954	0.959	0.945	0.938	0.942	0.955	0.949	0.96	0.946	0.94	0.942	0.951	0.947
0.545	0.969	**0.958**	0.947	0.942	**0.943**	0.97	0.966	**0.957**	0.948	0.942	**0.943**	0.963	0.969

(continued)

Table 5.25 (continued)

r_{gran}	Cov2	Cov3	Cov4	Cov5	Cov6	Cov7	Cov8	Cov11	Cov12	Cov13	Cov14	Cov15	Cov16
0.591	0.981	0.972	0.958	0.958	0.945	0.977	0.971	0.972	0.957	0.957	0.945	0.977	0.967
0.636	0.989	0.966	0.974	0.978	0.955	0.987	0.988	0.966	0.974	0.979	0.955	0.987	0.986
0.682	0.991	0.97	0.983	0.986	0.96	0.992	0.994	0.97	0.983	0.986	0.96	0.992	0.994
0.727	0.994	0.97	0.983	0.988	0.963	0.993	0.994	0.971	0.986	0.988	0.963	0.993	0.995
0.773	0.995	0.977	0.986	0.987	0.964	0.994	0.996	0.977	0.986	0.988	0.965	0.994	0.996
0.818	0.993	0.977	0.982	0.982	0.953	0.994	0.996	0.977	0.982	0.982	0.953	0.993	0.996
0.864	0.985	0.966	0.967	0.969	0.92	0.988	0.992	0.965	0.968	0.969	0.92	0.987	0.992
0.909	0.961	0.934	0.938	0.938	0.846	0.968	0.98	0.934	0.939	0.939	0.846	0.964	0.979
0.955	0.865	0.836	0.859	0.866	0.694	0.884	0.923	0.837	0.859	0.867	0.694	0.881	0.923
1	0.5	0.5	0.5	0.5	0.5	0.5	0.5	0.5	0.5	0.5	0.5	0.5	0.5
AccBias													
0	0.001	0.001	0.001	0.001	0.001	0.001	0.001	0	0	0.001	0.001	0.001	0.001
0.045	0.001	0.001	0.001	0.001	0.001	0.001	0.001	0	0	0.001	0.001	0.001	0.001
0.091	0.001	0.001	0.001	0.001	0.001	0.001	0.001	0	0	0.001	0.001	0.001	0.001
0.136	0.001	0.001	0.001	0.001		0.001	0.001	0	0	0.001	0.001	0.001	0.001
0.182	0.001	0.001	0.001	0.001	0	0.001	0.001	0.001	0	0.001	0.001	0.001	0.001
0.227	0.001	0	0.001	0.001	0	0.001	0.001	0.001	0	0.001	0.001	0.001	0.001
0.273	0.001	0.001	0.001	0.001	0.001	0.002	0.001	0.001	0.002	0.001	0.001	0.001	0.001
0.318	0.001	0	0.001	0.001	0.003	0.003	0.001	0.004	0.001	0.001	0.005	0.003	0.001
0.364	0.002	0.003	0.002	0.001	0.004	0.003	0.001	0.001	0.001	0.001	0.003	0.003	0.001
0.409	0.003	0.001	0.002	0.001	0.001	0.005		0.001	0.001	0.001	0	0.003	0.003

(continued)

Table 5.25 (continued)

r_{gran}	Cov2	Cov3	Cov4	Cov5	Cov6	Cov7	Cov8	Cov11	Cov12	Cov13	Cov14	Cov15	Cov16
0.455	0.002	0.002	0.001	0.003	0.002	0.005	0.008	0.002	0.002	0.003	0.002	0.004	0.005
0.5	0.008	0.002	0.002	0.003	0.008	0.008	0.004	0.002	0.002	0.003	0.005	0.004	0.007
0.545	0.006	**0.002**	0.002	0.001	**0.004**	0.009	0.008	**0.001**	0.004	0.001	**0.004**	0.008	0.005
0.591	**0.007**	**0.001**	0.003	0.003	**0.001**	**0.005**	0.004	**0.002**	0.002	0.002	**0.002**	**0.009**	0.005
0.636	**0.002**	**0.002**	**0.001**	**0.002**	**0.001**	**0.004**	**0.003**	**0.001**	**0.001**	**0.001**	0	**0.005**	**0.003**
0.682	**0.002**	**0.002**	**0.002**	**0.001**	**0.001**	**0.002**	**0.001**	**0.001**	**0.001**	**0.002**	0	**0.003**	**0.002**
0.727	**0.001**	0	**0.001**	0	**0.001**	**0.002**	**0.003**	0	**0.001**	0	0	**0.001**	**0.003**
0.773	**0.001**	0	**0.001**	0	0	**0.001**	**0.001**	0	**0.001**	0	**0.001**	**0.001**	**0.001**
0.818	0	0	**0.001**	0	0	0	**0.001**	**0.001**	0	0	**0.001**	0	0
0.864	0	0	0	0	0	0	0	0	0	0	0	0	0
0.909	0	0	0	0	0	0	0	0	0	0	0	0	0
0.955	0	0	0	0	0	0	0	0	0	0	0	0	0
1	0	0	0	0	0	0	0	0	0	0	0	0	0

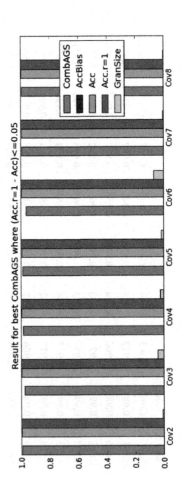

Result for best CombAGS where (Acc.r=1 - Acc)<=0.05

Table 5.26 5 *times* CV-5; Combination of accuracy and percentage reduction of objects in training data set (CombAGS) and Bias of accuracy form 5 × CV5 (AccBias); Result of experiments for Covering finding methods of concept-dependent granulation with use of kNN classifier; data set: **nursery**; r_{gran} = Granulation radius, Cov_i = Average accuracy for method Cov_i

r_{gran}	Cov2	Cov3	Cov4	Cov5	Cov6	Cov7	Cov8	Cov11	Cov12	Cov13	Cov14	Cov15	Cov16
CombAGS													
0	0.699	0.706	0.695	0.695	0.686	0.699	0.695	0.686	0.681	0.696	0.7	0.696	0.685
0.125	0.709	0.72	0.709	0.705	0.71	0.713	0.704	0.697	0.689	0.707	0.731	0.717	0.697
0.25	0.749	0.755	0.753	0.76	0.765	0.755	0.75	0.736	0.728	0.756	0.787	0.763	0.737
0.375	0.763	0.762	0.77	0.771	0.745	0.769	0.759	0.744	0.747	0.772	0.765	0.773	0.744
0.5	0.772	0.737	0.783	0.788	0.725	0.78	0.768	0.719	0.761	0.785	0.748	0.775	0.756
0.625	0.785	0.724	0.797	0.791	0.686	0.797	0.795	0.704	0.773	0.793	0.709	0.774	0.784
0.75	0.785	0.733	0.798	0.773	0.604	0.797	0.803	0.71	0.772	0.775	0.63	0.787	0.791
0.875	0.72	0.669	0.71	0.677	0.415	0.734	0.757	0.644	0.683	0.677	0.434	0.727	0.744
1	0.285	0.292	0.285	0.285	0.278	0.285	0.285	0.278	0.274	0.285	0.29	0.285	0.279

(continued)

Table 5.26 (continued)

r_{gran}	Cov2	Cov3	Cov4	Cov5	Cov6	Cov7	Cov8	Cov11	Cov12	Cov13	Cov14	Cov15	Cov16
AccBias													
0	0.025	0.011	0.04	0.036	0.036	0.034	0.034	0.045	0.056	0.029	0.01	0.021	0.034
0.125	0.031	0.008	0.027	0.031	0.044	0.029	0.022	0.044	0.054	0.022	0.013	0.023	0.036
0.25	0.035	0.015	0.035	**0.023**	**0.058**	0.031	0.028	0.047	0.061	0.031	**0.014**	**0.041**	0.053
0.375	**0.031**	0.008	**0.025**	**0.026**	**0.054**	**0.031**	0.022	0.049	**0.061**	**0.027**	**0.011**	**0.025**	0.046
0.5	**0.031**	0.007	**0.027**	**0.028**	**0.052**	**0.027**	**0.025**	0.044	**0.08**	**0.027**	**0.004**	**0.028**	**0.048**
0.625	**0.035**	0.005	**0.032**	**0.033**	**0.049**	**0.029**	**0.031**	**0.045**	**0.081**	**0.03**	**0.004**	**0.033**	**0.052**
0.75	**0.029**	**0.007**	**0.03**	**0.031**	**0.049**	**0.031**	**0.03**	**0.049**	**0.084**	**0.031**	**0.007**	**0.033**	**0.058**
0.875	**0.03**	**0.003**	**0.033**	**0.028**	0.043	**0.034**	**0.029**	**0.052**	**0.082**	**0.03**	**0.001**	**0.03**	**0.058**
1	**0.014**	**0.004**	**0.014**	**0.017**	**0.027**	**0.013**	**0.015**	**0.027**	**0.037**	**0.017**	**0.002**	**0.015**	**0.026**

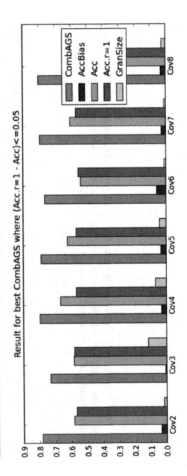

Result for best CombAGS where (Acc.r=1 - Acc)<=0.05

Table 5.27 5 *times* CV-5: Combination of accuracy and percentage reduction of objects in training data set (CombAGS) and Bias of accuracy form 5 × CV5 (AccBias); Result of experiments for Covering finding methods of concept-dependent granulation with use of kNN classifier; data set: **soybean-large-all-filled**; r_{gran} = Granulation radius, Cov_i = Average accuracy for method Cov_i

r_{gran}	Cov2	Cov3	Cov4	Cov5	Cov6	Cov7	Cov8	Cov11	Cov12	Cov13	Cov14	Cov15	Cov16
CombAGS													
0	0.822	0.822	0.823	0.824	0.823	0.823	0.823	0.823	0.817	0.822	0.826	0.824	0.82
…	…	…	…	…	…	…	…	…	…	…	…	…	…
0.4	0.822	0.822	0.823	0.824	0.823	0.823	0.823	0.823	0.817	0.822	0.826	0.824	0.82
0.429	0.822	0.821	0.822	0.824	0.822	0.822	0.823	0.822	0.816	0.822	0.826	0.824	0.82
0.457	0.822	0.822	0.823	0.824	0.821	0.822	0.823	0.822	0.816	0.822	0.825	0.824	0.82
0.486	0.822	0.826	0.822	0.824	0.825	0.822	0.823	0.825	0.816	0.822	0.828	0.825	0.82
0.514	0.822	0.826	0.822	0.824	0.824	0.822	0.823	0.826	0.814	0.822	0.828	0.825	0.82
0.543	0.823	0.828	0.82	0.824	0.824	0.821	0.823	0.83	0.814	0.822	0.83	0.826	0.82
0.571	0.823	0.834	0.824	0.825	0.826	0.826	0.823	0.831	0.814	0.822	0.83	0.826	0.82
0.6	0.826	0.838	0.824	0.825	0.826	0.829	0.824	0.835	0.818	0.823	0.833	0.825	0.821
0.629	0.828	0.85	0.831	0.828	0.838	0.842	0.831	0.85	0.826	0.824	0.839	0.831	0.823
0.657	0.83	0.83	0.839	0.828	0.811	0.84	0.83	0.827	0.831	0.824	0.819	0.83	0.827
0.686	0.839	0.812	0.826	0.826	0.801	0.84	0.833	0.813	0.818	0.824	0.803	0.83	0.828
0.714	0.839	0.807	0.82	0.825	0.78	0.85	0.838	0.8	0.815	0.827	0.781	0.833	0.84
0.743	0.846	0.807	0.817	0.819	0.78	0.854	0.854	0.807	0.815	0.82	0.779	0.839	0.843
0.771	0.856	0.817	0.838	0.834	0.775	0.861	0.871	0.814	0.836	0.831	0.777	0.855	0.869
0.8	0.851	0.813	0.84	0.837	0.774	0.86	0.874	0.81	0.84	0.838	0.775	0.845	0.867
0.829	0.84	**0.81**	0.839	0.839	**0.765**	0.849	0.86	0.804	0.841	0.839	**0.765**	**0.839**	0.862
0.857	0.816	**0.78**	0.816	**0.821**	**0.73**	**0.838**	0.846	**0.781**	**0.817**	0.816	**0.732**	0.807	0.844

(continued)

Table 5.27 (continued)

r_{gran}	Cov2	Cov3	Cov4	Cov5	Cov6	Cov7	Cov8	Cov11	Cov12	Cov13	Cov14	Cov15	Cov16
0.886	**0.78**	**0.744**	**0.781**	**0.781**	**0.689**	**0.803**	**0.806**	**0.742**	**0.778**	**0.782**	**0.688**	**0.776**	**0.811**
0.914	**0.726**	**0.691**	**0.727**	**0.728**	**0.638**	**0.74**	**0.756**	**0.691**	**0.725**	**0.732**	**0.637**	**0.715**	**0.758**
0.943	**0.646**	**0.625**	**0.645**	**0.659**	**0.584**	**0.654**	**0.677**	**0.627**	**0.643**	**0.661**	**0.582**	**0.639**	**0.677**
0.971	**0.561**	**0.55**	**0.552**	**0.572**	**0.535**	**0.548**	**0.577**	**0.548**	**0.55**	**0.574**	**0.534**	**0.549**	**0.577**
1	**0.459**	**0.457**	**0.46**	**0.459**	**0.459**	**0.458**	**0.457**	**0.457**	**0.458**	**0.458**	**0.458**	**0.456**	**0.456**
AccBias													
0	0.021	0.01	0.014	0.008	0.023	0.007	0.012	0.01	0.02	0.005	0.021	0.022	0.007
...	...	...	...	...	...	...	...	...	...	...	...	...	...
0.4	0.021	0.01	0.014	0.008	0.023	0.007	0.012	0.01	0.02	0.005	0.021	0.022	0.007
0.429	0.021	0.011	0.014	0.008	0.023	0.008	0.012	0.01	0.02	0.005	0.021	0.022	0.007
0.457	0.021	0.012	0.013	0.008	0.024	0.008	0.012	0.01	0.02	0.005	0.02	0.022	0.007
0.486	0.021	0.014	0.014	0.008	0.019	0.008	0.012	0.007	0.02	0.005	0.018	0.023	0.007
0.514	0.022	0.014	0.013	0.008	0.019	0.009	0.012	0.009	0.02	0.005	0.019	0.024	0.007
0.543	0.019	0.023	0.01	0.008	0.016	0.008	0.012	0.015	0.019	0.005	0.015	0.02	0.007
0.571	0.016	0.015	0.016	0.007	0.013	0.006	0.013	0.018	0.02	0.005	0.012	0.022	0.007
0.6	0.011	0.017	0.008	0.007	0.03	0.004	0.012	0.016	0.015	0.005	0.016	0.023	0.008
0.629	0.006	0.011	0.007	0.009	0.017	0.009	0.021	0.011	0.006	0.011	0.007	0.012	0.014
0.657	0.023	0.008	0.009	0.009	0.017	0.017	0.017	0.009	0.007	0.008	0.012	0.009	0.014
0.686	0.015	0.03	0.025	0.007	0.022	0.01	0.012	0.023	0.006	0.009	0.028	0.018	0.023
0.714	0.022	0.006	0.016	0.006	0.008	0.014	0.023	0.012	0.009	0.004	0.011	0.028	0.022

(continued)

Table 5.27 (continued)

r_{gran}	Cov2	Cov3	Cov4	Cov5	Cov6	Cov7	Cov8	Cov11	Cov12	Cov13	Cov14	Cov15	Cov16
0.743	0.015	0.01	0.02	0.02	0.024	0.011	0.02	0.015	0.014	0.023	0.006	0.022	0.012
0.771	0.01	0.013	0.022	0.011	0.009	0.008	0.012	0.004	0.01	0.01	0.018	0.014	0.011
0.8	0.009	0.017	0.012	0.013	0.008	0.016	0.012	0.007	0.013	0.009	0.013	0.009	0.012
0.829	0.015	**0.016**	0.01	0.012	**0.007**	0.015	0.014	0.017	0.007	0.004	**0.006**	**0.005**	0.024
0.857	0.004	**0.009**	0.023	**0.012**	**0.011**	**0.006**	0.013	**0.008**	**0.013**	0.011	**0.004**	0.02	0.011
0.886	**0.006**	**0.006**	**0.008**	**0.013**	0.014	**0.012**	**0.02**	**0.008**	**0.008**	**0.009**	**0.018**	**0.01**	**0.009**
0.914	**0.014**	**0.009**	**0.009**	**0.004**	0.016	**0.004**	**0.006**	0.021	**0.002**	**0.008**	**0.009**	**0.01**	**0.009**
0.943	**0.021**	**0.005**	**0.009**	**0.009**	**0.008**	**0.013**	**0.015**	**0.014**	**0.01**	**0.014**	**0.005**	**0.007**	**0.014**
0.971	**0.012**	**0.002**	**0.009**	**0.006**	**0.008**	**0.004**	**0.015**	**0.005**	**0.008**	**0.006**	**0.005**	**0.005**	**0.007**
1	**0.01**	**0.008**	**0.006**	**0.003**	**0.008**	**0.003**	**0.01**	**0.005**	**0.007**	**0.003**	**0.006**	**0.003**	**0.006**

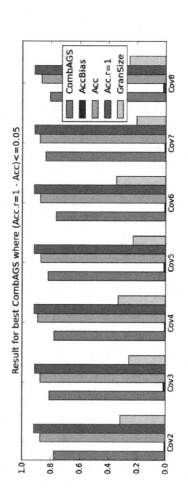

Result for best CombAGS where (Acc.r=1 - Acc)<=0.05

Table 5.28 5 *times* CV-5; Combination of accuracy and percentage reduction of objects in training data set (CombAGS) and Bias of accuracy form 5 × CV5 (AccBias); Result of experiments for Covering finding methods of concept-dependent granulation with use of kNN classifier; data set: **SPECT – all – modif**; r_{gran} = Granulation radius, Cov_i = Average accuracy for method Cov_i

r_{gran}	$Cov2$	$Cov3$	$Cov4$	$Cov5$	$Cov6$	$Cov7$	$Cov8$	$Cov11$	$Cov12$	$Cov13$	$Cov14$	$Cov15$	$Cov16$
CombAGS													
0	0.759	0.773	0.747	0.77	0.779	0.756	0.743	0.766	0.779	0.767	0.773	0.76	0.759
0.045	0.759	0.774	0.754	0.77	0.777	0.756	0.743	0.775	0.786	0.767	0.775	0.76	0.759
0.091	0.759	0.795	0.778	0.77	0.785	0.771	0.743	0.792	0.794	0.767	0.78	0.76	0.759
0.136	0.759	0.808	0.791	0.77	0.789	0.784	0.743	0.805	0.808	0.767	0.793	0.76	0.759
0.182	0.763	0.808	0.795	0.77	0.793	0.784	0.743	0.8	0.804	0.767	0.801	0.761	0.759
0.227	0.761	0.81	0.809	0.77	0.797	0.783	0.743	0.803	0.804	0.767	0.801	0.766	0.759
0.273	0.776	0.804	0.805	0.77	0.797	0.791	0.743	0.802	0.799	0.767	0.799	0.791	0.759
0.318	0.793	0.795	0.821	0.77	0.792	0.803	0.743	0.786	0.828	0.767	0.793	0.796	0.759
0.364	0.812	0.825	0.828	0.769	0.799	0.825	0.782	0.824	0.824	0.788	0.799	0.817	0.776
0.409	0.822	0.835	0.839	0.784	0.816	0.839	0.808	0.838	0.836	0.796	0.809	0.825	0.797
0.455	0.824	0.832	0.838	0.802	0.812	0.814	0.82	0.83	0.839	0.804	**0.806**	0.835	0.818
0.5	0.835	**0.826**	0.83	0.803	**0.789**	0.844	0.833	**0.821**	0.831	0.806	0.782	0.85	0.833
0.545	0.834	**0.815**	0.851	0.818	**0.764**	0.822	0.826	**0.814**	0.847	0.816	0.757	**0.851**	0.816
0.591	0.847	**0.799**	**0.854**	0.817	**0.75**	0.825	0.843	**0.796**	**0.849**	**0.818**	**0.749**	**0.849**	0.852
0.636	**0.847**	**0.78**	**0.831**	**0.804**	**0.708**	0.823	**0.857**	**0.783**	**0.839**	**0.807**	**0.715**	**0.847**	**0.864**
0.682	**0.827**	**0.764**	**0.807**	**0.78**	**0.68**	0.806	0.835	**0.768**	**0.812**	**0.786**	**0.685**	**0.829**	**0.842**
0.727	**0.799**	**0.729**	**0.783**	**0.751**	**0.655**	**0.796**	**0.823**	**0.719**	**0.781**	**0.749**	**0.653**	**0.8**	**0.833**
0.773	**0.761**	**0.685**	**0.746**	**0.735**	**0.614**	**0.773**	**0.802**	**0.682**	**0.743**	**0.734**	**0.613**	**0.759**	**0.812**
0.818	**0.699**	**0.637**	**0.689**	**0.696**	**0.559**	**0.73**	**0.751**	**0.637**	**0.69**	**0.695**	**0.559**	**0.699**	**0.752**
0.864	**0.624**	**0.578**	**0.623**	**0.639**	**0.527**	**0.651**	**0.67**	**0.58**	**0.624**	**0.637**	**0.525**	**0.625**	**0.67**
0.909	**0.539**	**0.507**	**0.533**	**0.552**	**0.484**	**0.55**	**0.569**	**0.508**	**0.531**	**0.551**	**0.485**	**0.531**	**0.569**

(continued)

Table 5.28 (continued)

r_{gran}	Cov2	Cov3	Cov4	Cov5	Cov6	Cov7	Cov8	Cov11	Cov12	Cov13	Cov14	Cov15	Cov16
0.955	**0.456**	**0.45**	**0.452**	**0.464**	**0.439**	**0.453**	**0.465**	**0.449**	**0.45**	**0.464**	**0.438**	**0.444**	**0.466**
1	**0.402**	**0.4**	**0.398**	**0.398**	**0.396**	**0.396**	**0.398**	**0.4**	**0.397**	**0.399**	**0.396**	**0.399**	**0.398**
AccBias													
0	0.022	0.047	0.056	0.047	0.02	0.029	0.068	0.05	0.042	0.052	0.036	0.036	0.027
0.045	0.022	0.041	0.039	0.047	0.02	0.044	0.068	0.028	0.032	0.052	0.04	0.036	0.027
0.091	0.022	0.031	0.048	0.047	0.031	0.038	0.068	0.02	0.028	0.052	0.049	0.036	0.027
0.136	0.021	0.013	0.044	0.047	0.038	0.027	0.068	0.014	0.017	0.052	0.021	0.036	0.027
0.182	0.043	0.011	0.033	0.047	0.005	0.023	0.068	0.016	0.023	0.052	0.019	0.028	0.027
0.227	0.049	0.016	0.015	0.047	0.01	0.014	0.068	0.021	0.027	0.052	0.035	0.076	0.027
0.273	0.036	0.028	0.026	0.047	0.015	0.03	0.068	0.01	0.026	0.052	0.006	0.05	0.027
0.318	0.03	0.021	0.032	0.047	0.025	0.037	0.068	0.017	0.034	0.052	0.018	0.041	0.027
0.364	0.033	0.013	0.014	0.045	0.023	0.031	0.054	0.011	0.038	0.024	0.009	0.026	0.044
0.409	0.06	0.016	0.027	0.041	0.009	0.016	0.076	0.016	0.035	0.02	0.006	0.059	0.081
0.455	0.036	0.013	0.044	0.005	**0.012**	0.043	0.088	**0.014**	0.028	0.011	**0.031**	0.047	0.026
0.5	0.02	**0.01**	0.036	0.015	**0.021**	0.058	0.031	0.014	0.047	0.022	0.027	0.038	0.038
0.545	0.012	**0.009**	0.011	0.013	**0.012**	0.057	0.034	**0.005**	0.021	0.013	0.016	**0.02**	0.028
0.591	0.018	**0.02**	**0.005**	**0.007**	**0.019**	0.049	0.017	**0.013**	**0.016**	**0.01**	**0.012**	**0.016**	0.047

(continued)

Table 5.28 (continued)

r_{gran}	Cov2	Cov3	Cov4	Cov5	Cov6	Cov7	Cov8	Cov11	Cov12	Cov13	Cov14	Cov15	Cov16
0.636	0.032	0.009	0.016	0.019	0.011	0.033	0.022	0.025	0.017	0.009	0.008	0.013	0.019
0.682	0.042	0.013	0.024	0.021	0.006	0.023	0.025	0.025	0.024	0.011	0.006	0.015	0.05
0.727	0.017	0.01	0.018	0.017	0.012	0.047	0.026	0.011	0.017	0.006	0.023	0.024	0.028
0.773	0.004	0.01	0.04	0.002	0.018	0.017	0.013	0.006	0.015	0.016	0.018	0.027	0.047
0.818	0.002	0	0	0	0.005	0.05	0	0.012	0.006	0.002	0.01	0	0.01
0.864	0	0	0	0	0.015	0.006	0	0	0	0	0	0	0.001
0.909	0	0	0	0	0	0	0	0	0	0	0	0	0.001
0.955	0.001	0	0	0	0.004	0.003	0	0	0	0	0.016	0	0.004
1	0.009	0.017	0.009	0.011	0.009	0.002	0.01	0.006	0.005	0.008	0.005	0.005	0.006

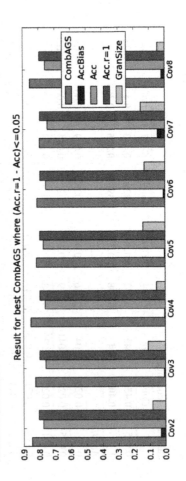

Result for best CombAGS where (Acc.r=1 - Acc)<=0.05

Table 5.29 5 *times* CV-5; Combination of accuracy and percentage reduction of objects in training data set (CombAGS) and Bias of accuracy form $5 \times$ CV5 (AccBias); Result of experiments for Covering finding methods of concept-dependent granulation with use of kNN classifier; data set: **SPECTF – all – modif**; r_{gran} = Granulation radius, Cov_i = Average accuracy for method Cov_i

r_{gran}	Cov2	Cov3	Cov4	Cov5	Cov6	Cov7	Cov8	Cov11	Cov12	Cov13	Cov14	Cov15	Cov16
CombAGS													
0	0.712	0.719	0.708	0.721	0.715	0.702	0.711	0.718	0.718	0.713	0.71	0.703	0.715
0.023	0.775	0.802	0.794	0.76	0.781	0.749	0.736	0.815	0.789	0.752	0.788	0.778	0.744
0.045	0.803	0.81	0.82	0.791	**0.791**	0.764	0.794	**0.823**	0.815	0.802	0.785	0.794	0.792
0.068	0.806	**0.783**	**0.807**	0.804	**0.74**	0.771	0.829	**0.785**	**0.81**	0.805	**0.741**	**0.804**	0.828
0.091	**0.773**	**0.729**	**0.77**	**0.779**	0.664	0.752	**0.812**	**0.733**	**0.77**	**0.782**	0.663	0.761	**0.809**
0.114	**0.698**	**0.647**	**0.686**	**0.723**	0.544	**0.674**	**0.759**	0.647	**0.689**	**0.719**	0.554	0.678	**0.757**
0.136	**0.597**	**0.569**	**0.582**	**0.616**	0.462	**0.573**	**0.659**	0.573	0.578	0.629	0.462	0.572	0.652
0.159	**0.505**	0.49	0.492	0.521	0.453	0.478	0.534	0.491	0.495	0.525	0.455	0.482	0.533
0.182	**0.439**	0.433	0.436	0.441	0.436	0.433	0.446	0.438	0.438	0.443	0.434	0.434	0.443
0.205	**0.406**	0.406	0.402	0.401	0.399	0.403	0.407	0.407	0.406	0.405	0.403	0.404	0.404
0.227	**0.398**	**0.397**	**0.393**	**0.393**	**0.392**	**0.395**	**0.398**	**0.399**	**0.397**	**0.395**	**0.395**	**0.396**	**0.397**
0.25	**0.398**	**0.396**	**0.392**	**0.393**	**0.392**	**0.395**	**0.398**	**0.399**	**0.397**	**0.394**	**0.394**	**0.396**	**0.396**
…	…	…	…	…	…	…	…	…	…	…	…	…	…
1	**0.398**	**0.396**	**0.392**	**0.393**	**0.392**	**0.395**	**0.398**	**0.399**	**0.397**	**0.394**	**0.394**	**0.396**	**0.396**
AccBias													
0	0.023	0.026	0.02	0.039	0.052	0.037	0.029	0.031	0.016	0.025	0.017	0.022	0.024
0.023	0.018	0.019	0.012	0.023	0.03	0.017	0.034	0.022	0.022	0.017	0.029	0.044	0.018
0.045	0.026	0.025	0.037	0.036	**0.011**	0.014	0.017	**0.023**	0.022	0.019	0.033	0.021	0.023

(continued)

Table 5.29 (continued)

r_{gran}	Cov2	Cov3	Cov4	Cov5	Cov6	Cov7	Cov8	Cov11	Cov12	Cov13	Cov14	Cov15	Cov16
0.068	0.035	0.016	0.017	0.027	0.012	0.032	0.041	0.007	0.017	0.007	0.03	0.014	0.026
0.091	0.011	0.006	0.011	0.013	0.026	0.025	0.027	0.007	0.008	0.018	0.013	0.009	0.014
0.114	0.011	0.011	0.007	0.006	0.008	0.008	0.01	0.02	0.011	0.005	0.014	0.011	0.006
0.136	0.015	0.01	0.002	0.009	0.016	0.015	0.006	0.009	0.008	0.003	0.019	0.009	0.011
0.159	0.013	0.011	0.009	0.014	0.015	0.01	0.019	0.011	0.011	0.008	0.009	0.011	0.01
0.182	0.004	0.011	0.011	0.01	0.006	0.008	0.009	0.009	0.007	0.011	0.009	0.019	0.015
0.205	0.013	0.012	0.017	0.016	0.012	0.011	0.016	0.011	0.006	0.022	0.011	0.003	0.012
0.227	0.009	0.007	0.014	0.013	0.012	0.011	0.019	0.013	0.007	0.019	0.014	0.005	0.011
0.25	0.007	0.006	0.014	0.016	0.011	0.009	0.018	0.012	0.009	0.018	0.014	0.003	0.01
...	...	...	...	...	...	...	...	...	...	...	...	...	...
1	0.007	0.006	0.014	0.016	0.011	0.009	0.018	0.012	0.009	0.018	0.014	0.003	0.01

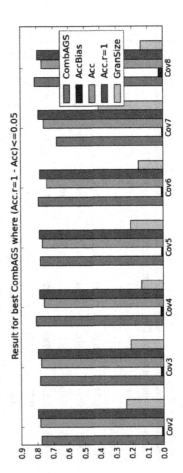

Result for best CombAGS where (Acc.r=1 − Acc)<=0.05

5.7 Best Result Based on CombAGS and the Error $(acc_{r=1} - acc) \leq 0.02$

The result of experiments are shown in Tables 5.30, 5.31, 5.32, 5.33, 5.34, 5.35, 5.36, 5.37, 5.38, 5.39, 5.40 and 5.41.

Considering results evaluated by parameter *CombAGS* chosen for *Acc* fulfilling inequality $(acc_{r=1} - acc) \leq 0.02$, where *AccBias* helps with tie resolution in case of the same value of *CombAGS*, the most impressive results for examined data sets in Table A.1 are as follows.

1. For **Australian Credit** data set, $Covr3$ wins over other methods; for radius $r_{gran} = 0.286$: $ComAGS = 0.9$ with $AccBias = 0.001$, error in accuracy $acc_{r=1} - acc = 0.007$, reduction of training objects in range of 94.7 %. See the result in Tables 5.4 and 5.30.
2. For **Car Evaluation** data set, $Covr3$ wins over other methods; for radius $r_{gran} = 0.833$: $ComAGS = 0.781$ with $AccBias = 0.005$, error in accuracy $acc_{r=1} - acc = 0.02$, reduction of training objects in range of 63.79 %. See the result in Tables 5.5 and 5.31.
3. For **Pima Indians Diabetes** data set, $Cov4$ wins over other methods; for radius $r_{gran} = 0.125$: $ComAGS = 0.774$ with $AccBias = 0.004$, error in accuracy $acc_{r=1} - acc = 0.001$, reduction of training objects in range of 90.89 %. See the result in Tables 5.6 and 5.32.
4. For **Fertility Diagnosis** data set, Cov4 wins over other method;, for radius $r_{gran} = 0.556$: $ComAGS = 0.814$ with $AccBias = 0.014$, error in accuracy $acc_{r=1} - acc = 0.004$, reduction of training objects in range of 75.75 %. See the result in Tables 5.7 and 5.33.
5. For **German Credit** data set, Cov8 wins over other methods; for radius $r_{gran} = 0.6$: $ComAGS = 0.789$ with $AccBias = 0.007$, error in accuracy $acc_{r=1} - acc = 0.011$, reduction of training objects in range of 85.93 %. See the result in Table 5.8 and Table 5.34.
6. For **Heart Disease** data set, Cov6 wins over other methods; for radius $r_{gran} = 0$: $ComAGS = 0.897$ with $AccBias = 0.021$, error in accuracy $acc_{r=1} - acc = 0.019$, reduction of training objects in range of 99.07 %. See the result in Table 9.3 and Table 5.35.
7. For **Hepatitis** data set, Cov8 wins over other methods; for radius $r_{gran} = 0.526$: $ComAGS = 0.901$ with $AccBias = 0.015$, error in accuracy $acc_{r=1} - acc = 0.009$, reduction of training objects in range of 92.13 %. See the result in Tables 5.10 and 5.36.
8. For **Congressional Voting Records** data set, Cov8 wins over other methods; for radius $r_{gran} = 0.688$: $ComAGS = 0.948$ with $AccBias = 0.006$, error in accuracy $acc_{r=1} - acc = 0.01$, reduction of training objects in range of 95.78 %. See the result in Tables 5.11 and 5.37.
9. For **Mushroom** data set, Cov8 wins over other methods; for radius $r_{gran} = 0.773$: $ComAGS = 0.996$ with $AccBias = 0.001$, error in accuracy $acc_{r=1} -$

$acc = 0.003$, reduction of training objects in range of 99.25 %. See the result in Tables 5.12 and 5.38.

10. For **Nursery** data set, Cov8 wins over other methods; for radius $r_{gran} = 0.75$: $ComAGS = 0.803$ with $AccBias = 0.03$, gain in accuracy $acc - acc_{r=1} = 0.062$, eduction of training objects in range of 97.43 %. See the result in Tables 5.13 and 5.39.

11. For **Soybean Large** data set, Cov4 wins over other methods; for radius $r_{gran} = 0.914$: $ComAGS = 0.727$ with $AccBias = 0.008$, error in accuracy $acc_{r=1} - acc = 0.019$, reduction of training objects in range of 55.3 %. See the result in Tables 5.14 and 5.40.

12. For **SPECT** data set, Cov8 wins over other methods; for radius $r_{gran} = 0.773$: $ComAGS = 0.802$ with $AccBias = 0.013$, error in accuracy $acc_{r=1} - acc = 0.011$, reduction of training objects in range of 81.87 %. See the result in Tables 5.15 and 5.41.

13. For **SPECTF** data set, Cov3 wins over other methods; for radius $r_{gran} = 0.068$: $ComAGS = 0.783$ with $AccBias = 0.016$, error in accuracy $acc_{r=1} - acc = 0.018$, reduction of training objects in range of 79.29 %. See the result in Tables 5.16 and 5.42.

Table 5.30 5 *times* CV-5; Combination of accuracy and percentage reduction of objects in training data set (CombAGS) and Bias of accuracy form $5 \times$ CV5 (AccBias); Result of experiments for Covering finding methods of concept-dependent granulation with use of kNN classifier; data set: **australian**; $r_{gran} =$ Granulation radius, Cov_i = Average accuracy for method Cov_i

r_{gran}	$Cov2$	$Cov3$	$Cov4$	$Cov5$	$Cov6$	$Cov7$	$Cov8$	$Cov11$	$Cov12$	$Cov13$	$Cov14$	$Cov15$	$Cov16$
CombAGS													
0	0.885	0.887	0.885	0.884	0.885	0.885	0.886	0.884	0.884	0.885	0.884	0.886	0.884
0.071	0.884	0.892	0.884	0.884	0.895	0.893	0.886	0.893	0.884	0.885	0.894	0.885	0.884
0.143	0.886	0.895	0.883	0.884	0.89	0.891	0.886	0.894	0.882	0.885	0.892	0.884	0.884
0.214	0.888	0.895	0.886	0.884	0.88	0.891	0.887	0.895	0.885	0.884	0.878	0.889	0.885
0.286	0.894	0.9	0.885	0.882	0.867	0.898	0.888	0.898	0.886	0.882	0.866	0.891	0.888
0.357	0.892	0.863	0.88	0.874	0.822	0.892	0.904	0.861	0.882	0.872	0.819	0.89	0.902
0.429	0.885	0.84	0.874	0.862	0.768	0.891	0.904	0.843	0.875	0.861	0.771	0.89	0.909
0.5	0.858	0.765	0.848	0.84	0.694	0.863	0.889	0.776	0.848	0.838	0.695	0.856	0.888
0.571	0.776	0.707	0.773	0.768	0.607	0.787	0.82	0.709	0.771	0.77	0.607	0.777	0.823
0.643	0.634	0.602	0.627	0.643	0.54	0.622	0.674	0.601	0.627	0.644	0.542	0.62	0.673
0.714	0.503	0.492	0.493	0.512	0.483	0.487	0.518	0.494	0.494	0.512	0.482	0.492	0.519
0.786	0.444	0.442	0.441	0.443	0.444	0.441	0.443	0.443	0.44	0.443	0.442	0.442	0.446
0.857	0.435	0.434	0.434	0.433	0.434	0.433	0.433	0.434	0.431	0.433	0.434	0.433	0.436
0.929	0.434	0.433	0.434	0.432	0.434	0.433	0.432	0.433	0.431	0.433	0.433	0.432	0.435
1	0.431	0.43	0.431	0.429	0.43	0.431	0.429	0.43	0.427	0.429	0.43	0.43	0.433
AccBias													
0	0.003	0.009	0.002	0.004	0.008	0.004	0.002	0.002	0.002	0.004	0.003	0.002	0.006
0.071	0.001	0.003	0.002	0.004	0.006	0.007	0.002	0.001	0.004	0.004	0.003	0.002	0.006
0.143	0.011	0.004	0.005	0.004	0.007	0.011	0.002	0.003	0.002	0.004	0.013	0.004	0.006
0.214	0.003	0.004	0.005	0.005	0.01	0.016	0.005	0.015	0.002	0.004	0.008	0.008	0.003

(continued)

Table 5.30 (continued)

r_{gran}	Cov2	Cov3	Cov4	Cov5	Cov6	Cov7	Cov8	Cov11	Cov12	Cov13	Cov14	Cov15	Cov16
0.286	0.008	**0.001**	0.008	0.012	0.008	0.005	0.019	**0.008**	0.01	0.014	0.012	0.008	0.019
0.357	0.011	0.016	0.006	0.014	0.008	0.025	0.014	0.003	0.006	0.005	0.011	0.017	0.011
0.429	0.013	0.012	0.009	0.003	0.008	0.017	0.009	0.007	0.009	0.013	0.007	**0.022**	0.017
0.5	**0.02**	0.016	0.01	0.006	**0.009**	0.016	0.017	0.018	**0.006**	0.008	**0.014**	0.008	0.005
0.571	0.008	0.019	**0.005**	0.011	0.006	0.006	0.011	0.012	**0.009**	**0.008**	0.01	0.006	0.006
0.643	**0.01**	**0.007**	**0.008**	**0.007**	0.015	0.006	**0.006**	**0.01**	0.005	0.009	0.01	**0.007**	**0.005**
0.714	**0.004**	**0.017**	**0.004**	**0.007**	0.003	**0.009**	**0.007**	**0.009**	0.005	0.011	0.011	**0.009**	**0.006**
0.786	**0.009**	**0.009**	**0.001**	0.003	**0.009**	0.006	**0.011**	**0.007**	0.001	0.01	0.002	0.01	**0.002**
0.857	**0.008**	**0.015**	**0.005**	**0.007**	**0.009**	**0.004**	**0.011**	**0.005**	0.002	0.01	**0.004**	**0.011**	**0.003**
0.929	**0.008**	**0.013**	**0.003**	0.006	**0.008**	**0.004**	**0.012**	**0.006**	**0.002**	**0.013**	**0.004**	**0.012**	**0.003**
1	**0.007**	**0.013**	**0.003**	**0.008**	0.009	**0.004**	0.012	0.005	**0.002**	**0.008**	0.003	0.014	**0.003**

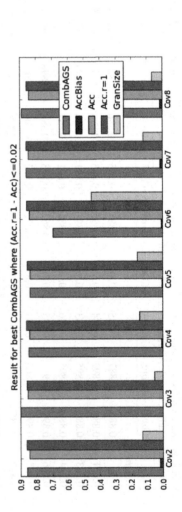

Result for best CombAGS where (Acc.r=1 - Acc)<=0.02

Table 5.31 5 *times* CV-5; Combination of accuracy and percentage reduction of objects in training data set (CombAGS) and Bias of accuracy form 5 × CV5 (AccBias); Result of experiments for Covering finding methods of concept-dependent granulation with use of kNN classifier; data set: **car evaluation**; r_{gran} = Granulation radius, Cov_i = Average accuracy for method Cov_i

r_{gran}	$Cov2$	$Cov3$	$Cov4$	$Cov5$	$Cov6$	$Cov7$	$Cov8$	$Cov11$	$Cov12$	$Cov13$	$Cov14$	$Cov15$	$Cov16$
CombAGS													
0	0.662	0.658	0.66	0.663	0.663	0.667	0.659	0.665	0.658	0.664	0.662	0.661	0.661
0.167	0.695	0.692	0.714	0.663	0.7	0.686	0.665	0.695	0.71	0.664	0.709	0.697	0.666
0.333	0.763	0.772	0.757	0.716	0.766	0.715	0.722	0.771	0.761	0.712	0.773	0.756	0.721
0.5	0.827	0.771	0.783	0.766	0.769	0.773	0.803	0.769	0.784	0.765	0.766	0.824	0.801
0.667	0.863	0.794	0.806	0.788	0.739	0.859	0.883	0.791	0.799	0.789	0.738	0.868	0.887
0.833	0.798	**0.781**	0.746	0.764	0.612	0.808	0.858	0.783	0.747	0.767	0.615	0.801	0.859
1	**0.472**	**0.472**	**0.472**	**0.472**	**0.472**	**0.471**	**0.472**	**0.473**	**0.472**	**0.474**	**0.472**	**0.473**	**0.473**
AccBias													
0	0.004	0.024	0.017	0.008	0.009	0.01	0.01	0.013	0.016	0.009	0.012	0.012	0.005
0.167	0.012	0.012	0.032	0.018	0.021	0.007	0.045	0.029	0.025	0.024	0.012	0.024	0.021
0.333	0.008	0.021	0.01	0.014	0.023	0.025	0.016	0.054	0.007	0.014	0.009	0.005	0.017

(continued)

Table 5.31 (continued)

r_{gran}	Cov2	Cov3	Cov4	Cov5	Cov6	Cov7	Cov8	Cov11	Cov12	Cov13	Cov14	Cov15	Cov16
0.5	0.013	0.01	0.009	0.021	0.019	0.047	0.017	0.008	0.011	0.01	0.009	0.014	0.026
0.667	0.024	0.014	0.007	0.022	0.015	0.011	0.015	0.006	0.009	0.007	0.012	0.007	0.019
0.833	0.014	**0.005**	0.003	0.002	0.006	0.016	0.006	0.002	0.01	0.007	0.004	0.01	0.003
1	**0.005**	**0.003**	**0.003**	**0.003**	**0.003**	**0.003**	**0.002**	**0.003**	**0.006**	**0.006**	**0.005**	**0.004**	**0.009**

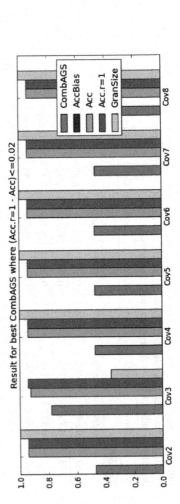

Result for best CombAGS where (Acc.r=1 - Acc)<=0.02

Table 5.32 5 *times* CV-5; Combination of accuracy and percentage reduction of objects in training data set (CombAGS) and Bias of accuracy form 5 × CV5 (AccBias); Result of experiments for Covering finding methods of concept-dependent granulation with use of kNN classifier; data set: **diabetes**; r_{gran} = Granulation radius, Cov_i = Average accuracy for method Cov_i

r_{gran}	$Cov2$	$Cov3$	$Cov4$	$Cov5$	$Cov6$	$Cov7$	$Cov8$	$Cov11$	$Cov12$	$Cov13$	$Cov14$	$Cov15$	$Cov16$
CombAGS													
0	0.798	0.797	0.801	0.796	0.798	0.799	0.8	0.8	0.797	0.798	0.801	0.797	0.799
0.125	0.775	0.749	**0.774**	0.757	0.707	0.787	0.778	0.746	**0.77**	0.76	0.703	0.777	0.789
0.25	0.683	**0.623**	**0.681**	0.666	**0.542**	0.702	0.728	**0.621**	0.682	0.663	**0.542**	0.698	0.733
0.375	**0.523**	**0.477**	0.513	0.521	**0.437**	0.522	0.551	**0.483**	0.519	0.521	0.436	**0.515**	**0.555**
0.5	**0.388**	**0.374**	0.372	0.39	0.366	0.368	0.399	0.377	0.376	0.39	0.367	0.375	**0.4**
0.625	0.332	**0.328**	0.324	0.329	0.327	0.327	0.327	0.328	0.328	0.33	0.329	0.331	0.332
0.75	**0.328**	**0.324**	**0.32**	0.325	0.324	0.324	0.323	0.324	0.325	0.326	0.326	0.327	0.328
0.875	**0.328**	**0.324**	**0.32**	0.325	0.324	0.324	0.323	0.324	0.325	0.326	0.326	0.327	0.328
1	**0.328**	**0.324**	**0.32**	0.325	0.324	0.324	0.323	0.324	0.325	0.326	0.326	0.327	0.328
AccBias													
0	0.007	0.012	0.009	0.007	0.016	0.002	0.008	0.006	0.009	0.012	0.008	0.014	0.007
0.125	0.015	0.018	**0.004**	0.016	0.009	0.019	0.019	0.016	**0.011**	0.024	0.025	0.007	0.025
0.25	0.018	**0.006**	**0.013**	0.01	**0.01**	0.012	0.021	**0.018**	0.016	0.012	**0.012**	0.017	0.013

(continued)

Table 5.32 (continued)

r_{gran}	Cov2	Cov3	Cov4	Cov5	Cov6	Cov7	Cov8	Cov11	Cov12	Cov13	Cov14	Cov15	Cov16
0.375	0.015	0.02	0.019	0.012	0.007	0.015	0.011	0.012	0.015	0.013	0.01	0.015	0.021
0.5	0.011	0.017	0.015	0.004	0.012	0.011	0.009	0.004	0.008	0.005	0.017	0.003	0.01
0.625	0.006	0.004	0.007	0.008	0.004	0.008	0.009	0.008	0.016	0.014	0.004	0.008	0.003
0.75	0.008	0.005	0.006	0.008	0.006	0.01	0.011	0.008	0.014	0.014	0.006	0.007	0.003
0.875	0.008	0.005	0.006	0.008	0.006	0.01	0.011	0.008	0.014	0.014	0.006	0.007	0.003
1	0.008	0.005	0.006	0.008	0.006	0.01	0.011	0.008	0.014	0.014	0.006	0.007	0.003

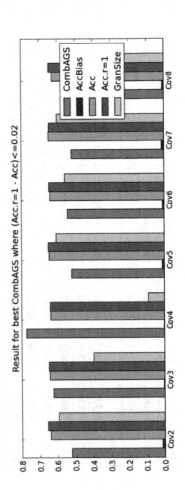

Result for best CombAGS where (Acc.r=1 - Acc)<=0.02

Table 5.33 5 *times* CV-5; Combination of accuracy and percentage reduction of objects in training data set (CombAGS) and Bias of accuracy form $5 \times$ CV5 (AccBias); Result of experiments for Covering finding methods of concept-dependent granulation with use of kNN classifier; data set: **fertility diagnosis**; r_{gran} = Granulation radius, Cov_i = Average accuracy for method Cov_i

r_{gran}	Cov2	Cov3	Cov4	Cov5	Cov6	Cov7	Cov8	Cov11	Cov12	Cov13	Cov14	Cov15	Cov16
CombAGS													
0	0.691	0.693	0.702	0.699	0.688	0.688	0.692	0.71	0.685	0.699	0.7	0.669	0.676
0.111	0.702	0.715	0.718	0.699	0.718	0.723	0.692	0.736	0.705	0.699	0.71	0.667	0.676
0.222	0.715	0.714	0.778	0.702	0.748	0.752	0.703	0.724	0.753	0.695	0.731	0.707	0.693
0.333	0.747	0.791	0.797	0.741	0.777	0.775	0.747	0.809	0.791	0.72	0.769	0.776	0.761
0.444	0.793	0.827	0.802	0.792	0.741	0.799	0.746	0.823	0.798	0.788	0.745	0.807	0.771
0.556	**0.815**	0.768	**0.814**	**0.807**	**0.669**	0.802	0.838	0.767	0.809	0.8	0.663	0.799	0.82
0.667	0.694	**0.656**	**0.696**	0.694	**0.564**	**0.708**	0.75	**0.655**	**0.69**	0.696	**0.557**	0.693	0.748
0.778	**0.573**	**0.554**	**0.56**	**0.582**	**0.532**	**0.547**	**0.597**	**0.558**	**0.556**	**0.575**	**0.533**	**0.552**	**0.595**
0.889	**0.472**	**0.474**	**0.474**	**0.479**	**0.475**	**0.467**	**0.48**	**0.473**	**0.474**	**0.472**	**0.469**	**0.471**	**0.477**
1	**0.436**	**0.437**	**0.435**	**0.439**	**0.439**	**0.434**	**0.44**	**0.437**	**0.437**	**0.431**	**0.438**	**0.435**	**0.434**

(continued)

Table 5.33 (continued)

r_{gran}	Cov2	Cov3	Cov4	Cov5	Cov6	Cov7	Cov8	Cov11	Cov12	Cov13	Cov14	Cov15	Cov16
AccBias													
0	0.044	0.02	0.032	0.028	0.04	0.07	0.022	0.046	0.046	0.028	0.026	0.068	0.074
0.111	0.046	0.016	0.02	0.028	0.06	0.066	0.022	0.054	0.036	0.028	0.026	0.066	0.074
0.222	0.082	0.054	0.05	0.058	0.062	0.018	0.026	0.056	0.056	0.032	0.042	0.056	0.056
0.333	0.072	0.056	0.08	0.062	0.036	0.062	0.046	0.05	0.064	0.076	0.028	0.058	0.06
0.444	0.024	0.016	0.078	0.038	0.032	0.04	0.072	0.026	0.074	0.066	0.05	0.092	0.012
0.556	**0.038**	0.026	**0.014**	**0.04**	**0.018**	0.066	0.03	0.016	0.026	0.018	0.028	0.024	0.054
0.667	0.014	**0.034**	0.016	0.012	**0.024**	**0.024**	0.01	**0.016**	**0.02**	0.016	**0.008**	0.04	0.032
0.778	**0.012**	**0.014**	**0.006**	**0.022**	**0.008**	**0.008**	**0.016**	**0.02**	**0.014**	**0.01**	**0.018**	**0.012**	**0.016**
0.889	**0.01**	**0.016**	**0.008**	**0.012**	**0.008**	**0.004**	**0.01**	**0.016**	**0.008**	**0.02**	**0.01**	**0.016**	**0.018**
1	**0.008**	**0.016**	**0.01**	**0.012**	**0.002**	**0.012**	**0.01**	**0.016**	**0.006**	**0.018**	**0.014**	**0.01**	**0.012**

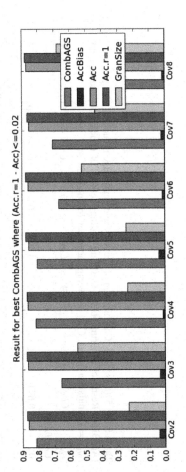

Result for best CombAGS where (Acc.r=1 - Acc)<=0.02

Table 5.34 5 *times* CV-5; Combination of accuracy and percentage reduction of objects in training data set (CombAGS) and Bias of accuracy form 5 × CV5 (AccBias); Result of experiments for Covering finding methods of concept-dependent granulation with use of kNN classifier; data set: **german credit**; $r_{gran} =$ Granulation radius, $Cov_i = $ Average accuracy for method Cov_i

r_{gran}	$Cov2$	$Cov3$	$Cov4$	$Cov5$	$Cov6$	$Cov7$	$Cov8$	$Cov11$	$Cov12$	$Cov13$	$Cov14$	$Cov15$	$Cov16$
CombAGS													
0	0.779	0.779	0.78	0.782	0.782	0.778	0.779	0.779	0.777	0.78	0.781	0.778	0.784
0.05	0.779	0.781	0.78	0.782	0.781	0.778	0.779	0.78	0.777	0.78	0.782	0.778	0.784
0.1	0.779	0.781	0.782	0.782	0.782	0.782	0.779	0.781	0.778	0.78	0.784	0.778	0.784
0.15	0.779	0.792	0.797	0.782	0.792	0.787	0.779	0.795	0.793	0.78	0.789	0.777	0.784
0.2	0.782	0.807	0.811	0.782	0.802	0.789	0.779	0.807	0.81	0.78	0.804	0.779	0.784
0.25	0.791	0.82	0.812	0.782	0.816	0.794	0.779	0.819	0.811	0.78	0.815	0.784	0.784
0.3	0.799	0.816	0.823	0.785	0.813	0.783	0.79	0.82	0.822	0.783	0.815	0.797	0.793
0.35	0.814	0.811	0.812	0.789	0.801	0.763	0.801	0.81	0.814	0.786	0.801	0.813	0.801
0.4	0.816	0.824	0.809	0.809	0.778	0.763	0.821	0.825	0.805	0.808	0.776	0.827	0.818
0.45	0.819	0.806	0.795	0.812	0.742	0.77	0.829	0.806	0.799	0.815	0.737	0.821	0.827
0.5	0.811	0.784	0.788	0.797	0.711	0.773	0.832	0.78	0.791	0.796	0.715	0.808	0.832
0.55	0.787	0.743	0.778	0.771	0.77	0.77	0.822	0.74	0.778	0.772	0.642	0.79	0.819
0.6	0.741	0.682	0.735	0.733	0.58	0.732	0.789	0.687	0.745	0.733	0.579	0.74	0.79
0.65	0.659	0.6	0.658	0.661	0.502	0.647	0.714	0.602	0.658	0.661	0.501	0.645	0.712
0.7	0.562	0.525	0.543	0.571	0.446	0.535	0.607	0.526	0.544	0.57	0.446	0.535	0.602
0.75	0.459	0.45	0.451	0.469	0.423	0.44	0.48	0.45	0.45	0.471	0.419	0.437	0.481
0.8	0.397	0.394	0.396	0.399	0.391	0.391	0.402	0.394	0.395	0.403	0.389	0.389	0.401
0.85	0.371	0.371	0.372	0.372	0.372	0.373	0.371	0.37	0.372	0.372	0.369	0.372	0.37
0.9	0.366	0.368	0.367	0.368	0.367	0.368	0.368	0.365	0.368	0.368	0.365	0.368	0.366
0.95	0.365	0.366	0.366	0.367	0.366	0.366	0.366	0.363	0.366	0.366	0.364	0.366	0.364
1	0.364	0.366	0.365	0.366	0.365	0.366	0.365	0.363	0.365	0.365	0.363	0.365	0.364

(continued)

Table 5.34 (continued)

r_{gran}	Cov2	Cov3	Cov4	Cov5	Cov6	Cov7	Cov8	Cov11	Cov12	Cov13	Cov14	Cov15	Cov16
AccBias													
0	0.018	0.008	0.015	0.01	0.012	0.009	0.005	0.011	0.018	0.015	0.011	0.002	0.007
0.05	0.018	0.015	0.015	0.01	0.012	0.009	0.005	0.021	0.017	0.015	0.011	0.002	0.007
0.1	0.018	0.018	0.014	0.01	0.017	0.018	0.005	0.021	0.023	0.015	0.025	0.002	0.007
0.15	0.018	0.01	0.009	0.01	0.01	0.018	0.005	0.01	0.023	0.015	0.01	0.002	0.007
0.2	0.02	0.019	0.011	0.01	0.015	0.007	0.005	0.007	0.018	0.015	0.011	0.007	0.007
0.25	0.023	0.011	0.014	0.01	0.015	0.014	0.005	0.01	0.013	0.015	0.013	0.015	0.007
0.3	0.007	0.012	0.003	0.011	0.014	0.02	0.014	0.016	0.022	0.017	0.009	0.009	0.014
0.35	0.019	0.011	0.025	0.017	0.008	0.019	0.012	0.01	0.018	0.017	0.008	0.009	0.019
0.4	0.023	0.016	0.015	0.011	0.014	0.033	0.009	0.014	0.031	0.015	0.016	0.023	0.006
0.45	0.023	0.018	0.009	0.009	0.009	0.039	0.015	0.011	0.006	0.008	0.027	0.009	0.041
0.5	0.016	0.009	0.008	0.014	0.012	0.028	0.028	0.018	0.014	0.004	0.005	0.008	0.005
0.55	0.012	**0.01**	0.013	0.011	**0.004**	0.041	0.014	0.016	0.003	0.004	0.013	0.008	0.013
0.6	**0.005**	**0.015**	0.007	**0.01**	**0.013**	**0.01**	**0.007**	**0.003**	**0.011**	**0.01**	**0.009**	**0.015**	**0.015**
0.65	**0.011**	**0.004**	**0.01**	0.016	**0.007**	**0.01**	**0.02**	**0.011**	**0.014**	**0.011**	**0.014**	**0.008**	**0.011**
0.7	**0.009**	**0.007**	**0.016**	**0.009**	**0.013**	**0.013**	**0.016**	**0.009**	**0.025**	**0.011**	**0.004**	**0.002**	**0.01**

(continued)

Table 5.34 (continued)

r_{gran}	Cov2	Cov3	Cov4	Cov5	Cov6	Cov7	Cov8	Cov11	Cov12	Cov13	Cov14	Cov15	Cov16
0.75	0.003	0.005	0.01	0.003	0.008	0.01	0.005	0.007	0.009	0.003	0.008	0.01	0.009
0.8	0.004	0.007	0.013	0.009	0.006	0.007	0.008	0.008	0.004	0.01	0.005	0.01	0.003
0.85	0.009	0.009	0.011	0.005	0.006	0.007	0.01	0.007	0.007	0.007	0.006	0.004	0.004
0.9	0.009	0.007	0.008	0.003	0.004	0.008	0.012	0.006	0.007	0.006	0.006	0.009	0.005
0.95	0.009	0.007	0.007	0.003	0.004	0.007	0.012	0.006	0.007	0.006	0.006	0.009	0.005
1	0.009	0.007	0.007	0.004	0.003	0.007	0.012	0.006	0.007	0.007	0.006	0.008	0.004

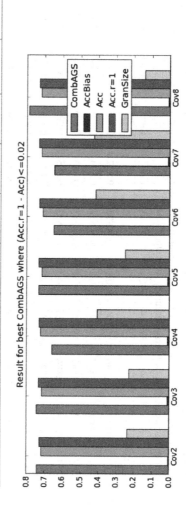

Result for best CombAGS where (Acc.r=1 - Acc)<=0.02

Table 5.35 5 *times* CV-5; Combination of accuracy and percentage reduction of objects in training data set (CombAGS) and Bias of accuracy form 5 × CV5 (AccBias); Result of experiments for Covering finding methods of concept-dependent granulation with use of kNN classifier; data set: **heart disease**; r_{gran} = Granulation radius, Cov_i = Average accuracy for method Cov_i

r_{gran}	Cov2	Cov3	Cov4	Cov5	Cov6	Cov7	Cov8	Cov11	Cov12	Cov13	Cov14	Cov15	Cov16
CombAGS													
0	0.897	0.896	0.895	0.895	0.897	0.897	0.895	0.893	0.893	0.894	0.897	0.896	0.891
0.077	0.897	0.887	0.893	0.895	0.892	0.894	0.895	0.884	0.891	0.894	0.891	0.896	0.891
0.154	0.892	0.884	0.892	0.895	0.882	0.885	0.895	0.892	0.889	0.894	0.884	0.893	0.891
0.231	0.888	0.886	0.887	0.895	0.873	0.885	0.895	0.89	0.887	0.894	0.87	0.884	0.891
0.308	0.881	0.871	0.88	0.883	0.837	0.88	0.891	0.873	0.88	0.88	0.842	0.885	0.887
0.385	0.871	0.841	0.869	0.861	0.782	0.873	0.886	0.838	0.866	0.86	0.779	0.865	0.883
0.462	0.831	0.779	0.828	0.819	0.709	0.832	0.861	0.776	0.824	0.821	0.701	0.827	0.866
0.538	0.746	0.693	0.753	0.746	0.598	0.752	0.802	0.687	0.759	0.747	0.597	0.745	0.801
0.615	0.616	0.589	0.611	0.623	0.524	0.601	0.65	0.583	0.61	0.622	0.517	0.594	0.648
0.692	0.495	0.487	0.484	0.493	0.468	0.473	0.503	0.484	0.486	0.499	0.468	0.476	0.506
0.769	0.425	0.424	0.425	0.427	0.425	0.424	0.429	0.421	0.424	0.427	0.426	0.424	0.425
0.846	0.413	0.413	0.412	0.412	0.412	0.412	0.414	0.409	0.412	0.414	0.415	0.413	0.413
0.923	0.413	0.413	0.412	0.412	0.412	0.412	0.414	0.409	0.412	0.414	0.415	0.413	0.413
1	0.413	0.413	0.412	0.412	0.412	0.412	0.414	0.409	0.412	0.414	0.415	0.413	0.413

(continued)

Table 5.35 (continued)

r_{gran}	Cov2	Cov3	Cov4	Cov5	Cov6	Cov7	Cov8	Cov11	Cov12	Cov13	Cov14	Cov15	Cov16
AccBias													
0	0.011	0.025	0.011	0.016	0.021	0.022	0.009	0.011	0.016	0.01	0.012	0.009	0.013
0.077	0.011	0.016	0.011	0.016	0.016	0.027	0.009	0.02	0.015	0.01	0.014	0.009	0.013
0.154	0.017	0.017	0.013	0.016	0.011	0.022	0.009	0.013	0.019	0.01	0.004	0.007	0.013
0.231	0.009	0.019	0.024	0.016	0.007	0.013	0.009	0.008	0.014	0.01	0.008	0.01	0.013
0.308	0.023	0.011	0.015	0.013	0.01	0.036	0.019	0.006	0.02	0.006	0.011	0.022	0.011
0.385	0.007	0.01	0.014	0.009	0.015	0.012	0.013	0.016	0.013	0.014	0.009	0.018	0.013
0.462	0.025	0.015	0.007	0.015	0.019	0.01	0.021	0.009	0.007	0.011	0.011	0.009	0.01
0.538	0.005	0.006	0.007	0.011	0.005	0.018	0.013	0.005	0.004	0.011	0.017	0.009	0.019
0.615	0.02	0.013	0.013	0.007	0.009	0.013	0.013	0.01	0.006	0.007	0.007	0.01	0.012

(continued)

Table 5.35 (continued)

r_{gran}	Cov2	Cov3	Cov4	Cov5	Cov6	Cov7	Cov8	Cov11	Cov12	Cov13	Cov14	Cov15	Cov16
0.692	0.01	0.008	0.004	0.003	0.01	0.007	0.013	0.007	0.007	0.004	0.004	0.012	0.007
0.769	0.004	0.006	0.008	0.009	0.008	0.006	0.005	0.009	0.005	0.001	0.005	0.005	0.006
0.846	0.004	0.004	0.007	0.01	0.007	0.005	0.009	0.008	0.009	0.006	0.004	0.004	0.004
0.923	0.004	0.004	0.007	0.01	0.007	0.005	0.009	0.008	0.009	0.006	0.004	0.004	0.004
1	0.004	0.004	0.007	0.01	0.007	0.005	0.009	0.008	0.009	0.006	0.004	0.004	0.004

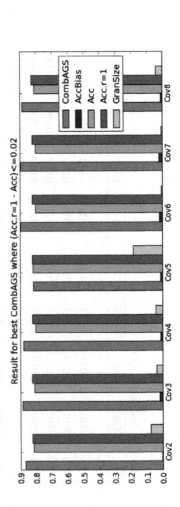

Result for best CombAGS where (Acc.r=1 - Acc) <=0.02

Table 5.36 5 *times* CV-5; Combination of accuracy and percentage reduction of objects in training data set (CombAGS) and Bias of accuracy form 5 × CV5 (AccBias); Result of experiments for Covering finding methods of concept-dependent granulation with use of kNN classifier; data set: **hepatitis-filled**; r_{gran} = Granulation radius, Cov_i = Average accuracy for method Cov_i

r_{gran}	Cov2	Cov3	Cov4	Cov5	Cov6	Cov7	Cov8	Cov11	Cov12	Cov13	Cov14	Cov15	Cov16
CombAGS													
0	0.893	0.901	0.899	0.893	0.897	0.896	0.896	0.899	0.895	0.899	0.897	0.895	0.895
0.053	0.893	0.901	0.899	0.893	0.897	0.896	0.896	0.899	0.895	0.899	0.897	0.895	0.895
0.105	0.893	0.901	0.899	0.893	0.897	0.896	0.896	0.899	0.895	0.899	0.897	0.895	0.895
0.158	0.893	0.898	0.895	0.893	0.893	0.893	0.896	0.896	0.892	0.899	0.893	0.895	0.895
0.211	0.893	0.896	0.896	0.893	0.89	0.892	0.896	0.893	0.891	0.899	0.89	0.894	0.895
0.263	0.892	0.901	0.895	0.893	0.886	0.902	0.896	0.889	0.896	0.899	0.886	0.891	0.895
0.316	0.892	0.899	0.906	0.893	0.888	0.9	0.896	0.897	0.905	0.899	0.892	0.89	0.895
0.368	0.894	**0.897**	0.898	0.891	**0.875**	0.891	0.893	**0.905**	0.895	0.895	**0.874**	0.886	0.895
0.421	0.897	**0.877**	0.891	0.881	**0.845**	0.885	0.904	**0.878**	0.893	0.888	**0.843**	0.885	0.899
0.474	0.891	0.842	0.879	0.871	**0.809**	0.881	0.91	**0.841**	0.886	0.875	**0.806**	0.884	0.913
0.526	**0.862**	**0.802**	0.864	0.852	0.74	0.859	**0.901**	**0.803**	0.871	0.855	**0.74**	0.854	**0.908**
0.579	**0.815**	**0.755**	**0.815**	0.807	**0.686**	0.814	0.852	**0.756**	**0.813**	0.816	**0.683**	0.804	0.853
0.632	**0.759**	**0.692**	**0.75**	0.756	**0.631**	0.745	**0.802**	**0.698**	**0.754**	0.762	**0.631**	**0.744**	**0.802**
0.684	**0.658**	0.617	**0.66**	0.671	**0.59**	0.649	**0.699**	0.621	**0.665**	0.677	**0.59**	**0.638**	**0.702**
0.737	**0.583**	**0.563**	**0.568**	**0.593**	**0.54**	0.565	**0.599**	0.557	**0.578**	**0.59**	**0.532**	**0.562**	**0.602**
0.789	**0.503**	**0.497**	**0.495**	**0.516**	**0.489**	0.493	**0.51**	0.493	**0.498**	**0.511**	**0.484**	**0.491**	**0.516**
0.842	**0.475**	**0.475**	**0.472**	**0.481**	**0.471**	0.471	**0.474**	0.471	**0.476**	**0.477**	**0.467**	**0.469**	**0.478**
0.895	**0.46**	**0.463**	**0.459**	**0.464**	**0.462**	0.463	**0.457**	0.459	**0.464**	**0.464**	**0.458**	**0.458**	**0.462**
0.947	**0.456**	**0.459**	**0.455**	**0.46**	**0.458**	0.459	**0.453**	0.455	**0.46**	**0.46**	**0.454**	**0.454**	**0.458**
1	**0.448**	**0.451**	**0.447**	**0.452**	**0.45**	0.451	**0.445**	0.447	**0.452**	**0.452**	**0.446**	**0.446**	**0.45**

(continued)

Table 5.36 (continued)

r_{gran}	Cov2	Cov3	Cov4	Cov5	Cov6	Cov7	Cov8	Cov11	Cov12	Cov13	Cov14	Cov15	Cov16
AccBias													
0	0.01	0.021	0.03	0.01	0.015	0.012	0.01	0.012	0.006	0.012	0.015	0.013	0.006
0.053	0.01	0.021	0.03	0.01	0.015	0.012	0.01	0.012	0.006	0.012	0.015	0.013	0.006
0.105	0.01	0.021	0.03	0.01	0.015	0.012	0.01	0.012	0.006	0.012	0.015	0.013	0.006
0.158	0.01	0.021	0.031	0.01	0.015	0.012	0.01	0.012	0.006	0.012	0.015	0.013	0.006
0.211	0.009	0.021	0.028	0.01	0.014	0.012	0.01	0.014	0.005	0.012	0.014	0.013	0.006
0.263	0.019	0.012	0.023	0.01	0.008	0.01	0.01	0.009	0.014	0.012	0.034	0.013	0.006
0.316	0.025	0.004	0.018	0.01	0.014	0.019	0.01	0.005	0.021	0.012	0.019	0.008	0.006
0.368	0.015	**0.017**	0.018	0.006	**0.021**	0.01	0.01	**0.022**	0.001	0.012	**0.009**	0.005	0.014
0.421	0.036	**0.021**	0.014	0.018	**0.01**	0.009	0.026	**0.01**	0.013	0.018	**0.022**	0.012	0.034
0.474	0.022	0.009	0.013	0.017	**0.023**	0.023	0.013	**0.009**	0.026	0.009	**0.012**	0.018	0.031
0.526	**0.034**	**0.017**	**0.015**	0.036	0.026	0.048	**0.015**	**0.034**	0.023	0.004	**0.019**	0.031	**0.023**
0.579	**0.015**	**0.018**	**0.015**	0.018	**0.01**	0.012	0.023	**0.01**	**0.026**	0.015	**0.015**	0.026	0.009
0.632	**0.01**	**0.01**	**0.018**	0.028	**0.019**	0.014	**0.006**	**0.017**	**0.01**	**0.017**	**0.01**	**0.01**	**0.027**
0.684	**0.026**	0.025	**0.019**	0.015	**0.006**	**0.01**	**0.014**	**0.019**	**0.021**	**0.018**	**0.008**	**0.01**	**0.009**

(continued)

Table 5.36 (continued)

r_{gran}	Cov2	Cov3	Cov4	Cov5	Cov6	Cov7	Cov8	Cov11	Cov12	Cov13	Cov14	Cov15	Cov16
0.737	0.022	0.019	0.026	0.012	0.01	0.01	0.01	0.015	0.014	0.036	0.006	0.009	0.026
0.789	0.018	0.023	0.009	0.018	0.01	0.012	0.019	0.018	0.015	0.013	0.006	0.008	0.013
0.842	0.018	0.023	0.009	0.013	0.012	0.012	0.012	0.017	0.014	0.012	0.006	0.01	0.014
0.895	0.014	0.022	0.01	0.013	0.01	0.015	0.013	0.015	0.013	0.013	0.005	0.012	0.01
0.947	0.014	0.022	0.01	0.013	0.01	0.015	0.013	0.015	0.013	0.013	0.005	0.012	0.01
1	0.014	0.022	0.01	0.013	0.01	0.015	0.013	0.015	0.013	0.013	0.005	0.012	0.01

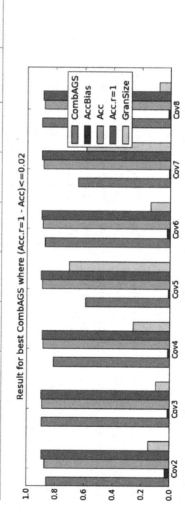

Result for best CombAGS where (Acc.r=1 - Acc)<=0.02

Table 5.37 5 *times* CV-5: Combination of accuracy and percentage reduction of objects in training data set (CombAGS) and Bias of accuracy form 5 × CV5 (AccBias); Result of experiments for Covering finding methods of concept-dependent granulation with use of kNN classifier; data set: congressional voting records; r_{gran} = Granulation radius, Cov_i = Average accuracy for method Cov_i

r_{gran}	Cov2	Cov3	Cov4	Cov5	Cov6	Cov7	Cov8	Cov11	Cov12	Cov13	Cov14	Cov15	Cov16
CombAGS													
0	0.946	0.946	0.945	0.945	0.946	0.946	0.944	0.945	0.946	0.946	0.945	0.944	0.945
0.063	0.946	0.944	0.944	0.945	0.944	0.944	0.944	0.944	0.944	0.946	0.943	0.944	0.945
0.125	0.946	0.943	0.942	0.945	0.942	0.943	0.944	0.942	0.941	0.946	0.941	0.944	0.945
0.188	0.945	0.938	0.943	0.945	0.936	0.943	0.944	0.939	0.941	0.946	0.935	0.942	0.945
0.25	0.944	0.937	0.943	0.945	0.936	0.944	0.944	0.937	0.943	0.946	0.936	0.94	0.945
0.313	0.942	0.927	0.942	0.945	0.926	0.936	0.944	0.928	0.942	0.946	0.926	0.936	0.945
0.375	0.94	**0.924**	0.941	0.945	**0.919**	0.935	0.944	**0.924**	0.942	0.946	**0.917**	0.932	0.945
0.438	0.937	**0.915**	0.941	0.943	**0.908**	0.934	0.943	**0.914**	0.94	0.944	**0.908**	0.931	0.944
0.5	0.937	**0.898**	0.937	0.939	**0.888**	0.935	0.943	**0.899**	0.939	0.939	**0.886**	0.932	0.944
0.563	0.935	**0.886**	0.927	0.922	**0.862**	0.933	0.947	**0.886**	0.928	0.92	**0.859**	0.932	0.948
0.625	**0.932**	0.874	0.914	0.9	0.837	0.934	0.945	0.87	0.915	0.898	0.835	0.933	0.941
0.688	**0.925**	0.867	**0.902**	**0.893**	0.803	**0.931**	**0.948**	0.864	**0.898**	**0.891**	0.802	**0.928**	**0.951**
0.75	**0.899**	0.824	**0.888**	**0.871**	0.758	**0.907**	**0.935**	0.821	**0.886**	**0.874**	0.757	**0.902**	**0.934**
0.813	**0.851**	0.773	**0.832**	**0.819**	0.707	**0.866**	**0.894**	0.772	**0.835**	**0.821**	0.705	**0.845**	**0.893**
0.875	**0.76**	0.709	**0.759**	**0.753**	0.655	**0.786**	**0.808**	0.708	**0.761**	**0.752**	0.657	**0.77**	**0.807**
0.938	**0.636**	0.594	**0.636**	**0.642**	0.558	**0.658**	**0.676**	0.593	**0.631**	**0.643**	0.555	**0.651**	**0.676**
1	**0.474**	0.473	**0.474**	**0.474**	0.474	**0.475**	**0.475**	0.473	**0.474**	**0.474**	0.472	**0.475**	**0.475**

(continued)

Table 5.37 (continued)

r_{gran}	Cov2	Cov3	Cov4	Cov5	Cov6	Cov7	Cov8	Cov11	Cov12	Cov13	Cov14	Cov15	Cov16
AccBias													
0	0	0	0.001	0.001	0	0	0.003	0.001	0	0	0.002	0.003	0.001
0.063	0	0	0	0.001	0	0	0.003	0	0	0	0.002	0.003	0.001
0.125	0	0.001	0.002	0.001	0		0.003	0.002	0.001	0	0.003	0.003	0.001
0.188	0	0.005	0.002	0.001	0.004	0	0.003	0.004	0.002	0	0.003	0.003	0.001
0.25	0	0.004	0.003	0.001	0.004	0.004	0.003	0.003	0.002	0	0.001	0.003	0.001
0.313	0	0.008	0.004	0.001	0.006	0.002	0.003	0.003	0.003	0	0.003	0.003	0.001
0.375	0.001	**0.005**	0.003	0.001	**0.006**	0.008	0.003	**0.008**	0.003	0	**0.007**	0.002	0.001
0.438	0.006	**0.003**	0.001	0.001	**0.004**	0.006	0.003	**0.002**	0.006	0	**0.004**	0.005	0.004
0.5	0.008	**0.007**	0.006	0	**0.005**	0.005	0.01	**0.007**	0.004	0	**0.006**	0	0.017
0.563	0.009	**0.012**	0.003	0.004	**0.006**	0.012	0.002	**0.005**	0.005	0.003	**0.01**	0.004	0.011
0.625	**0.006**	**0.008**	0.004	0.004	**0.006**	0.005	0.02	**0.001**	0.003	0.004	**0.008**	**0.003**	0.034
0.688	**0.012**	**0.003**	**0.008**	**0.011**	**0.011**	**0.013**	**0.006**	**0.006**	**0.006**	**0.006**	**0.004**	**0.005**	**0.005**

(continued)

Table 5.37 (continued)

r_{gran}	Cov2	Cov3	Cov4	Cov5	Cov6	Cov7	Cov8	Cov11	Cov12	Cov13	Cov14	Cov15	Cov16
0.75	0.014	0.003	0.002	0.003	0.008	0.006	0.013	0.006	0.003	0.006	0.003	0.002	0.012
0.813	0.004	0.008	0.002	0.004	0.007	0.006	0.008	0.009	0.009	0.005	0.011	0.006	0.003
0.875	0.003	0.006	0.007	0.005	0.007	0.009	0.006	0.005	0.006	0.002	0.005	0.011	0.005
0.938	0.003	0.006	0.007	0.006	0.01	0.006	0.008	0.007	0.007	0.006	0.011	0.006	0.005
1	0.009	0.008	0.007	0.004	0.005	0.004	0.008	0.001	0.004	0.004	0.005	0.006	0.003

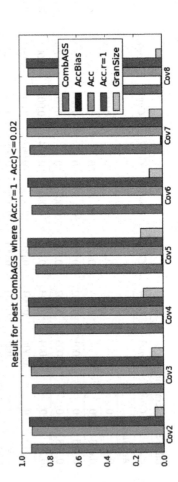

Result for best CombAGS where (Acc.r=1 - Acc)<=0.02

Table 5.38 5 *times* CV-5; Combination of accuracy and percentage reduction of objects in training data set (CombAGS) and Bias of accuracy form 5 × CV5 (AccBias); Result of experiments for Covering finding methods of concept-dependent granulation with use of kNN classifier; data set: mushroom-modified-filled; r_{gran} = Granulation radius, Cov_i = Average accuracy for method Cov_i

r_{gran}	Cov2	Cov3	Cov4	Cov5	Cov6	Cov7	Cov8	Cov11	Cov12	Cov13	Cov14	Cov15	Cov16
CombAGS													
0	0.944	0.943	0.944	0.943	0.944	0.943	0.943	0.943	0.944	0.944	0.944	0.944	0.943
0.045	0.944	0.943	0.944	0.943	0.944	0.943	0.943	0.943	0.944	0.944	0.944	0.944	0.943
0.091	0.944	0.943	0.944	0.943	0.944	0.943	0.943	0.943	0.944	0.944	0.944	0.944	0.943
0.136	0.944	0.943	0.944	0.943	0.944	0.943	0.943	0.943	0.944	0.944	0.944	0.944	0.943
0.182	0.944	0.94	0.944	0.943	0.94	0.943	0.943	0.94	0.944	0.944	0.94	0.944	0.943
0.227	0.943	0.938	0.944	0.943	0.939	0.943	0.943	0.938	0.944	0.944	0.939	0.944	0.943
0.273	0.944	0.94	0.944	0.943	0.938	0.942	0.943	0.941	0.945	0.944	0.938	0.944	0.943
0.318	0.944	0.939	0.946	0.943	0.937	0.94	0.943	0.94	0.946	0.944	0.936	0.942	0.943
0.364	0.943	0.949	0.945	0.943	0.943	0.942	0.943	0.949	0.946	0.944	0.942	0.943	0.943
0.409	0.942	0.945	0.943	0.943	0.94	0.944	0.94	0.945	0.943	0.944	0.939	0.943	0.94
0.455	0.946	0.959	0.946	0.94	0.928	0.945	0.943	0.958	0.946	0.94	0.928	0.942	0.941
0.5	0.954	0.959	0.945	0.938	0.942	0.955	0.949	0.96	0.946	0.94	0.942	0.951	0.947
0.545	0.969	0.958	0.947	0.942	0.943	0.97	0.966	0.957	0.948	0.942	0.943	0.963	0.969
0.591	0.981	**0.972**	0.958	0.958	**0.945**	0.977	0.971	**0.972**	0.957	0.957	**0.945**	0.977	0.967
0.636	**0.989**	0.966	0.974	0.978	**0.955**	0.987	0.988	**0.966**	0.974	0.979	**0.955**	0.987	0.986
0.682	**0.991**	0.97	**0.983**	**0.986**	0.96	0.992	**0.994**	0.97	0.983	0.986	0.96	**0.992**	**0.994**
0.727	**0.994**	0.97	**0.983**	**0.988**	0.963	0.993	**0.994**	0.971	0.983	0.988	0.963	**0.993**	**0.995**
0.773	**0.995**	0.977	**0.986**	0.987	0.964	0.994	**0.996**	0.977	0.986	0.988	0.965	**0.994**	**0.996**
0.818	0.993	0.977	0.982	0.982	0.953	0.994	**0.996**	0.977	0.982	0.982	0.953	0.993	**0.996**
0.864	0.985	0.966	0.967	0.969	0.92	0.988	0.992	0.965	0.968	0.969	0.92	0.987	0.992

(continued)

Table 5.38 (continued)

r_{gran}	Cov2	Cov3	Cov4	Cov5	Cov6	Cov7	Cov8	Cov11	Cov12	Cov13	Cov14	Cov15	Cov16
0.909	**0.961**	**0.934**	**0.938**	**0.938**	**0.846**	**0.968**	**0.98**	**0.934**	**0.939**	**0.939**	**0.846**	**0.964**	**0.979**
0.955	**0.865**	**0.836**	**0.859**	**0.866**	**0.694**	**0.884**	**0.923**	**0.837**	**0.859**	**0.867**	**0.694**	**0.881**	**0.923**
1	**0.5**	**0.5**	**0.5**	**0.5**	**0.5**	**0.5**	**0.5**	**0.5**	**0.5**	**0.5**	**0.5**	**0.5**	**0.5**
AccBias													
0	0.001	0.001	0.001	0.001	0.001	0.001	0.001	0	0	0.001	0.001	0.001	0.001
0.045	0.001	0.001	0.001	0.001	0.001	0.001	0.001	0	0	0.001	0.001	0.001	0.001
0.091	0.001	0.001	0.001	0.001	0.001	0.001	0.001	0	0	0.001	0.001	0.001	0.001
0.136	0.001	0.001	0.001	0.001	0.001	0.001	0.001	0	0	0.001	0.001	0.001	0.001
0.182	0.001	0.001	0.001	0.001	0	0.001	0.001	0.001	0	0.001	0.001	0.001	0.001
0.227	0.001	0	0.001	0.001	0	0.001	0.001	0.001	0	0.001	0.001	0.001	0.001
0.273	0.001	0.001	0.001	0.001	0.001	0.002	0.001	0.001	0.002	0.001	0.001	0.001	0.001
0.318	0.001	0	0.001	0.001	0.003	0.003	0.001	0.004	0.001	0.001	0.005	0.003	0.001
0.364	0.002	0.003	0.002	0.001	0.004	0.003	0.001	0.001	0.001	0.001	0.003	0.003	0.003
0.409	0.003	0.001	0.002	0.001	0.001	0.005	0	0.001	0.001	0.001	0	0.003	0.005
0.455	0.002	0.002	0.001	0.003	0.002	0.005	0.008	0.002	0.002	0.003	0.002	0.004	0.005
0.5	0.008	0.002	0.002	0.003	0.008	0.008	0.004	0.002	0.002	0.003	0.005	0.004	0.007
0.545	0.006	0.002	0.002	0.001	0.004	0.009	0.008	0.001	0.004	0.001	0.004	0.008	0.005

(continued)

Table 5.38 (continued)

r_{gran}	Cov2	Cov3	Cov4	Cov5	Cov6	Cov7	Cov8	Cov11	Cov12	Cov13	Cov14	Cov15	Cov16
0.591	0.007	**0.001**	0.003	0.003	**0.001**	0.005	0.004	**0.002**	0.002	0.002	**0.002**	0.009	0.005
0.636	**0.002**	**0.002**	0.001	0.002	**0.001**	0.004	0.003	**0.001**	0.001	0.001	0	0.005	0.003
0.682	**0.002**	**0.002**	**0.002**	**0.001**	**0.001**	**0.002**	**0.001**	**0.001**	**0.001**	**0.002**	0	**0.003**	**0.002**
0.727	**0.001**	0	**0.001**	0	**0.001**	**0.002**	**0.003**	**0.001**	**0.001**		0	**0.001**	**0.003**
0.773	**0.001**	0	**0.001**	0		**0.001**	**0.001**	0	**0.001**		**0.001**	**0.001**	**0.001**
0.818	0	0	**0.001**	0	0	0	**0.001**	**0.001**	0	0	**0.001**	0	0
0.864	0	0	0	0	0	0	0	0	0	0	0	0	0
0.909	0	0	0	0	0	0	0	0	0	0	0	0	0
0.955	0	0	0	0	0	0	0	0	0	0	0	0	0
1	0	0	0	0	0	0	0	0	0	0	0	0	0

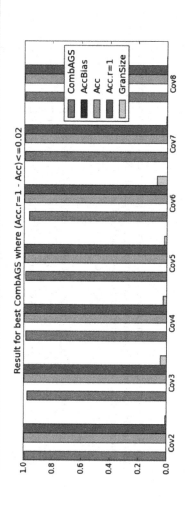

Result for best CombAGS where (Acc.r=1 - Acc) <=0.02

CombAGS   AccBias   Acc   Acc.r=1   GranSize

Table 5.39 5 *times* CV-5; Combination of accuracy and percentage reduction of objects in training data set (CombAGS) and Bias of accuracy form 5 × CV5 (AccBias); Result of experiments for Covering finding methods of concept-dependent granulation with use of kNN classifier; data set: **nursery**; r_{gran} = Granulation radius, Cov_i = Average accuracy for method Cov_i

r_{gran}	$Cov2$	$Cov3$	$Cov4$	$Cov5$	$Cov6$	$Cov7$	$Cov8$	$Cov11$	$Cov12$	$Cov13$	$Cov14$	$Cov15$	$Cov16$
CombAGS													
0	0.699	0.706	0.695	0.695	0.686	0.699	0.695	0.686	0.681	0.696	0.7	0.696	0.685
0.125	0.709	0.72	0.709	0.705	0.71	0.713	0.704	0.697	0.689	0.707	0.731	0.717	0.697
0.25	0.749	0.755	0.753	0.76	**0.765**	0.755	0.75	0.736	0.728	0.756	**0.787**	0.763	0.737
0.375	0.763	0.762	0.77	0.771	0.745	0.769	0.759	0.744	0.747	0.772	**0.765**	0.773	0.744
0.5	**0.772**	0.737	**0.783**	**0.788**	0.725	**0.78**	0.768	0.719	**0.761**	**0.785**	**0.748**	**0.775**	0.756
0.625	**0.785**	0.724	**0.797**	**0.791**	**0.686**	**0.797**	**0.795**	0.704	**0.773**	**0.793**	**0.709**	**0.774**	**0.784**
0.75	**0.785**	**0.733**	**0.798**	**0.773**	**0.604**	**0.797**	**0.803**	**0.71**	**0.772**	**0.775**	**0.63**	**0.787**	**0.791**
0.875	**0.72**	**0.669**	**0.71**	**0.677**	**0.415**	**0.734**	**0.757**	**0.644**	**0.683**	**0.677**	**0.434**	**0.727**	**0.744**
1	**0.285**	**0.292**	**0.285**	**0.285**	**0.278**	**0.285**	**0.285**	**0.278**	**0.274**	**0.285**	**0.29**	**0.285**	**0.279**
AccBias													
0	0.025	0.011	0.04	0.036	0.036	0.034	0.034	0.045	0.056	0.029	0.01	0.021	0.034
0.125	0.031	0.008	0.027	0.031	0.044	0.029	0.022	0.044	0.054	0.022	0.013	0.023	0.036
0.25	0.035	0.015	0.035	0.023	**0.058**	0.031	0.028	0.047	0.061	0.031	**0.014**	0.041	0.053

(continued)

Table 5.39 (continued)

r_{gran}	Cov2	Cov3	Cov4	Cov5	Cov6	Cov7	Cov8	Cov11	Cov12	Cov13	Cov14	Cov15	Cov16
0.375	0.031	0.008	0.025	0.026	0.054	0.031	0.022	0.049	0.061	0.027	0.011	0.025	0.046
0.5	0.031	0.007	0.027	0.028	0.052	0.027	0.025	0.044	0.08	0.027	0.004	0.028	0.048
0.625	0.035	0.005	0.032	0.033	0.049	0.029	0.031	0.045	0.081	0.03	0.004	0.033	0.052
0.75	0.029	0.007	0.03	0.031	0.049	0.031	0.03	0.049	0.084	0.031	0.007	0.033	0.058
0.875	0.03	0.003	0.033	0.028	0.043	0.034	0.029	0.052	0.082	0.03	0.001	0.03	0.058
1	0.014	0.004	0.014	0.017	0.027	0.013	0.015	0.027	0.037	0.017	0.002	0.015	0.026

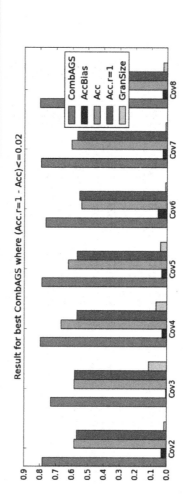

Result for best CombAGS where (Acc.r=1 - Acc)<=0.02

Table 5.40 5 *times* CV-5; Combination of accuracy and percentage reduction of objects in training data set (CombAGS) and Bias of accuracy form 5 × CV5 (AccBias); Result of experiments for Covering finding methods of concept-dependent granulation with use of kNN classifier; data set: **soybean-large-all-filled**; r_{gran} = Granulation radius, Cov_i = Average accuracy for method Cov_i

CombAGS

r_{gran}	$Cov2$	$Cov3$	$Cov4$	$Cov5$	$Cov6$	$Cov7$	$Cov8$	$Cov11$	$Cov12$	$Cov13$	$Cov14$	$Cov15$	$Cov16$
0	0.822	0.822	0.823	0.824	0.823	0.823	0.823	0.823	0.817	0.822	0.826	0.824	0.82
...	...	...	...	...	...	...	...	...	...	...	...	...	...
0.4	0.822	0.822	0.823	0.824	0.823	0.823	0.823	0.823	0.817	0.822	0.826	0.824	0.82
0.429	0.822	0.821	0.822	0.824	0.822	0.822	0.823	0.822	0.816	0.822	0.826	0.824	0.82
0.457	0.822	0.822	0.823	0.824	0.821	0.822	0.823	0.822	0.816	0.822	0.825	0.824	0.82
0.486	0.822	0.826	0.822	0.824	0.825	0.822	0.823	0.825	0.816	0.822	0.828	0.825	0.82
0.514	0.822	0.826	0.822	0.824	0.824	0.822	0.823	0.826	0.814	0.822	0.828	0.825	0.82
0.543	0.823	0.828	0.82	0.824	0.824	0.821	0.823	0.83	0.814	0.822	0.83	0.826	0.82
0.571	0.823	0.834	0.824	0.825	0.826	0.826	0.823	0.831	0.818	0.823	0.83	0.826	0.82
0.6	0.826	0.838	0.824	0.825	0.826	0.829	0.824	0.835	0.826	0.823	0.833	0.825	0.821
0.629	0.828	0.85	0.831	0.828	0.838	0.842	0.831	0.85	0.831	0.824	0.839	0.831	0.823
0.657	0.83	0.83	0.839	0.828	0.811	0.84	0.83	0.827	0.831	0.824	0.819	0.83	0.827
0.686	0.839	0.812	0.826	0.826	0.801	0.84	0.833	0.813	0.818	0.824	0.803	0.83	0.828
0.714	0.839	0.807	0.82	0.825	0.78	0.85	0.838	0.8	0.815	0.827	0.781	0.833	0.84
0.743	0.846	0.807	0.817	0.819	0.78	0.854	0.854	0.807	0.815	0.82	0.779	0.839	0.843
0.771	0.856	0.817	0.838	0.834	0.775	0.861	0.871	0.814	0.836	0.831	0.777	0.855	0.869
0.8	0.851	0.813	0.84	0.837	0.774	0.86	0.874	0.81	0.84	0.838	0.775	0.845	0.867
0.829	0.84	0.81	0.839	0.839	0.765	0.849	0.86	0.804	0.841	0.839	0.765	0.839	0.862
0.857	0.816	0.78	0.816	0.821	0.73	0.838	0.846	0.781	0.817	0.816	0.732	0.807	0.844

(continued)

Table 5.40 (continued)

r_{gran}	Cov2	Cov3	Cov4	Cov5	Cov6	Cov7	Cov8	Cov11	Cov12	Cov13	Cov14	Cov15	Cov16
0.886	0.78	0.744	0.781	0.781	0.689	0.803	0.806	0.742	0.778	0.782	0.688	0.776	0.811
0.914	0.726	0.691	**0.727**	0.728	0.638	0.74	0.756	0.691	0.725	**0.732**	0.637	0.715	0.758
0.943	0.646	0.625	0.645	0.659	0.584	0.654	0.677	0.627	0.643	0.661	0.582	0.639	0.677
0.971	**0.561**	**0.55**	**0.552**	**0.572**	**0.535**	**0.548**	**0.577**	**0.548**	**0.55**	**0.574**	**0.534**	**0.549**	**0.577**
1	**0.459**	**0.457**	**0.46**	**0.459**	**0.459**	**0.458**	**0.457**	**0.457**	**0.458**	**0.458**	**0.458**	**0.456**	**0.456**
AccBias													
0	0.021	0.01	0.014	0.008	0.023	0.007	0.012	0.01	0.02	0.005	0.021	0.022	0.007
...	...	...	...	...	...	...	...	...	...	...	...	...	...
0.4	0.021	0.01	0.014	0.008	0.023	0.007	0.012	0.01	0.02	0.005	0.021	0.022	0.007
0.429	0.021	0.011	0.014	0.008	0.023	0.008	0.012	0.01	0.02	0.005	0.021	0.022	0.007
0.457	0.021	0.012	0.013	0.008	0.024	0.008	0.012	0.01	0.02	0.005	0.02	0.022	0.007
0.486	0.021	0.014	0.014	0.008	0.019	0.008	0.012	0.007	0.02	0.005	0.018	0.023	0.007
0.514	0.022	0.014	0.013	0.008	0.019	0.009	0.012	0.009	0.02	0.005	0.019	0.024	0.007
0.543	0.019	0.023	0.01	0.008	0.016	0.008	0.012	0.015	0.019	0.005	0.015	0.02	0.007
0.571	0.016	0.015	0.016	0.007	0.013	0.006	0.013	0.018	0.02	0.005	0.012	0.022	0.007
0.6	0.011	0.017	0.008	0.007	0.03	0.004	0.012	0.016	0.015	0.005	0.016	0.023	0.008
0.629	0.006	0.011	0.007	0.009	0.017	0.009	0.021	0.011	0.006	0.011	0.007	0.012	0.014
0.657	0.023	0.008	0.009	0.009	0.017	0.017	0.017	0.009	0.007	0.008	0.012	0.009	0.014
0.686	0.015	0.03	0.025	0.007	0.022	0.01	0.012	0.023	0.006	0.009	0.028	0.018	0.023
0.714	0.022	0.006	0.016	0.006	0.008	0.014	0.023	0.012	0.009	0.004	0.011	0.028	0.022
0.743	0.015	0.01	0.02	0.02	0.024	0.011	0.02	0.015	0.014	0.023	0.006	0.022	0.012
0.771	0.01	0.013	0.022	0.011	0.009	0.008	0.012	0.004	0.01	0.01	0.018	0.014	0.011

(continued)

Table 5.40 (continued)

r_{gran}	Cov2	Cov3	Cov4	Cov5	Cov6	Cov7	Cov8	Cov11	Cov12	Cov13	Cov14	Cov15	Cov16
0.8	0.009	0.017	0.012	0.013	0.008	0.016	0.012	0.007	0.013	0.009	0.013	0.009	0.012
0.829	0.015	0.016	0.01	0.012	0.007	0.015	0.014	0.017	0.007	0.004	0.006	0.005	0.024
0.857	0.004	0.009	0.023	0.012	0.011	0.006	0.013	0.008	0.013	0.011	0.004	0.02	0.011
0.886	0.006	0.006	0.008	0.013	0.014	0.012	0.02	0.008	0.008	**0.008**	0.018	0.01	0.009
0.914	0.014	0.009	**0.008**	0.004	0.016	0.004	0.006	0.021	0.002	**0.008**	0.009	0.01	0.009
0.943	0.021	0.005	0.009	0.009	0.008	0.013	0.015	0.014	0.01	0.014	0.005	0.007	0.014
0.971	**0.012**	**0.002**	**0.009**	**0.006**	**0.008**	**0.004**	**0.015**	**0.005**	**0.008**	**0.006**	**0.005**	**0.005**	**0.007**
1	**0.01**	**0.008**	**0.006**	**0.003**	**0.008**	**0.003**	**0.01**	**0.005**	**0.007**	**0.003**	**0.006**	**0.003**	**0.006**

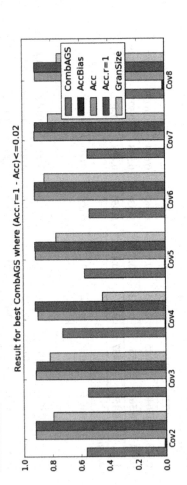

Result for best CombAGS where (Acc.r=1 - Acc)<=0.02

Table 5.41 5 *times* CV-5: Combination of accuracy and percentage reduction of objects in training data set (CombAGS) and Bias of accuracy form 5 × CV5 (AccBias); Result of experiments for Covering finding methods of concept-dependent granulation with use of kNN classifier; data set: **SPECT-all-modif**; r_{gran} = Granulation radius, Cov_i = Average accuracy for method Cov_i

r_{gran}	$Cov2$	$Cov3$	$Cov4$	$Cov5$	$Cov6$	$Cov7$	$Cov8$	$Cov11$	$Cov12$	$Cov13$	$Cov14$	$Cov15$	$Cov16$
CombAGS													
0	0.759	0.773	0.747	0.77	0.779	0.756	0.743	0.766	0.779	0.767	0.773	0.76	0.759
0.045	0.759	0.774	0.754	0.77	0.777	0.756	0.743	0.775	0.786	0.767	0.775	0.76	0.759
0.091	0.759	0.795	0.778	0.77	0.785	0.771	0.743	0.792	0.794	0.767	0.78	0.76	0.759
0.136	0.759	0.808	0.791	0.77	0.789	0.784	0.743	0.805	0.808	0.767	0.793	0.76	0.759
0.182	0.763	0.808	0.795	0.77	0.793	0.784	0.743	0.8	0.804	0.767	0.801	0.761	0.759
0.227	0.761	0.81	0.809	0.77	0.797	0.783	0.743	0.803	0.804	0.767	0.801	0.766	0.759
0.273	0.776	0.804	0.805	0.77	0.797	0.791	0.743	0.802	0.799	0.767	0.799	0.791	0.759
0.318	0.793	0.795	0.821	0.77	0.792	0.803	0.743	0.786	0.828	0.767	0.793	0.796	0.759
0.364	0.812	0.825	0.828	0.769	0.799	0.825	0.782	0.824	0.824	0.788	0.799	0.817	0.776
0.409	0.822	0.835	0.839	0.784	0.816	0.839	0.808	0.838	0.836	0.796	0.809	0.825	0.797
0.455	0.824	0.832	0.838	0.802	0.812	0.814	0.82	0.83	0.839	0.804	0.806	0.835	0.818
0.5	0.835	0.826	0.83	0.803	0.789	0.844	0.833	0.821	0.831	0.806	0.782	0.85	0.833
0.545	0.834	0.815	0.851	0.818	0.764	0.822	0.826	0.814	0.847	0.816	0.757	0.851	0.816
0.591	0.847	0.799	0.854	0.817	**0.75**	0.825	0.843	0.796	0.849	0.818	**0.749**	0.849	0.852
0.636	0.847	0.78	0.831	**0.804**	0.708	0.823	0.857	**0.783**	0.839	**0.807**	**0.715**	**0.847**	0.864
0.682	0.827	**0.764**	0.807	**0.78**	**0.68**	0.806	0.835	**0.768**	**0.812**	**0.786**	**0.685**	**0.829**	0.842
0.727	**0.799**	**0.729**	0.783	**0.751**	**0.655**	0.796	0.823	0.719	**0.781**	**0.749**	**0.653**	**0.8**	0.833
0.773	**0.761**	**0.685**	**0.746**	**0.735**	**0.614**	**0.773**	**0.802**	**0.682**	**0.743**	**0.734**	**0.613**	**0.759**	**0.812**

(continued)

Table 5.41 (continued)

r_{gran}	Cov2	Cov3	Cov4	Cov5	Cov6	Cov7	Cov8	Cov11	Cov12	Cov13	Cov14	Cov15	Cov16
0.818	**0.699**	**0.637**	**0.689**	**0.696**	**0.559**	**0.73**	**0.751**	**0.637**	**0.69**	**0.695**	**0.559**	**0.699**	**0.752**
0.864	**0.624**	**0.578**	**0.623**	**0.639**	**0.527**	**0.651**	**0.67**	**0.58**	**0.624**	**0.637**	**0.525**	**0.625**	**0.67**
0.909	**0.539**	**0.507**	**0.533**	**0.552**	**0.484**	**0.55**	**0.569**	**0.508**	**0.531**	**0.551**	**0.485**	**0.531**	**0.569**
0.955	**0.456**	**0.45**	**0.452**	**0.464**	**0.439**	**0.453**	**0.465**	**0.449**	**0.45**	**0.464**	**0.438**	**0.444**	**0.466**
1	**0.402**	**0.4**	**0.398**	**0.398**	**0.396**	**0.396**	**0.398**	**0.4**	**0.397**	**0.399**	**0.396**	**0.399**	**0.398**
AccBias													
0	0.022	0.047	0.056	0.047	0.02	0.029	0.068	0.05	0.042	0.052	0.036	0.036	0.027
0.045	0.022	0.041	0.039	0.047	0.02	0.044	0.068	0.028	0.032	0.052	0.04	0.036	0.027
0.091	0.022	0.031	0.048	0.047	0.031	0.038	0.068	0.02	0.028	0.052	0.049	0.036	0.027
0.136	0.021	0.013	0.044	0.047	0.038	0.027	0.068	0.014	0.017	0.052	0.021	0.036	0.027
0.182	0.043	0.011	0.033	0.047	0.005	0.023	0.068	0.016	0.023	0.052	0.019	0.028	0.027
0.227	0.049	0.016	0.015	0.047	0.01	0.014	0.068	0.021	0.027	0.052	0.035	0.076	0.027
0.273	0.036	0.028	0.026	0.047	0.015	0.03	0.068	0.01	0.026	0.052	0.006	0.05	0.027
0.318	0.03	0.021	0.032	0.047	0.025	0.037	0.068	0.017	0.034	0.052	0.018	0.041	0.027
0.364	0.033	0.013	0.014	0.045	0.023	0.031	0.054	0.011	0.038	0.024	0.009	0.026	0.044
0.409	0.06	0.016	0.027	0.041	0.009	0.016	0.076	0.016	0.035	0.02	0.006	0.059	0.081
0.455	0.036	0.013	0.044	0.005	0.012	0.043	0.088	0.025	0.028	0.011	0.031	0.047	0.026
0.5	0.02	0.01	0.036	0.015	0.021	0.058	0.031	0.014	0.047	0.022	0.027	0.038	0.038
0.545	0.012	0.009	0.011	0.013	0.012	0.057	0.034	0.005	0.021	0.013	0.016	0.02	0.028
0.591	0.018	0.02	0.005	0.007	**0.019**	0.049	0.017	0.013	0.016	0.01	**0.012**	0.016	0.047
0.636	0.032	0.009	0.016	**0.019**	0.011	0.033	0.022	**0.025**	0.017	**0.009**	**0.008**	**0.013**	0.019
0.682	0.042	**0.013**	0.024	**0.021**	**0.006**	0.023	0.025	**0.025**	**0.024**	**0.011**	**0.006**	**0.015**	0.05

(continued)

Table 5.41 (continued)

r_{gran}	Cov2	Cov3	Cov4	Cov5	Cov6	Cov7	Cov8	Cov11	Cov12	Cov13	Cov14	Cov15	Cov16
0.727	0.017	0.01	0.018	0.017	0.012	0.047	0.026	0.011	0.017	0.006	0.023	0.024	0.028
0.773	0.004	0.01	0.04	0.002	0.018	0.017	0.013	0.006	0.015	0.016	0.018	0.027	0.047
0.818	0.002	0	0	0	0.005	0.05	0	0.012	0.006	0.002	0.01	0	0.01
0.864	0	0	0	0	0.015	0.006	0	0	0	0	0	0	0.001
0.909	0	0	0	0	0	0	0	0	0	0	0	0	0.001
0.955	0.001	0	0	0	0.004	0.003	0	0	0	0	0.016	0	0.004
1	0.009	0.017	0.009	0.011	0.009	0.002	0.01	0.006	0.005	0.008	0.005	0.005	0.006

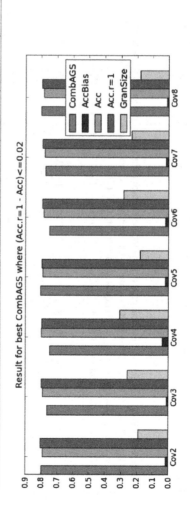

Result for best CombAGS where (Acc.r=1 - Acc) <=0.02

Table 5.42 5 *times* CV-5; Combination of accuracy and percentage reduction of objects in training data set (CombAGS) and Bias of accuracy form $5 \times$ CV5 (AccBias); Result of experiments for Covering finding methods of concept-dependent granulation with use of kNN classifier; data set: **SPECTF-all-modif**; r_{gran} = Granulation radius, Cov_i = Average accuracy for method Cov_i

r_{gran}	$Cov2$	$Cov3$	$Cov4$	$Cov5$	$Cov6$	$Cov7$	$Cov8$	$Cov11$	$Cov12$	$Cov13$	$Cov14$	$Cov15$	$Cov16$
CombAGS													
0	0.712	0.719	0.708	0.721	0.715	0.702	0.711	0.718	0.718	0.713	0.71	0.703	0.715
0.023	0.775	0.802	0.794	0.76	0.781	0.749	0.736	0.815	0.789	0.752	0.788	0.778	0.744
0.045	0.803	0.81	0.82	0.791	0.791	0.764	0.794	0.823	0.815	0.802	0.785	0.794	0.792
0.068	0.806	0.783	0.807	0.804	0.74	0.771	0.829	0.785	0.81	0.805	0.741	0.804	0.828
0.091	0.773	0.729	0.77	0.779	0.664	0.752	0.812	0.733	0.77	0.782	0.663	0.761	0.809
0.114	0.698	0.647	0.686	0.723	0.544	0.674	0.759	0.647	0.689	0.719	0.554	0.678	0.757
0.136	0.597	0.569	0.582	0.616	0.462	0.573	0.659	0.573	0.578	0.629	0.462	0.572	0.652
0.159	0.505	0.49	0.492	0.521	0.453	0.478	0.534	0.491	0.495	0.525	0.455	0.482	0.533
0.182	0.439	0.433	0.436	0.441	0.436	0.433	0.446	0.438	0.438	0.443	0.434	0.434	0.443

(continued)

Table 5.42 (continued)

r_{gran}	Cov2	Cov3	Cov4	Cov5	Cov6	Cov7	Cov8	Cov11	Cov12	Cov13	Cov14	Cov15	Cov16
0.205	**0.406**	**0.406**	**0.402**	**0.401**	**0.399**	**0.403**	**0.407**	**0.407**	**0.406**	**0.405**	**0.403**	**0.404**	**0.404**
0.227	**0.398**	**0.397**	**0.393**	**0.393**	**0.392**	**0.395**	**0.398**	**0.399**	**0.397**	**0.395**	**0.395**	**0.396**	**0.397**
0.25	**0.398**	**0.396**	**0.392**	**0.393**	**0.392**	**0.395**	**0.398**	**0.399**	**0.397**	**0.394**	**0.394**	**0.396**	**0.396**
...	...	...	...	...	...	...	...	...	...	...	...	...	...
1	**0.398**	**0.396**	**0.392**	**0.393**	**0.392**	**0.395**	**0.398**	**0.399**	**0.397**	**0.394**	**0.394**	**0.396**	**0.396**
AccBias													
0	0.023	0.026	0.02	0.039	0.052	0.037	0.029	0.031	0.016	0.025	0.017	0.022	0.024
0.023	0.018	0.019	0.012	0.023	0.03	0.017	0.034	0.022	0.022	0.017	0.029	0.044	0.018
0.045	0.026	0.025	0.037	0.036	0.011	0.014	0.017	0.023	0.022	0.019	0.033	0.021	0.023
0.068	0.035	**0.016**	0.017	0.027	0.012	0.032	0.041	**0.007**	0.017	0.007	0.03	0.014	0.026
0.091	**0.011**	**0.006**	**0.011**	0.013	**0.026**	0.025	0.027	**0.007**	**0.008**	**0.018**	**0.013**	**0.009**	0.014
0.114	**0.011**	0.011	**0.007**	**0.006**	0.008	0.008	**0.01**	0.02	**0.011**	**0.005**	**0.014**	**0.011**	**0.006**
0.136	**0.015**	0.01	0.002	**0.009**	**0.016**	**0.015**	**0.006**	0.009	0.008	**0.003**	0.019	0.009	**0.011**

(continued)

Table 5.42 (continued)

r_{gran}	Cov2	Cov3	Cov4	Cov5	Cov6	Cov7	Cov8	Cov11	Cov12	Cov13	Cov14	Cov15	Cov16
0.159	0.013	0.011	0.009	0.014	0.015	0.01	0.019	0.011	0.011	0.008	0.009	0.011	0.01
0.182	0.004	0.011	0.011	0.01	0.006	0.008	0.009	0.009	0.007	0.011	0.009	0.019	0.015
0.205	0.013	0.012	0.017	0.016	0.012	0.011	0.016	0.011	0.006	0.022	0.011	0.003	0.012
0.227	0.009	0.007	0.014	0.013	0.012	0.011	0.019	0.013	0.007	0.019	0.014	0.005	0.011
0.25	0.007	0.006	0.014	0.016	0.011	0.009	0.018	0.012	0.009	0.018	0.014	0.003	0.01
...	...	...	...	...	...	...	...	...	...	...	...	...	...
1	0.007	0.006	0.014	0.016	0.011	0.009	0.018	0.012	0.009	0.018	0.014	0.003	0.01

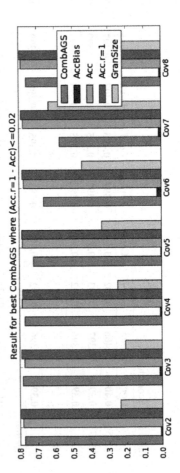

Result for best CombAGS where (Acc.r=1 - Acc)<=0.02

Chapter 6
Layered Granulation

Five times, Hermann, five times ...
[David Hilbert to Hermann Weyl on teaching. From Constance Reid: David Hilbert.]

In this chapter we study the method of layered (multiple) granulation, see Sect. 4.7.1, which can be regarded as a variant of the general paradigm of *layered learning*, see, Stone and Veloso [15] and Stone [14]. We show results of the layered classification performed on real data sets already studied in Chap. 5. Layered granulation leads to a sequence of granular reflections of decreasing sizes, which stabilizes after a finite number of steps, usually about five steps are sufficient. The final granular reflection produces a compact set of rules which may be identified with the core of the decision algorithm, i.e., its most effective rules, as this small set of rules yields the classifier of accuracy on par with that using the whole realm of rules.

Another development which may be stressed here is the heuristic rule for finding the optimal granulation radius giving the highest accuracy. It may be stated on the basis of results in the following sections that

the optimal granulation radius is located around the value which yields the maximal decrease in size of the granular reflection between the first and the second granulation layers.

6.1 Introduction

As a basic granulation algorithm we use standard granulation, see Polkowski [7–9, 12], Polkowski and Artiemjew [10, 11], iterated in accordance with the idea of layered (multiple) granulation coupled with the concept-dependent constraint of granulation, see Artiemjew [1–5], mentioned already in Chap. 4.

We consider a sequence $\{(U_{l_i}, A, d) : l = 0, 1, 2, \ldots\}$ of decision systems [6], where U_{l_i} is the universe of data samples in the layer l_i^{th}, A is the set of conditional attributes, and d is the decision attribute.

Clearly, (U_{l_0}, A, d) is the original decision system before granulation, and, $(U_{l_{i+1}}, A, d)$ for $i > 0$ is the granular reflection of the system (U_{l_i}, A, d).

© Springer International Publishing Switzerland 2015
L. Polkowski and P. Artiemjew, *Granular Computing in Decision Approximation*,
Intelligent Systems Reference Library 77, DOI 10.1007/978-3-319-12880-1_6

We apply the variant of concept-dependent granulation, i.e., the granule about an object u of the radius r is defined as follows:

$$g_{r,l_i}^{cd}(u) = \{v \in U_{l_{i-1}} : \frac{|IND(u,v)|}{|A|} \leq r \ and \ d(u) = d(v)\}$$

for $i = 1, 2, \ldots$.

6.1.1 An Example of Multiple Granulation

Exemplary multiple granulation of Quinlan's data set [13], see Table 6.1, for the granulation radius of 0.5 runs as follows.

For decision system from Table 6.1, granules in the first layer are ($r_{gran} = 0.5$):

$$g_{0.5,l_1}^{cd}(D_1) = \{D_1, D_2, D_8\},$$

$$g_{0.5,l_1}^{cd}(D_2) = \{D_1, D_2, D_8, D_{14}\},$$

$$g_{0.5,l_1}^{cd}(D_3) = \{D_3, D_4, D_{12}, D_{13}\},$$

$$g_{0.5,l_1}^{cd}(D_4) = \{D_3, D_4, D_5, D_{10}, D_{12}\},$$

$$g_{0.5,l_1}^{cd}(D_5) = \{D_4, D_5, D_7, D_9, D_{10}, D_{13}\},$$

$$g_{0.5,l_1}^{cd}(D_6) = \{D_6, D_{14}\},$$

$$g_{0.5,l_1}^{cd}(D_7) = \{D_5, D_7, D_9, D_{11}, D_{12}, D_{13}\},$$

$$g_{0.5,l_1}^{cd}(D_8) = \{D_1, D_2, D_8, D_{14}\},$$

Table 6.1 Exemplary decision system (U_{l_0}, A, d) by Quinlan [13]

Day	Outlook	Temperature	Humidity	Wind	Play golf
D_1	Sunny	Hot	High	Weak	No
D_2	Sunny	Hot	High	Strong	No
D_3	Overcast	Hot	High	Weak	Yes
D_4	Rainy	Mild	High	Weak	Yes
D_5	Rainy	Cool	Normal	Weak	Yes
D_6	Rainy	Cool	Normal	Strong	No
D_7	Overcast	Cool	Normal	Strong	Yes
D_8	Sunny	Mild	High	Weak	No
D_9	Sunny	Cool	Normal	Weak	Yes
D_{10}	Rainy	Mild	Normal	Weak	Yes
D_{11}	Sunny	Mild	Normal	Strong	Yes
D_{12}	Overcast	Mild	High	Strong	Yes
D_{13}	Overcast	Hot	Normal	Weak	Yes
D_{14}	Rainy	Mild	High	Strong	No

$$g_{0.5,l_1}^{cd}(D_9) = \{D_5, D_7, D_9, D_{10}, D_{11}, D_{13}\},$$
$$g_{0.5,l_1}^{cd}(D_{10}) = \{D_4, D_5, D_9, D_{10}, D_{11}, D_{13}\},$$
$$g_{0.5,l_1}^{cd}(D_{11}) = \{D_7, D_9, D_{10}, D_{11}, D_{12}\},$$
$$g_{0.5,l_1}^{cd}(D_{12}) = \{D_3, D_4, D_7, D_{11}, D_{12}\},$$
$$g_{0.5,l_1}^{cd}(D_{13}) = \{D_3, D_5, D_7, D_9, D_{10}, D_{13}\},$$
$$g_{0.5,l_1}^{cd}(D_{14}) = \{D_2, D_6, D_8, D_{14}\}.$$

Covering process of U_{l_0} with usage of order-preserving strategy Cov3, cf., Chap. 5, yields us the covering:

$U_{l_0,Cover} \leftarrow \emptyset,$

Step 1 $g_{0.5,l_1}^{cd}(D_1) \rightarrow U_{l_0,Cover}, U_{l_0,Cover} = \{D_1, D_2, D_8\},$

Step 2 $g_{0.5,l_1}^{cd}(D_2) \rightarrow U_{l_0,Cover}, U_{l_0,Cover} = \{D_1, D_2, D_8, D_{14}\},$

Step 3 $g_{0.5,l_1}^{cd}(D_3) \rightarrow U_{l_0,Cover}, U_{l_0,Cover} = \{D_1, D_2, D_3, D_4, D_8, D_{12}, D_{13}, D_{14}\},$

Step 4 $g_{0.5,l_1}^{cd}(D_4) \rightarrow U_{l_0,Cover},$

$\quad U_{l_0,Cover} = \{D_1, D_2, D_3, D_4, D_5, D_8, D_{10}, D_{12}, D_{13}, D_{14}\},$

Step 5 $g_{0.5,l_1}^{cd}(D_5) \rightarrow U_{l_0,Cover},$

$\quad U_{l_0,Cover} = \{D_1, D_2, D_3, D_4, D_5, D_7, D_8, D_9, D_{10}, D_{12}, D_{13}, D_{14}\},$

Step 6 $g_{0.5,l_1}^{cd}(D_6) \rightarrow U_{l_0,Cover},$

$\quad U_{l_0,Cover} = \{D_1, D_2, D_3, D_4, D_5, D_6, D_7, D_8, D_9, D_{10}, D_{12}, D_{13}, D_{14}\},$

Step 7 $g_{0.5,l_1}^{cd}(D_7) \rightarrow U_{l_0,Cover}, U_{l_0,Cover} = U_{l_0}.$

The granular reflection of (U_{l_0}, A, d) based on granules from $U_{l_0,Cover}$, with use of Majority Voting, where ties are resolved according to the ordering of granules is shown in Table 6.2.

Table 6.2 The decision system (U_{l_1}, A, d)

Day	Outlook	Temperature	Humidity	Wind	Play golf
$MV(g_{0.5,l_1}^{cd}(D_1))$	Sunny	Hot	High	Weak	No
$MV(g_{0.5,l_1}^{cd}(D_2))$	Sunny	Hot	High	Weak	No
$MV(g_{0.5,l_1}^{cd}(D_3))$	Overcast	Mild	High	Weak	Yes
$MV(g_{0.5,l_1}^{cd}(D_4))$	Rainy	Mild	High	Weak	Yes
$MV(g_{0.5,l_1}^{cd}(D_5))$	Rainy	Cool	Normal	Weak	Yes
$MV(g_{0.5,l_1}^{cd}(D_6))$	Rainy	Cool	Normal	Strong	No
$MV(g_{0.5,l_1}^{cd}(D_7))$	Overcast	Cool	Normal	Strong	Yes

Exemplary granular reflection formation based on Majority Voting looks as follows. In case, e.g., of the granule $g_{0.5,l_1}^{cd}(D_1)$ we have,

$$MV(g_{0.5,l_1}^{cd}(D_1)) = \left\{ \begin{array}{llll} \underline{Sunny} & \underline{Hot} & \underline{High} & \underline{Weak} \\ \underline{Sunny} & \underline{Hot} & \underline{High} & \underline{Strong} \\ \underline{Sunny} & \underline{Mild} & \underline{High} & \underline{Weak} \end{array} \right\} \rightarrow \underline{Sunny}\ \underline{Hot}\ \underline{High}\ \underline{Weak}$$

Treating all other granules in the same way, we obtain the granular reflection (U_{l_1}, A, d) shown in Table 6.2.

Granulation performed in the same manner with the granular reflection (U_{l_1}, A, d) from Table 6.2, yields the granule set in the second layer.

$$g_{0.5,l_2}^{cd}(MV(g_{0.5,l_1}^{cd}(D_1))) = \{MV(g_{0.5,l_1}^{cd}(D_1)), MV(g_{0.5,l_1}^{cd}(D_2))\}$$

$$g_{0.5,l_2}^{cd}(MV(g_{0.5,l_1}^{cd}(D_2))) = \{MV(g_{0.5,l_1}^{cd}(D_1)), MV(g_{0.5,l_1}^{cd}(D_2))\}$$

$$g_{0.5,l_2}^{cd}(MV(g_{0.5,l_1}^{cd}(D_3))) = \{MV(g_{0.5,l_1}^{cd}(D_3)), MV(g_{0.5,l_1}^{cd}(D_4))\}$$

$$g_{0.5,l_2}^{cd}(MV(g_{0.5,l_1}^{cd}(D_4))) = \{MV(g_{0.5,l_1}^{cd}(D_3)), MV(g_{0.5,l_1}^{cd}(D_4)), MV(g_{0.5,l_1}^{cd}(D_5))\}$$

$$g_{0.5,l_2}^{cd}(MV(g_{0.5,l_1}^{cd}(D_5))) = \{MV(g_{0.5,l_1}^{cd}(D_4)), MV(g_{0.5,l_1}^{cd}(D_5)), MV(g_{0.5,l_1}^{cd}(D_7))\}$$

$$g_{0.5,l_2}^{cd}(MV(g_{0.5,l_1}^{cd}(D_6))) = \{MV(g_{0.5,l_1}^{cd}(D_6))\}$$

$$g_{0.5,l_2}^{cd}(MV(g_{0.5,l_1}^{cd}(D_7))) = \{MV(g_{0.5,l_1}^{cd}(D_5)), MV(g_{0.5,l_1}^{cd}(D_7))\}$$

Covering process of $U_{l_1,Cover}$, runs in the following steps:

Step 1 $g_{0.5,l_2}^{cd}(MV(g_{0.5,l_1}^{cd}(D_1))) \rightarrow U_{l_1,Cover}$,

$\quad U_{l_1,Cover} = \{MV(g_{0.5,l_1}^{cd}(D_1)), MV(g_{0.5,l_1}^{cd}(D_2))\}$

Step 2 $g_{0.5,l_2}^{cd}(MV(g_{0.5,l_1}^{cd}(D_2))) \nrightarrow U_{l_1,Cover}$,

Step 3 $g_{0.5,l_2}^{cd}(MV(g_{0.5,l_1}^{cd}(D_3))) \rightarrow U_{l_1,Cover}$,

$\quad U_{l_1,Cover} = \{MV(g_{0.5,l_1}^{cd}(D_1)), MV(g_{0.5,l_1}^{cd}(D_2)), MV(g_{0.5,l_1}^{cd}(D_3)), MV(g_{0.5,l_1}^{cd}(D_4))\}$

Step 4 $g_{0.5,l_2}^{cd}(MV(g_{0.5,l_1}^{cd}(D_4))) \rightarrow U_{l_1,Cover}$,

$\quad U_{l_1,Cover} = \{MV(g_{0.5,l_1}^{cd}(D_1)), MV(g_{0.5,l_1}^{cd}(D_2)), MV(g_{0.5,l_1}^{cd}(D_3)), MV(g_{0.5,l_1}^{cd}(D_4)),$

$\quad\quad MV(g_{0.5,l_1}^{cd}(D_5))\}$

Step 5 $g_{0.5,l_2}^{cd}(MV(g_{0.5,l_1}^{cd}(D_5))) \rightarrow U_{l_1,Cover}$,

$\quad U_{l_1,Cover} = \{MV(g_{0.5,l_1}^{cd}(D_1)), MV(g_{0.5,l_1}^{cd}(D_2)), MV(g_{0.5,l_1}^{cd}(D_3)), MV(g_{0.5,l_1}^{cd}(D_4)),$

$\quad\quad MV(g_{0.5,l_1}^{cd}(D_5)), MV(g_{0.5,l_1}^{cd}(D_7))\}$

Step 6 $g_{0.5,l_2}^{cd}(MV(g_{0.5,l_1}^{cd}(D_6))) \rightarrow U_{l_1,Cover}$,

$\quad U_{l_1,Cover} = U_{l_1}$

Table 6.3 The decision system (U_{l_2}, A, d)

Day	Outlook	Temperature	Humidity	Wind	Play golf
$MV(g_{0.5,l_2}^{cd}(MV(g_{0.5,l_1}^{cd}(D_1))))$	Sunny	Hot	High	Weak	No
$MV(g_{0.5,l_2}^{cd}(MV(g_{0.5,l_1}^{cd}(D_3))))$	Overcast	Mild	High	Weak	Yes
$MV(g_{0.5,l_2}^{cd}(MV(g_{0.5,l_1}^{cd}(D_4))))$	Rainy	Mild	High	Weak	Yes
$MV(g_{0.5,l_2}^{cd}(MV(g_{0.5,l_1}^{cd}(D_5))))$	Rainy	Cool	Normal	Weak	Yes
$MV(g_{0.5,l_2}^{cd}(MV(g_{0.5,l_1}^{cd}(D_6))))$	Rainy	Cool	Normal	Strong	No

Applying Majority Voting to granules in U_{l_1}, we obtain the second granular reflection shown in Table 6.3.

The third layer of granulation based on system (U_{l_2}, A, d) from Table 6.3 is as follows,

$$g_{0.5,l_3}^{cd}(MV(g_{0.5,l_2}^{cd}(MV(g_{0.5,l_1}^{cd}(D_1))))) = \{MV(g_{0.5,l_2}^{cd}(MV(g_{0.5,l_1}^{cd}(D_1))))\}$$

$$g_{0.5,l_3}^{cd}(MV(g_{0.5,l_2}^{cd}(MV(g_{0.5,l_1}^{cd}(D_3))))) = \{MV(g_{0.5,l_2}^{cd}(MV(g_{0.5,l_1}^{cd}(D_3)))),$$
$$MV(g_{0.5,l_2}^{cd}(MV(g_{0.5,l_1}^{cd}(D_4))))\}$$

$$g_{0.5,l_3}^{cd}(MV(g_{0.5,l_2}^{cd}(MV(g_{0.5,l_1}^{cd}(D_4))))) = \{MV(g_{0.5,l_2}^{cd}(MV(g_{0.5,l_1}^{cd}(D_3)))),$$
$$MV(g_{0.5,l_2}^{cd}(MV(g_{0.5,l_1}^{cd}(D_4)))),$$
$$MV(g_{0.5,l_2}^{cd}(MV(g_{0.5,l_1}^{cd}(D_5))))\}$$

$$g_{0.5,l_3}^{cd}(MV(g_{0.5,l_2}^{cd}(MV(g_{0.5,l_1}^{cd}(D_5))))) = \{MV(g_{0.5,l_2}^{cd}(MV(g_{0.5,l_1}^{cd}(D_4)))),$$
$$MV(g_{0.5,l_2}^{cd}(MV(g_{0.5,l_1}^{cd}(D_5))))\}$$

$$g_{0.5,l_3}^{cd}(MV(g_{0.5,l_2}^{cd}(MV(g_{0.5,l_1}^{cd}(D_6))))) = \{MV(g_{0.5,l_2}^{cd}(MV(g_{0.5,l_1}^{cd}(D_6))))\}$$

Covering process for the third layer is as follows:

Step 1 $g_{0.5,l_3}^{cd}(MV(g_{0.5,l_2}^{cd}(MV(g_{0.5,l_1}^{cd}(D_1))))) \to U_{l_2,Cover}$,
$$U_{l_2,Cover} = \{MV(g_{0.5,l_2}^{cd}(MV(g_{0.5,l_1}^{cd}(D_1))))\}$$

Step 2 $g_{0.5,l_3}^{cd}(MV(g_{0.5,l_2}^{cd}(MV(g_{0.5,l_1}^{cd}(D_3))))) \to U_{l_2,Cover}$,
$$U_{l_2,Cover} = \{MV(g_{0.5,l_2}^{cd}(MV(g_{0.5,l_1}^{cd}(D_1)))), MV(g_{0.5,l_2}^{cd}(MV(g_{0.5,l_1}^{cd}(D_3)))),$$
$$MV(g_{0.5,l_2}^{cd}(MV(g_{0.5,l_1}^{cd}(D_4))))\}$$

Step 3 $g_{0.5,l_3}^{cd}(MV(g_{0.5,l_2}^{cd}(MV(g_{0.5,l_1}^{cd}(D_4))))) \to U_{l_2,Cover}$,
$$U_{l_2,Cover} = \{MV(g_{0.5,l_2}^{cd}(MV(g_{0.5,l_1}^{cd}(D_1)))), MV(g_{0.5,l_2}^{cd}(MV(g_{0.5,l_1}^{cd}(D_3)))),$$
$$MV(g_{0.5,l_2}^{cd}(MV(g_{0.5,l_1}^{cd}(D_4)))), MV(g_{0.5,l_2}^{cd}(MV(g_{0.5,l_1}^{cd}(D_5))))\}$$

Step 4 $g_{0.5,l_3}^{cd}(MV(g_{0.5,l_2}^{cd}(MV(g_{0.5,l_1}^{cd}(D_5))))) \not\to U_{l_2,Cover}$,

Step 5 $g_{0.5,l_3}^{cd}(MV(g_{0.5,l_2}^{cd}(MV(g_{0.5,l_1}^{cd}(D_6))))) \to U_{l_2,Cover}, U_{l_2,Cover} = U_{l_2}$

Table 6.4 The decision system (U_{l_3}, A, d)

Day	Outlook	Temperature	Humidity	Wind	Play golf
$MV(g_{0.5,l_3}^{cd}(MV(g_{0.5,l_2}^{cd}(MV(g_{0.5,l_1}^{cd}(D_1))))))$	Sunny	Hot	High	Weak	No
$MV(g_{0.5,l_3}^{cd}(MV(g_{0.5,l_2}^{cd}(MV(g_{0.5,l_1}^{cd}(D_3))))))$	Overcast	Mild	High	Weak	Yes
$MV(g_{0.5,l_3}^{cd}(MV(g_{0.5,l_2}^{cd}(MV(g_{0.5,l_1}^{cd}(D_4))))))$	Rainy	Mild	High	Weak	Yes
$MV(g_{0.5,l_3}^{cd}(MV(g_{0.5,l_2}^{cd}(MV(g_{0.5,l_1}^{cd}(D_6))))))$	Rainy	Cool	Normal	Strong	No

Using Majority voting we get the third layer of granular reflections shown in Table 6.4.

Granules of fourth layer computed from the third layer (U_{l_3}, A, d) (Table 6.4) are as follows:

$$g_{0.5,l_4}^{cd}(MV(g_{0.5,l_3}^{cd}(MV(g_{0.5,l_2}^{cd}(MV(g_{0.5,l_1}^{cd}(D_1)))))))$$
$$= \{MV(g_{0.5,l_3}^{cd}(MV(g_{0.5,l_2}^{cd}(MV(g_{0.5,l_1}^{cd}(D_1))))))\}$$
$$g_{0.5,l_4}^{cd}(MV(g_{0.5,l_3}^{cd}(MV(g_{0.5,l_2}^{cd}(MV(g_{0.5,l_1}^{cd}(D_3)))))))$$
$$= \{MV(g_{0.5,l_3}^{cd}(MV(g_{0.5,l_2}^{cd}(MV(g_{0.5,l_1}^{cd}(D_3)))))),$$
$$MV(g_{0.5,l_3}^{cd}(MV(g_{0.5,l_2}^{cd}(MV(g_{0.5,l_1}^{cd}(D_4))))))\}$$
$$g_{0.5,l_4}^{cd}(MV(g_{0.5,l_3}^{cd}(MV(g_{0.5,l_2}^{cd}(MV(g_{0.5,l_1}^{cd}(D_4)))))))$$
$$= \{MV(g_{0.5,l_3}^{cd}(MV(g_{0.5,l_2}^{cd}(MV(g_{0.5,l_1}^{cd}(D_3)))))),$$
$$MV(g_{0.5,l_3}^{cd}(MV(g_{0.5,l_2}^{cd}(MV(g_{0.5,l_1}^{cd}(D_4))))))\}$$
$$g_{0.5,l_4}^{cd}(MV(g_{0.5,l_3}^{cd}(MV(g_{0.5,l_2}^{cd}(MV(g_{0.5,l_1}^{cd}(D_6)))))))$$
$$= \{MV(g_{0.5,l_3}^{cd}(MV(g_{0.5,l_2}^{cd}(MV(g_{0.5,l_1}^{cd}(D_6))))))\}$$

Covering of U_{l_3} for the fourth layer is:

Step 1 $g_{0.5,l_4}^{cd}(MV(g_{0.5,l_3}^{cd}(MV(g_{0.5,l_2}^{cd}(MV(g_{0.5,l_1}^{cd}(D_1))))))) \rightarrow U_{l_3,Cover}$,

$\quad U_{l_3,Cover} = \{MV(g_{0.5,l_3}^{cd}(MV(g_{0.5,l_2}^{cd}(MV(g_{0.5,l_1}^{cd}(D_1))))))\}$

Step 2 $g_{0.5,l_4}^{cd}(MV(g_{0.5,l_3}^{cd}(MV(g_{0.5,l_2}^{cd}(MV(g_{0.5,l_1}^{cd}(D_3))))))) \rightarrow U_{l_3,Cover}$,

$\quad U_{l_3,Cover} = \{MV(g_{0.5,l_3}^{cd}(MV(g_{0.5,l_2}^{cd}(MV(g_{0.5,l_1}^{cd}(D_1)))))),$

$\quad\quad MV(g_{0.5,l_3}^{cd}(MV(g_{0.5,l_2}^{cd}(MV(g_{0.5,l_1}^{cd}(D_3)))))),$

$\quad\quad MV(g_{0.5,l_3}^{cd}(MV(g_{0.5,l_2}^{cd}(MV(g_{0.5,l_1}^{cd}(D_4))))))\}$

Step 3 $g_{0.5,l_4}^{cd}(MV(g_{0.5,l_3}^{cd}(MV(g_{0.5,l_2}^{cd}(MV(g_{0.5,l_1}^{cd}(D_4))))))) \nrightarrow U_{l_3,Cover}$,

Step 4 $g_{0.5,l_4}^{cd}(MV(g_{0.5,l_3}^{cd}(MV(g_{0.5,l_2}^{cd}(MV(g_{0.5,l_1}^{cd}(D_6))))))) \rightarrow U_{l_3,Cover}$,

$\quad U_{l_3,Cover} = U_{l_3}$

Table 6.5 The decision system (U_{l_4}, A, d)

Day	Outlook	Temperature	Humidity	Wind	Play golf
$MV(g^{cd}_{0.5,l_4}(MV(g^{cd}_{0.5,l_3}$ $(MV(g^{cd}_{0.5,l_2}(MV(g^{cd}_{0.5,l_1}(D_1))))))))$	Sunny	Hot	High	Weak	No
$MV(g^{cd}_{0.5,l_4}(MV(g^{cd}_{0.5,l_3}$ $(MV(g^{cd}_{0.5,l_2}(MV(g^{cd}_{0.5,l_1}(D_3))))))))$	Overcast	Mild	High	Weak	Yes
$MV(g^{cd}_{0.5,l_4}(MV(g^{cd}_{0.5,l_3}$ $(MV(g^{cd}_{0.5,l_2}(MV(g^{cd}_{0.5,l_1}(D_6))))))))$	Rainy	Cool	Normal	Strong	No

Using Majority Voting strategy on granules in the covering $U_{l_3,Cover}$, we get the granular reflection in Table 6.5.

For the radius of granulation equal to 0.5, the granular reflection in Table 6.5 is irreducible, as granules in the next layer contain only their centers:

$$g^{cd}_{0.5,l_5}(MV(g^{cd}_{0.5,l_4}(MV(g^{cd}_{0.5,l_3}(MV(g^{cd}_{0.5,l_2}(MV(g^{cd}_{0.5,l_1}(D_1)))))))))$$
$$= \{MV(g^{cd}_{0.5,l_4}(MV(g^{cd}_{0.5,l_3}(MV(g^{cd}_{0.5,l_2}(MV(g^{cd}_{0.5,l_1}(D_1)))))))) \}$$
$$g^{cd}_{0.5,l_5}(MV(g^{cd}_{0.5,l_4}(MV(g^{cd}_{0.5,l_3}(MV(g^{cd}_{0.5,l_2}(MV(g^{cd}_{0.5,l_1}(D_3)))))))))$$
$$= \{MV(g^{cd}_{0.5,l_4}(MV(g^{cd}_{0.5,l_3}(MV(g^{cd}_{0.5,l_2}(MV(g^{cd}_{0.5,l_1}(D_3)))))))) \}$$
$$g^{cd}_{0.5,l_5}(MV(g^{cd}_{0.5,l_4}(MV(g^{cd}_{0.5,l_3}(MV(g^{cd}_{0.5,l_2}(MV(g^{cd}_{0.5,l_1}(D_6)))))))))$$
$$= \{MV(g^{cd}_{0.5,l_4}(MV(g^{cd}_{0.5,l_3}(MV(g^{cd}_{0.5,l_2}(MV(g^{cd}_{0.5,l_1}(D_6)))))))) \}$$

6.1.2 Experiments with Real Data

We perform CV5 cross-validation test, where training data sets from all folds undergo multiple granulation. As a reference classifier, we use simple kNN method, which is described in detail in Chap. 5. In short, we use k nearest objects from all training classes with respect to the Hamming metric, and the nearest class with the smallest summary distance value passes the decision to the test object.

The results for real data sets in Table 6.6 are reported in Sects. 6.2 and 6.3.

In Table 6.6, we collect descriptions of examined data sets, which come from UCI Repository [16]. The successive columns in Table 6.6 give the name of the data *name*, the type of attributes *attr type*, the number of attributes *attr no.*, the number of objects *obj no.*, the number of decision classes *class no.*, the optimal value of the parameter k in kNN procedure, denoted *opt k*, found in the way described in Chap. 5.

Table 6.6 Data sets description

name	attr type	attr no.	obj no.	class no.	opt k
Adult	categorical, integer	15	48,842	2	79
Australian-credit	categorical, integer, real	15	690	2	5
Car Evaluation	categorical	7	1,728	4	8
Diabetes	categorical, integer	9	768	2	3
Fertility_Diagnosis	real	10	100	2	5
German-credit	categorical, integer	21	1,000	2	18
Heartdisease	categorical, real	14	270	2	19
Hepatitis	categorical, integer, real	20	155	2	3
Congressional Voting Records	categorical	17	435	2	3
Mushroom	categorical	23	8, 124	2	1
Nursery	categorical	9	12,960	5	4
Soybean-large	categorical	36	307	19	3
SPECT Heart	categorical	23	267	2	19
SPECTF Heart	integer	45	267	2	14

6.2 Results of Experiments for Symbolic Data from UCI Repository

Using the procedure of experimentation described in Sect. 6.1.1, we have performed a series of experiments for data from Table 6.6. In Tables 6.7, 6.8, 6.9, 6.10, 6.11, 6.12, 6.13, 6.14, 6.15, 6.16, 6.17, 6.18, 6.19, 6.20, 6.21, 6.22, 6.23, 6.24, 6.25, 6.26, 6.27, 6.28, 6.29, 6.30, 6.31, 6.32 and 6.33 accuracy results averaged over five folds are recorded as acc, respectively; the average number of granules is recorded in the column $GranSize$. Values are given in rows indexed by granulation radii values in the form of $r = \frac{i}{|A|}$ where $i = 1, \ldots, |A|$ (Table 6.34).

Let us point to the fact that values for non-granulated data are collected for the radius of 1.000 meaning the canonical indiscernibility granulation, i.e., the canonical rough set procedure. In columns, we have results for five layers of granulation.

Respective 2D visualizations of results in Tables 6.7, 6.8, 6.9, 6.10, 6.11, 6.12, 6.13, 6.14, 6.15, 6.16, 6.17, 6.18, 6.19, 6.20, 6.21, 6.22, 6.23, 6.24, 6.25, 6.26, 6.27, 6.28, 6.29, 6.30, 6.31, 6.32 and 6.33 are reported below tables. Respective results in 3D are shown in Figs. 6.1, 6.2, 6.3, 6.4, 6.5, 6.6, 6.7, 6.8, 6.9, 6.10, 6.11, 6.12, 6.13, and 6.14.

The results of experiments bring a quite surprising result: in most of experiments, the neighborhood of the optimal radius of granulation can be estimated based on double granulation. The radius for which the approximation between the first and the second layer is the best, in the sense of the maximal decrease in the number of objects, points quite precisely to the neighborhood of the optimal radius, and in many cases shows it exactly.

Table 6.7 CV-5; result of experiments for multi-layer c–d granulation with use of kNN classifier; data set **adult.all.filled**; r_{gran} = Granulation radius, acc = Average accuracy for considered layer, *TRNsize* the mean size of granular decision system for considered layer

r_{gran}	Layer1		Layer2		Layer3		Layer4		Layer5	
	acc	*GranSize*	*acc*	*GranSize*	*acc*	*GranSize*	*acc*	*GranSize*	*acc*	*GranSize*
0	0.635	2	0.635	2	0.635	2	0.635	2	0.635	2
0.071	0.636	2.2	0.636	2	0.636	2	0.636	2	0.636	2
0.143	0.636	3.6	0.636	2	0.636	2	0.636	2	0.636	2
0.214	0.636	9.4	0.636	2	0.636	2	0.636	2	0.636	2
0.286	0.687	21.2	0.635	2	0.635	2	0.635	2	0.635	2
0.357	0.72	44.6	0.636	2	0.636	2	0.636	2	0.636	2
0.429	0.736	104	0.635	2	0.635	2	0.635	2	0.635	2
0.5	0.765	257.4	0.64	7	0.63	3	0.63	3	0.63	3
0.571	0.798	640	0.715	39.2	0.522	13.6	0.32	11.2	0.32	11.2
0.643	0.814	1601.2	0.803	237.8	0.674	126.2	0.447	111	0.417	108.2
0.714	0.813	3955.2	0.818	1188.8	0.767	857.6	0.67	804.8	0.614	792.4
0.786	0.832	9654	0.828	5250.6	0.822	4497.8	0.803	4357.4	0.789	4333.2
0.857	0.843	20872	0.843	17491.8	0.84	16810.4	0.832	16692.4	0.829	16676
0.929	0.843	34238.4	0.843	34165.6	0.843	34160.4	0.843	34160.4	0.843	34160.4
1	0.841	39040	0.841	39040	0.841	39040	0.841	39040	0.841	39040

Table 6.8 Visualisation of results for adult

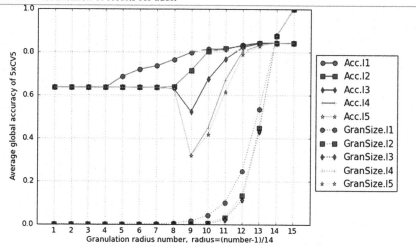

Table 6.9 CV-5; result of experiments for multi-layer c–d granulation with use of kNN classifier; data set **australian**; r_{gran} = Granulation radius, acc = Average accuracy for considered layer, *TRNsize* the mean size of granular decision system for considered layer

r_{gran}	Layer1		Layer2		Layer3		Layer4		Layer5	
	acc	GranSize	acc	GranSize	acc	GranSize	acc	GranSize	acc	GranSize
0	0.768	2	0.768	2	0.768	2	0.768	2	0.768	2
0.071	0.772	2	0.772	2	0.772	2	0.772	2	0.772	2
0.143	0.696	2.6	0.774	2	0.774	2	0.774	2	0.774	2
0.214	0.781	5.6	0.775	2	0.775	2	0.775	2	0.775	2
0.286	0.8	6.8	0.797	2	0.797	2	0.797	2	0.797	2
0.357	0.813	16.4	0.78	2	0.78	2	0.78	2	0.78	2
0.429	0.838	29.6	0.704	3.6	0.67	2.2	0.67	2.2	0.67	2.2
0.5	0.843	68.6	0.729	15.4	0.37	7.4	0.37	7.4	0.37	7.4
0.571	0.851	154.8	0.799	70.6	0.69	47.4	0.628	43.2	0.625	42.8
0.643	0.854	313.2	0.841	245.6	0.806	228.8	0.781	225.6	0.758	224.2
0.714	0.852	468.2	0.854	444.8	0.855	440	0.857	438.6	0.857	438.6
0.786	0.858	535.6	0.858	535.4	0.858	535.4	0.858	535.4	0.858	535.4
0.857	0.854	547.4	0.854	547.4	0.854	547.4	0.854	547.4	0.854	547.4
0.929	0.864	548.8	0.864	548.8	0.864	548.8	0.864	548.8	0.864	548.8
1	0.855	552	0.855	552	0.855	552	0.855	552	0.855	552

Table 6.10 Visualisation of results for australian credit

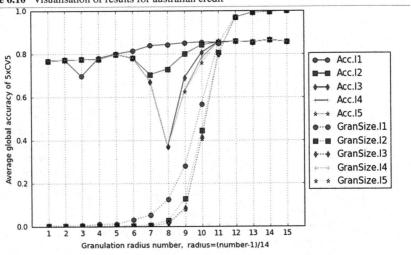

Table 6.11 CV-5; result of experiments for multi-layer c–d granulation with use of kNN classifier; data set **car evaluation**; r_{gran} = Granulation radius, acc = Average accuracy for considered layer, *TRNsize* the mean size of granular decision system for considered layer

r_{gran}	Layer1		Layer2		Layer3		Layer4		Layer5	
	acc	GranSize	acc	GranSize	acc	GranSize	acc	GranSize	acc	GranSize
0	0.315	4	0.315	4	0.315	4	0.315	4	0.315	4
0.167	0.395	8.6	0.296	4	0.296	4	0.296	4	0.296	4
0.333	0.484	16.4	0.351	6.2	0.326	4.6	0.326	4.6	0.326	4.6
0.5	0.668	44	0.477	16.2	0.374	9.4	0.296	7	0.289	6.8
0.667	0.811	102.8	0.723	47.4	0.632	29.8	0.601	25.4	0.59	24.8
0.833	0.865	370	0.841	199.8	0.832	147.2	0.833	137	0.833	134.8
1	0.944	1382.4	0.944	1382.4	0.944	1382.4	0.944	1382.4	0.944	1382.4

Table 6.12 Visualisation of results for car

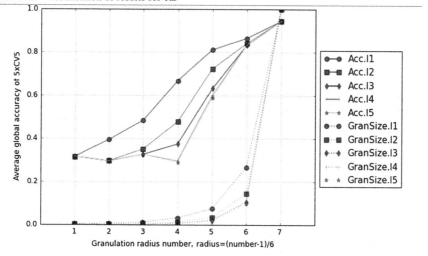

Table 6.13 CV-5; result of experiments for multi-layer c–d granulation with use of kNN classifier; data set **diabetes**; r_{gran} = Granulation radius, acc = Average accuracy for considered layer, *TRNsize* the mean size of granular decision system for considered layer

r_{gran}	Layer1		Layer2		Layer3		Layer4		Layer5	
	acc	GranSize	acc	GranSize	acc	GranSize	acc	GranSize	acc	GranSize
0	0.596	2	0.596	2	0.596	2	0.596	2	0.596	2
0.125	0.603	30.2	0.611	2	0.611	2	0.611	2	0.611	2
0.25	0.616	155.2	0.581	52.8	0.499	36	0.488	35.8	0.488	35.8
0.375	0.617	368	0.626	297.8	0.617	279.8	0.601	275.6	0.599	275
0.5	0.659	539.6	0.661	518.2	0.664	515.4	0.664	515.2	0.664	515.2
0.625	0.658	610	0.658	609.6	0.658	609.6	0.658	609.6	0.658	609.6
0.75	0.639	614.4	0.639	614.4	0.639	614.4	0.639	614.4	0.639	614.4
0.875	0.639	614.4	0.639	614.4	0.639	614.4	0.639	614.4	0.639	614.4
1	0.631	614.4	0.631	614.4	0.631	614.4	0.631	614.4	0.631	614.4

Table 6.14 Visualisation of results for diabetes

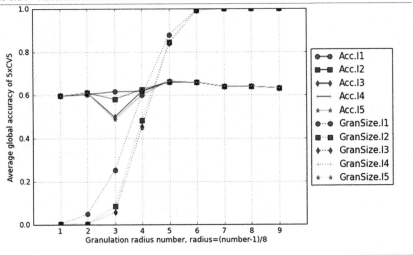

Table 6.15 CV-5; result of experiments for multi-layer c–d granulation with use of kNN classifier; data set **fertility diagnosis**; r_{gran} = Granulation radius, acc = Average accuracy for considered layer, *TRNsize* the mean size of granular decision system for considered layer

r_{gran}	Layer1		Layer2		Layer3		Layer4		Layer5	
	acc	GranSize	acc	GranSize	acc	GranSize	acc	GranSize	acc	GranSize
0	0.43	2	0.43	2	0.43	2	0.43	2	0.43	2
0.111	0.43	2	0.43	2	0.43	2	0.43	2	0.43	2
0.222	0.51	3.8	0.48	2	0.48	2	0.48	2	0.48	2
0.333	0.62	6.4	0.43	2.2	0.43	2.2	0.43	2.2	0.43	2.2
0.444	0.68	10	0.62	3.4	0.48	2.6	0.48	2.6	0.48	2.6
0.556	0.85	17.4	0.62	8.4	0.46	6.4	0.44	5.8	0.44	5.8
0.667	0.84	36	0.81	24.8	0.76	21.4	0.72	20.2	0.72	20.2
0.778	0.85	57.2	0.85	52.2	0.85	51	0.85	50.8	0.85	50.8
0.889	0.87	74.4	0.87	74	0.87	74	0.87	74	0.87	74
1	0.87	80	0.87	80	0.87	80	0.87	80	0.87	80

Table 6.16 Visualisation of results for fertility diagnosis

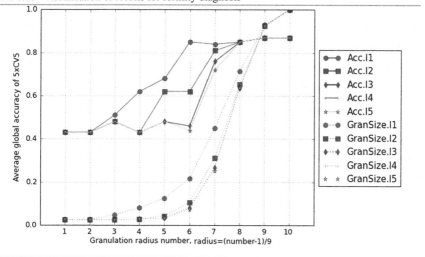

Table 6.17 CV-5; result of experiments for multi-layer c–d granulation with use of kNN classifier; data set **german credit**; r_{gran} = Granulation radius, acc = Average accuracy for considered layer, $TRNsize$ the mean size of granular decision system for considered layer

r_{gran}	Layer1		Layer2		Layer3		Layer4		Layer5	
	acc	GranSize	acc	GranSize	acc	GranSize	acc	GranSize	acc	GranSize
0	0.552	2	0.552	2	0.552	2	0.552	2	0.552	2
0.05	0.554	2	0.554	2	0.554	2	0.554	2	0.554	2
0.1	0.564	2.2	0.564	2	0.564	2	0.564	2	0.564	2
0.15	0.576	2.8	0.576	2	0.576	2	0.576	2	0.576	2
0.2	0.561	3	0.561	2	0.561	2	0.561	2	0.561	2
0.25	0.583	3.8	0.57	2	0.57	2	0.57	2	0.57	2
0.3	0.616	7	0.562	2	0.562	2	0.562	2	0.562	2
0.35	0.655	9.2	0.569	2	0.569	2	0.569	2	0.569	2
0.4	0.679	20.4	0.567	2	0.567	2	0.567	2	0.567	2
0.45	0.711	31.6	0.595	2	0.595	2	0.595	2	0.595	2
0.5	0.695	58.4	0.558	3.6	0.537	2.2	0.537	2.2	0.537	2.2
0.55	0.713	98	0.642	17.6	0.482	6.8	0.485	5.6	0.505	5.4
0.6	0.719	190.4	0.703	58.6	0.394	31.4	0.369	28.8	0.369	28.8
0.65	0.724	321.2	0.697	158.4	0.642	116.2	0.482	106.8	0.448	104.8
0.7	0.718	481.8	0.714	357.6	0.698	315.4	0.666	304.6	0.656	302.8
0.75	0.74	647.8	0.733	599.6	0.733	585	0.733	580.8	0.734	580
0.8	0.756	749.6	0.749	739.6	0.748	738.8	0.748	738.6	0.748	738.6
0.85	0.736	789.6	0.736	789.6	0.736	789.6	0.736	789.6	0.736	789.6
0.9	0.742	796.2	0.742	796	0.742	796	0.742	796	0.742	796
0.95	0.727	798.6	0.727	798.6	0.727	798.6	0.727	798.6	0.727	798.6
1	0.73	800	0.73	800	0.73	800	0.73	800	0.73	800

Table 6.18 Visualisation of results for german credit

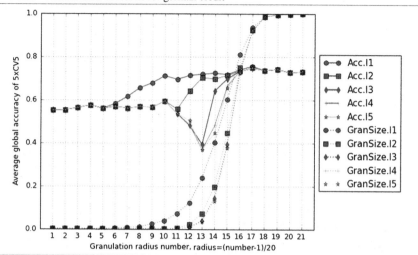

Table 6.19 CV-5; result of experiments for multi-layer c–d granulation with use of kNN classifier; data set **heart disease**; r_{gran} = Granulation radius, acc = Average accuracy for considered layer, *TRNsize* the mean size of granular decision system for considered layer

r_{gran}	Layer1		Layer2		Layer3		Layer4		Layer5	
	acc	GranSize	acc	GranSize	acc	GranSize	acc	GranSize	acc	GranSize
0	0.811	2	0.811	2	0.811	2	0.811	2	0.811	2
0.077	0.793	2	0.793	2	0.793	2	0.793	2	0.793	2
0.154	0.811	3	0.811	2	0.811	2	0.811	2	0.811	2
0.231	0.796	3.2	0.759	2	0.759	2	0.759	2	0.759	2
0.308	0.804	6.8	0.781	2	0.781	2	0.781	2	0.781	2
0.385	0.807	17	0.763	2.2	0.763	2	0.763	2	0.763	2
0.462	0.833	35.6	0.737	6.6	0.681	4	0.693	3.8	0.693	3.8
0.538	0.83	69.8	0.778	34.2	0.678	24.6	0.63	23	0.63	23
0.615	0.807	129.4	0.781	100.8	0.667	92.6	0.652	91.4	0.652	91.4
0.692	0.807	180.2	0.8	172.6	0.804	171	0.804	170.8	0.804	170.8
0.769	0.83	211	0.826	210.2	0.826	210	0.826	210	0.826	210
0.846	0.83	216	0.83	216	0.83	216	0.83	216	0.83	216
0.923	0.833	216	0.833	216	0.833	216	0.833	216	0.833	216
1	0.837	216	0.837	216	0.837	216	0.837	216	0.837	216

Table 6.20   Visualisation of results for heart disease

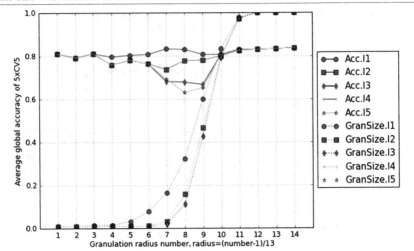

Table 6.21   CV-5; result of experiments for multi-layer c–d granulation with use of kNN classifier; data set **hepatitis-filled**; r_{gran} = Granulation radius, acc = Average accuracy for considered layer, $TRNsize$ the mean size of granular decision system for considered layer

r_{gran}	Layer1		Layer2		Layer3		Layer4		Layer5	
	acc	$GranSize$	acc	$GranSize$	acc	$GranSize$	acc	$GranSize$	acc	$GranSize$
0	0.8	2	0.8	2	0.8	2	0.8	2	0.8	2
0.053	0.806	2	0.806	2	0.806	2	0.806	2	0.806	2
0.105	0.813	2	0.813	2	0.813	2	0.813	2	0.813	2
0.158	0.826	2	0.826	2	0.826	2	0.826	2	0.826	2
0.211	0.826	2	0.826	2	0.826	2	0.826	2	0.826	2
0.263	0.813	3	0.813	2	0.813	2	0.813	2	0.813	2
0.316	0.806	2.8	0.806	2	0.806	2	0.806	2	0.806	2
0.368	0.819	7.2	0.819	2	0.819	2	0.819	2	0.819	2
0.421	0.832	6.8	0.806	2	0.806	2	0.806	2	0.806	2
0.474	0.871	12.4	0.8	2.2	0.8	2	0.8	2	0.8	2
0.526	0.877	20.2	0.794	4.8	0.703	2.8	0.703	2.8	0.703	2.8
0.579	0.865	32.2	0.658	10.6	0.652	7.4	0.652	7.4	0.652	7.4
0.632	0.884	49.6	0.806	27	0.703	22.4	0.69	21.8	0.69	21.8
0.684	0.89	67	0.865	54.6	0.865	52.6	0.845	52.2	0.845	52.2
0.737	0.89	88.4	0.877	79	0.871	77.8	0.871	77.6	0.871	77.6
0.789	0.91	108.6	0.91	104.4	0.91	103.8	0.91	103.8	0.91	103.8
0.842	0.903	117.4	0.903	114.6	0.903	114.6	0.903	114.6	0.903	114.6
0.895	0.89	121	0.89	120.2	0.89	120.2	0.89	120.2	0.89	120.2
0.947	0.916	122	0.916	122	0.916	122	0.916	122	0.916	122
1	0.89	124	0.89	124	0.89	124	0.89	124	0.89	124

Table 6.22 Visualisation of results for hepatitis

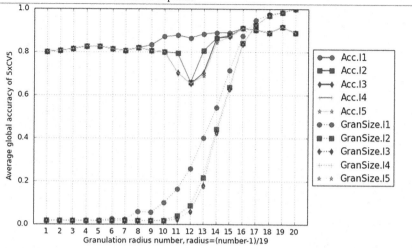

Table 6.23 CV-5; result of experiments for multi-layer c–d granulation with use of kNN classifier; data set **congressional voting records**; r_{gran} = Granulation radius, acc = Average accuracy for considered layer, *TRNsize* the mean size of granular decision system for considered layer

r_{gran}	Layer1		Layer2		Layer3		Layer4		Layer5	
	acc	*GranSize*	*acc*	*GranSize*	*acc*	*GranSize*	*acc*	*GranSize*	*acc*	*GranSize*
0	0.901	2	0.901	2	0.901	2	0.901	2	0.901	2
0.063	0.901	2	0.901	2	0.901	2	0.901	2	0.901	2
0.125	0.901	2.4	0.901	2	0.901	2	0.901	2	0.901	2
0.188	0.901	2.2	0.901	2	0.901	2	0.901	2	0.901	2
0.25	0.894	2.8	0.894	2	0.894	2	0.894	2	0.894	2
0.313	0.901	3.4	0.901	2	0.901	2	0.901	2	0.901	2
0.375	0.901	4	0.897	2	0.897	2	0.897	2	0.897	2
0.438	0.913	6.8	0.899	2	0.899	2	0.899	2	0.899	2
0.5	0.903	6.8	0.956	3	0.956	3	0.956	3	0.956	3
0.563	0.917	8.8	0.92	3	0.917	2.4	0.917	2.4	0.917	2.4
0.625	0.94	12.2	0.943	3.2	0.943	3.2	0.943	3.2	0.943	3.2
0.688	0.933	18.6	0.92	4.6	0.913	4	0.913	4	0.913	4
0.75	0.949	31	0.8	8.2	0.894	6	0.88	5.4	0.88	5.4
0.813	0.947	53.4	0.92	21.6	0.89	17.2	0.899	16.4	0.899	16.4
0.875	0.952	84.6	0.906	55	0.786	50.8	0.784	50.2	0.784	50.2
0.938	0.943	146	0.936	113.6	0.915	108.6	0.913	107.6	0.913	107.6
1	0.938	214.6	0.938	214.6	0.938	214.6	0.938	214.6	0.938	214.6

Table 6.24 Visualisation of results for congressional house votes

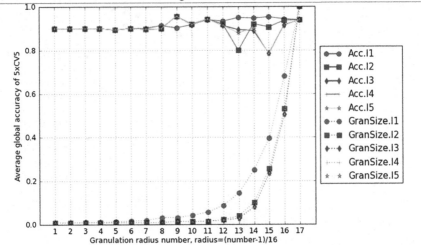

Table 6.25 CV-5; result of experiments for multi-layer c–d granulation with use of kNN classifier; data set **mushroom—modified—filled**; r_{gran} = Granulation radius, acc = Average accuracy for considered layer, *TRNsize* the mean size of granular decision system for considered layer

r_{gran}	Layer1		Layer2		Layer3		Layer4		Layer5	
	acc	*GranSize*	*acc*	*GranSize*	*acc*	*GranSize*	*acc*	*GranSize*	*acc*	*GranSize*
0	0.888	2	0.888	2	0.888	2	0.888	2	0.888	2
0.045	0.887	2	0.887	2	0.887	2	0.887	2	0.887	2
0.091	0.887	2	0.887	2	0.887	2	0.887	2	0.887	2
0.136	0.886	2	0.886	2	0.886	2	0.886	2	0.886	2
0.182	0.888	2	0.888	2	0.888	2	0.888	2	0.888	2
0.227	0.887	2	0.887	2	0.887	2	0.887	2	0.887	2
0.273	0.888	2.8	0.888	2	0.888	2	0.888	2	0.888	2
0.318	0.888	4	0.888	2	0.888	2	0.888	2	0.888	2
0.364	0.887	5.4	0.886	2	0.886	2	0.886	2	0.886	2
0.409	0.884	9.4	0.884	2	0.884	2	0.884	2	0.884	2
0.455	0.891	15.6	0.89	2	0.89	2	0.89	2	0.89	2
0.5	0.915	20.2	0.894	2.8	0.896	2.2	0.896	2.2	0.896	2.2
0.545	0.947	33.8	0.903	4.8	0.897	3.6	0.897	3.6	0.897	3.6
0.591	0.966	40.8	0.887	8.6	0.833	4.4	0.833	4.4	0.833	4.4
0.636	0.983	44.2	0.905	11.2	0.859	7.8	0.859	7.4	0.859	7.4
0.682	0.994	43.8	0.946	15.8	0.887	12.2	0.885	11.6	0.885	11.6
0.727	0.995	48	0.977	21.6	0.976	19	0.952	18.8	0.952	18.8
0.773	0.996	58	0.992	27	0.99	22	0.989	21.8	0.989	21.8
0.818	1	94.2	0.996	41.6	0.994	30	0.993	27.8	0.993	27.8

(continued)

Table 6.25 (continued)

r_{gran}	Layer1		Layer2		Layer3		Layer4		Layer5	
	acc	GranSize	acc	GranSize	acc	GranSize	acc	GranSize	acc	GranSize
0.864	1	200.4	1	82.8	0.998	51	0.997	44.8	0.997	44.6
0.909	1	504.8	1	226.6	1	147.4	1	128.6	1	126.8
0.955	1	1762.8	1	947.6	1	681.2	1	615.6	1	608.8
1	1	6499.2	1	6499.2	1	6499.2	1	6499.2	1	6499.2

Table 6.26 Visualisation of results for mushroom

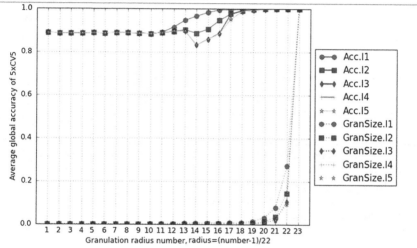

Table 6.27 CV-5; result of experiments for multi-layer c–d granulation with use of kNN classifier; data set **nursery**; r_{gran} = Granulation radius, acc = Average accuracy for considered layer, TRNsize the mean size of granular decision system for considered layer

r_{gran}	Layer1		Layer2		Layer3		Layer4		Layer5	
	acc	GranSize	acc	GranSize	acc	GranSize	acc	GranSize	acc	GranSize
0	0.371	5	0.371	5	0.371	5	0.371	5	0.371	5
0.125	0.428	7.4	0.419	5	0.419	5	0.419	5	0.419	5
0.25	0.525	14.2	0.441	5.4	0.441	5.4	0.441	5.4	0.441	5.4
0.375	0.556	26.4	0.477	9.2	0.431	6.4	0.431	6.2	0.431	6.2
0.5	0.575	57.4	0.525	19.8	0.476	10.4	0.455	8.4	0.455	8.4
0.625	0.491	155	0.458	57.2	0.442	30.6	0.43	23	0.429	22
0.75	0.654	550.4	0.637	219.6	0.602	129.6	0.589	103.4	0.585	100.2
0.875	0.696	2375	0.698	1223.6	0.674	854.6	0.666	760.2	0.664	745.8
1	0.578	10368	0.578	10368	0.578	10368	0.578	10368	0.578	10368

Table 6.28 Visualisation of results for nursery

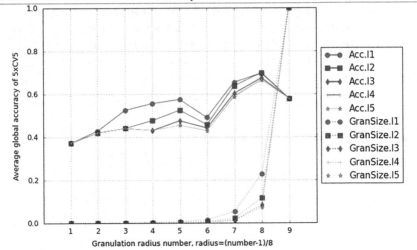

Table 6.29 CV-5; result of experiments for multi-layer c–d granulation with use of kNN classifier; data set **soybean—large—all—filled**; r_{gran} = Granulation radius, acc = Average accuracy for considered layer, *TRNsize* the mean size of granular decision system for considered layer

r_{gran}	Layer1		Layer2		Layer3		Layer4		Layer5	
	acc	GranSize	acc	GranSize	acc	GranSize	acc	GranSize	acc	GranSize
0	0.663	19	0.663	19	0.663	19	0.663	19	0.663	19
⋮	⋮	⋮	⋮	⋮	⋮	⋮	⋮	⋮	⋮	⋮
0.4	0.684	19	0.684	19	0.684	19	0.684	19	0.684	19
0.429	0.66	19.4	0.66	19	0.66	19	0.66	19	0.66	19
0.457	0.69	19.2	0.69	19	0.69	19	0.69	19	0.69	19
0.486	0.681	19.2	0.681	19	0.681	19	0.681	19	0.681	19
0.514	0.688	19.8	0.684	19	0.684	19	0.684	19	0.684	19
0.543	0.703	19.6	0.703	19	0.703	19	0.703	19	0.703	19
0.571	0.692	21.6	0.682	19	0.682	19	0.682	19	0.682	19
0.6	0.698	23.8	0.673	19	0.673	19	0.673	19	0.673	19
0.629	0.706	26	0.69	19.8	0.69	19.6	0.69	19.6	0.69	19.6
0.657	0.71	31	0.695	19.4	0.695	19.4	0.695	19.4	0.695	19.4
0.686	0.734	35.4	0.682	19.6	0.679	19.4	0.679	19.4	0.679	19.4
0.714	0.748	42	0.691	22.4	0.687	21.2	0.687	21.2	0.687	21.2
0.743	0.795	46.8	0.729	25.6	0.697	25	0.697	24.8	0.697	24.8
0.771	0.816	58.8	0.758	30	0.734	28	0.734	27.8	0.734	27.8
0.8	0.839	71.4	0.794	38	0.78	35.8	0.78	35.8	0.78	35.8
0.829	0.876	86.6	0.859	50.8	0.839	48.4	0.839	48.4	0.839	48.4
0.857	0.861	118.2	0.855	71.2	0.83	64.4	0.83	64.4	0.83	64.4

(continued)

Table 6.29 (continued)

r_{gran}	Layer1		Layer2		Layer3		Layer4		Layer5	
	acc	GranSize	acc	GranSize	acc	GranSize	acc	GranSize	acc	GranSize
0.886	0.871	164.4	0.852	113.8	0.855	105.4	0.854	105.2	0.854	105.2
0.914	0.884	220	0.852	170.4	0.845	162.4	0.845	161.8	0.845	161.8
0.943	0.903	302.6	0.889	256.8	0.877	249.6	0.877	249.2	0.877	249.2
0.971	0.914	402.8	0.912	379.8	0.914	378	0.914	378	0.914	378
1	0.928	510.2	0.928	510.2	0.928	510.2	0.928	510.2	0.928	510.2

Table 6.30 Visualisation of results for soybean large

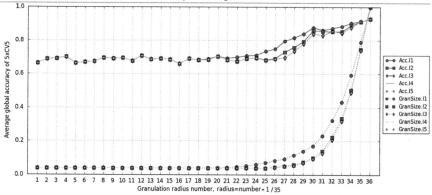

Table 6.31 CV-5; result of experiments for multi-layer c–d granulation with use of kNN classifier; data set **SPECT-all-modif**; r_{gran} = Granulation radius, acc = Average accuracy for considered layer, *TRNsize* the mean size of granular decision system for considered layer

r_{gran}	Layer1		Layer2		Layer3		Layer4		Layer5	
	acc	GranSize	acc	GranSize	acc	GranSize	acc	GranSize	acc	GranSize
0	0.529	2	0.529	2	0.529	2	0.529	2	0.529	2
0.045	0.513	2	0.513	2	0.513	2	0.513	2	0.513	2
0.091	0.52	2.2	0.52	2	0.52	2	0.52	2	0.52	2
0.136	0.498	2.4	0.498	2	0.498	2	0.498	2	0.498	2
0.182	0.522	2	0.522	2	0.522	2	0.522	2	0.522	2
0.227	0.566	3	0.517	2	0.517	2	0.517	2	0.517	2
0.273	0.629	3.2	0.546	2	0.546	2	0.546	2	0.546	2
0.318	0.577	3.2	0.569	2	0.569	2	0.569	2	0.569	2
0.364	0.614	4.8	0.46	2	0.46	2	0.46	2	0.46	2
0.409	0.629	4.4	0.588	2	0.588	2	0.588	2	0.588	2
0.455	0.745	7.2	0.555	2	0.555	2	0.555	2	0.555	2
0.5	0.621	6	0.498	2.4	0.494	2.2	0.494	2.2	0.494	2.2
0.545	0.756	9.6	0.632	3	0.621	2.8	0.621	2.8	0.621	2.8
0.591	0.779	12	0.67	3.6	0.621	2.6	0.621	2.6	0.621	2.6

(continued)

Table 6.31 (continued)

r_{gran}	Layer1		Layer2		Layer3		Layer4		Layer5	
	acc	GranSize	acc	GranSize	acc	GranSize	acc	GranSize	acc	GranSize
0.636	0.783	15.4	0.711	5.2	0.555	3	0.555	3	0.555	3
0.682	0.801	22.8	0.719	7.6	0.667	5.8	0.663	5.6	0.663	5.6
0.727	0.813	33	0.783	10.6	0.602	6.2	0.602	6	0.602	6
0.773	0.79	47.2	0.787	21.4	0.753	16.2	0.738	15.6	0.749	15.4
0.818	0.794	72.6	0.794	43.6	0.794	37.4	0.779	36.4	0.779	36.4
0.864	0.794	102	0.794	73.6	0.79	69.4	0.798	68.2	0.798	68.2
0.909	0.794	131	0.794	113.6	0.794	111.2	0.794	111	0.794	111
0.955	0.794	162.8	0.794	156	0.794	155.4	0.794	155.4	0.794	155.4
1	0.794	184.6	0.794	184.6	0.794	184.6	0.794	184.6	0.794	184.6

Table 6.32 Visualisation of results for spect

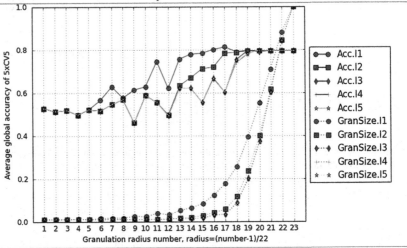

Table 6.33 CV-5; result of experiments for multi-layer c-d granulation with use of kNN classifier; data set **SPECTF—all—modif**; r_{gran} = Granulation radius, acc = Average accuracy for considered layer, TRNsize the mean size of granular decision system for considered layer

r_{gran}	Layer1		Layer2		Layer3		Layer4		Layer5	
	acc	GranSize	acc	GranSize	acc	GranSize	acc	GranSize	acc	GranSize
0	0.472	2	0.472	2	0.472	2	0.472	2	0.472	2
0.023	0.554	6.4	0.427	2	0.427	2	0.427	2	0.427	2
0.045	0.652	13.2	0.446	2	0.446	2	0.446	2	0.446	2
0.068	0.749	26.2	0.558	3.8	0.464	2	0.464	2	0.464	2
0.091	0.775	50.4	0.693	14.8	0.596	7.6	0.584	7.2	0.584	7.2
0.114	0.802	84.8	0.75	38.6	0.716	25.2	0.708	23.8	0.708	23.8
0.136	0.783	126.6	0.798	80.4	0.801	63.2	0.791	57.6	0.76	56.4

(continued)

Table 6.33 (continued)

r_{gran}	Layer1		Layer2		Layer3		Layer4		Layer5	
	acc	GranSize	acc	GranSize	acc	GranSize	acc	GranSize	acc	GranSize
0.159	0.794	166.2	0.812	147.2	0.812	142	0.812	141.2	0.812	141
0.182	0.809	195.4	0.805	192.2	0.805	191.8	0.805	191.8	0.805	191.8
0.205	0.779	209.6	0.783	208.2	0.783	208.2	0.783	208.2	0.783	208.2
0.227	0.786	213	0.786	213	0.786	213	0.786	213	0.786	213
0.25	0.798	213.6	0.798	213.6	0.798	213.6	0.798	213.6	0.798	213.6
⋮	⋮	⋮	⋮	⋮	⋮	⋮	⋮	⋮	⋮	⋮
1	0.779	213.6	0.779	213.6	0.779	213.6	0.779	213.6	0.779	213.6

Table 6.34 Visualisation of results for spectf

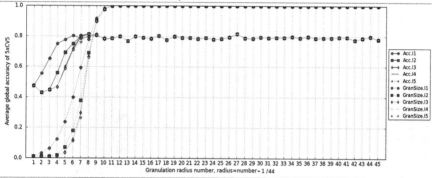

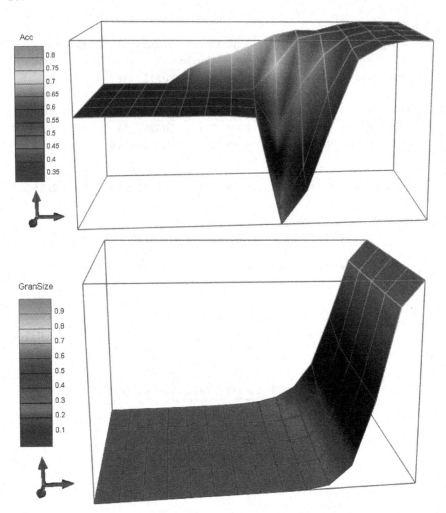

Fig. 6.1 Visualisation of accuracy and training set size for five layer granulation from *back* to *front*, and radii increasing from *left* to *right*; Data set: adult

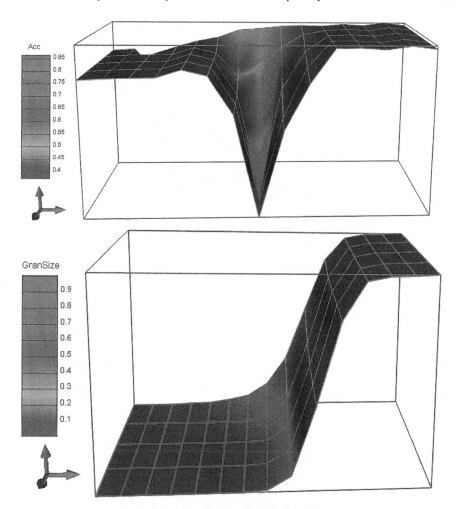

Fig. 6.2 Visualisation of accuracy and training set size for five layer granulation from *back* to *front*, and radii increasing from *left* to *right*; Data set: australian

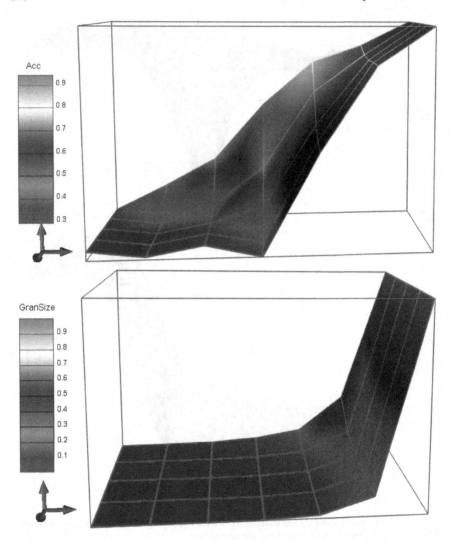

Fig. 6.3 Visualisation of accuracy and training set size for five layer granulation from *back* to *front*, and radii increasing from *left* to *right*; Data set: Car Evaluation

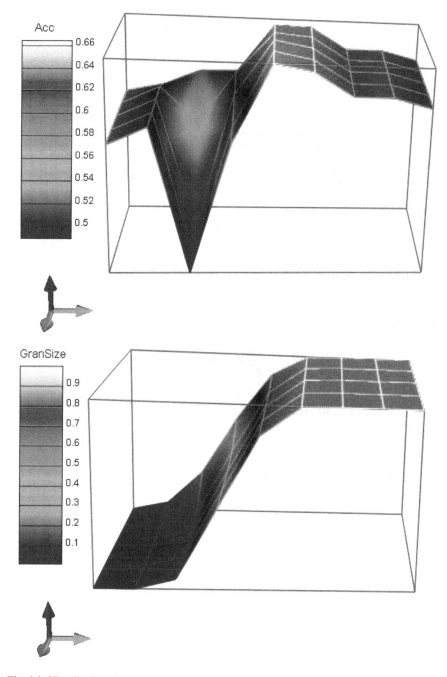

Fig. 6.4 Visualisation of accuracy and training set size for five layer granulation from *back* to *front*, and radii increasing from *left* to *right*; Data set: Pima Indians

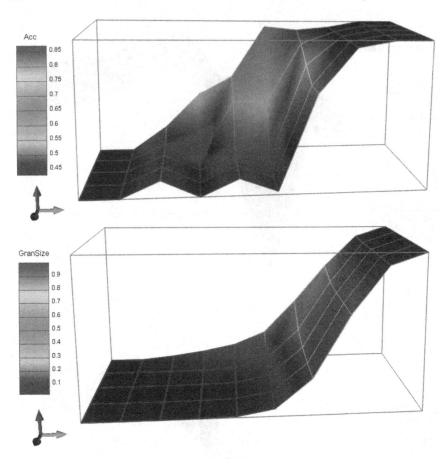

Fig. 6.5 Visualisation of accuracy and training set size for five layer granulation from *back* to *front*, and radii increasing from *left* to *right*; Data set: fertility

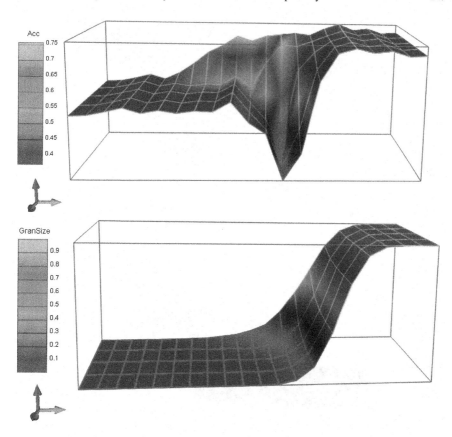

Fig. 6.6 Visualisation of accuracy and training set size for five layer granulation from *back* to *front*, and radii increasing from *left* to *right*; Data set: German Credit

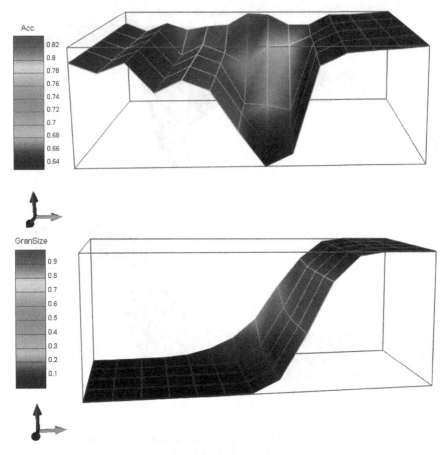

Fig. 6.7 Visualisation of accuracy and training set size for five layer granulation from *back* to *front*, and radii increasing from *left* to *right*; Data set: Heart disease

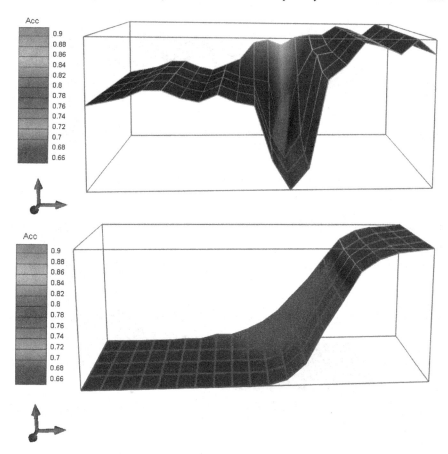

Fig. 6.8 Visualisation of accuracy and training set size for five layer granulation from *back* to *front*, and radii increasing from *left* to *right*; Data set: Hepatitis

By the optimal radius of granulation we mean the radius for which we get the highest approximation of training decision system size with maintenance of high accuracy of classification. Based on double granulation, optimal radius could be estimated without classification, based only on the size of data sets in the first and the second layers.

Considering original decision system (U_{l_0}, A, d), granular reflections (U_{l_1}, A, d), i.e. the first layer, and (U_{l_2}, A, d), i.e., the second layer, the value of the optimal r_{gran} (granulation radius) can be evaluated as

$$optimal(r_{gran}) = argmax_r |U_{r,l_1}| - |U_{r,l_2}|.$$

From classification results one infers that the value of the generalized optimal radius of granulation understood as the arg max of the difference in sizes between two

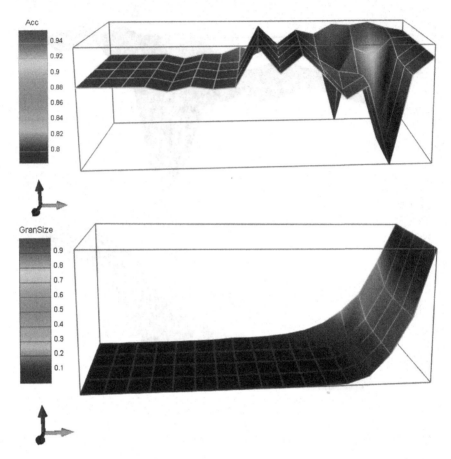

Fig. 6.9 Visualisation of accuracy and training set size for five layer granulation from *back* to *front*, and radii increasing from *left* to *right*; Data set: Congressional House Votes

consecutive layers seems to shift to the right, i.e., it does increase, in successive layers of granulation.

As a rule, only the first three layers show significant approximation in terms of the decrease in the number of objects, further layers give a slight decrease only, and, finally the process stops due to irreducibility. The first two layers give best approximations to training data.

To support and verify our hypothesis about the optimal granulation radius, we have performed additional experimental session, directed on numerical identification of best radii, based on the parameter which combines accuracy (acc) of classification and percentage size of decision systems $1 - \dfrac{GranSize}{GranSize_{r=1}}$, and, using difference between layer sizes in the form $GranSize_{l_{i-1}} - GranSize_{l_i}$. Discussion of results, including as well results from this section is given in Sect. 6.3.

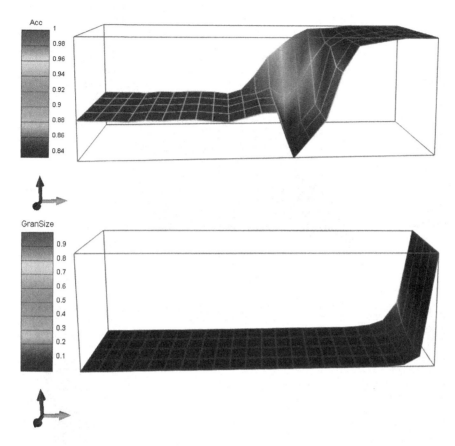

Fig. 6.10 Visualisation of accuracy and training set size for five layer granulation from *back* to *front*, and radii increasing from *left* to *right*; Data set: Mushroom

6.3 In Search for the Optimal Granulation Radius

In this section, we show additional results, aimed at the detecting the optimal radius of granulation. The results are shown in Tables 6.35, 6.36, 6.37, 6.38, 6.39, 6.40, 6.41, 6.42, 6.43, 6.44, 6.45, 6.46, 6.47 and 6.48. In these tables we give the two series of values, the first series is the result for the parameter *CombAGS*, defined as

$$CombAGS = \frac{acc + (1 - \frac{GranSize}{GranSize_{r=1}})}{2}.$$

The second series of values is the difference in layer size,

$$GranSize_{l_{i-1}} - GranSize_{l_i}.$$

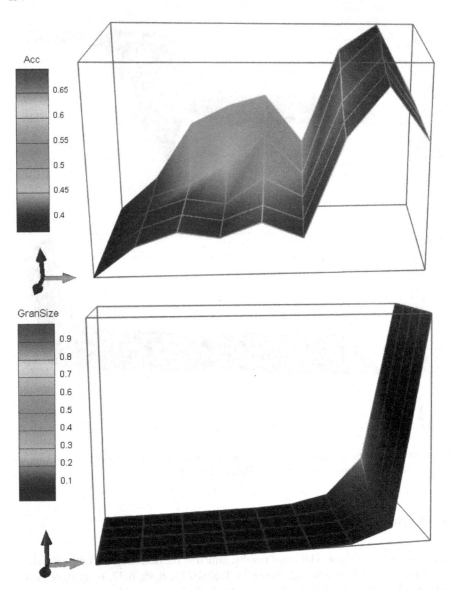

Fig. 6.11 Visualisation of accuracy and training set size for five layer granulation from *back* to *front*, and radii increasing from *left* to *right*; Data set: Nursery

As Best *CombAGS* parameter for considered data set, we understand the largest value of this parameter among radii for which the error of accuracy defined as $(acc_{r=1} - acc)$ is less than 0.01. If *CombAGS* points to the radius valued 1, then as the best radius we accept the radius with the value of $\frac{|A-1|}{|A|}$.

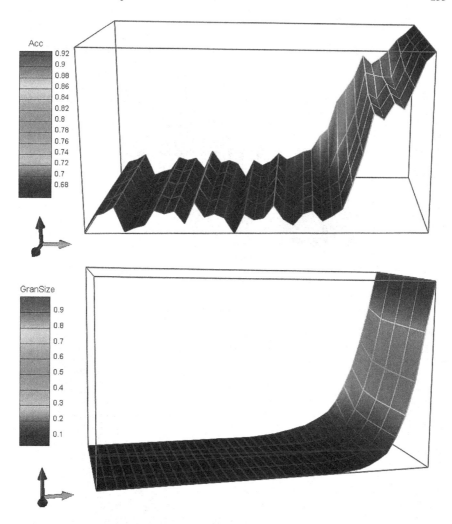

Fig. 6.12 Visualisation of accuracy and training set size for five layer granulation from *back* to *front*, and radii increasing from *left* to *right*; Data set: Soybean Large

6.3.1 Results Pointed to by the Two-layered Granulation

Considering the respective data sets with respect to the outlined above search for the optimal granulation radius between the first and the second layers of granulation, we put for the record the following results.

1. For Adult data set, for radius of 0.786, we have *CombAGS* = 0.792, with accuracy error of 0.009; we have reduction in training object number in range of 75.27 %.

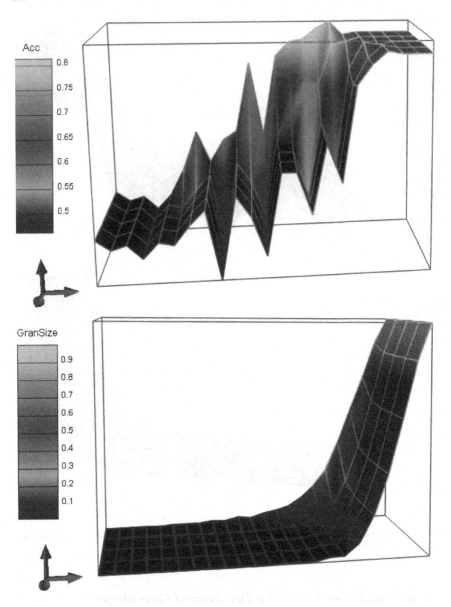

Fig. 6.13 Visualisation of accuracy and training set size for five layer granulation from *back* to *front*, and radii increasing from *left* to *right*; Data set: SPECT

2. For Australian Credit data set, for radius of 0.571, we have *CombAGS* = 0.785, with accuracy error of 0.004; we have reduction in training object number in range of 71.96 %.

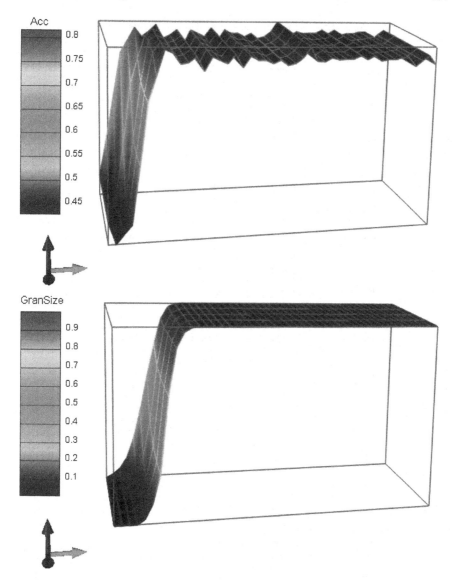

Fig. 6.14 Visualisation of accuracy and training set size for five layer granulation from *back* to *front*, and radii increasing from *left* to *right*; Data set: SPECTF

3. For Car Evaluation data set, for radius of 0.833, we have *CombAGS* = 0.799, with accuracy error of 0.079; we have reduction in training object number in range of 73.23 %.

Table 6.35 CV-5; CombAGS—Combination of Acc and GranSize; result of experiments for multiple concept-dependent granulation with use of kNN classifier; data set: **adult**; r_{gran} = Granulation radius

r_{gran}	Layer1	Layer2	Layer3	Layer4	Layer5
CombAGS					
0	0.817	0.817	0.817	0.817	0.817
0.071	0.818	0.818	0.818	0.818	0.818
0.143	0.818	0.818	0.818	0.818	0.818
0.214	0.818	0.818	0.818	0.818	0.818
0.286	0.843	0.817	0.817	0.817	0.817
0.357	0.859	0.818	0.818	0.818	0.818
0.429	0.867	0.817	0.817	0.817	0.817
0.5	0.879	0.82	0.815	0.815	0.815
0.571	0.891	0.857	0.761	0.66	0.66
0.643	0.886	0.898	0.835	0.722	0.707
0.714	0.856	0.894	0.873	0.825	0.797
0.786	**0.792**	0.847	0.853	0.846	0.839
0.857	**0.654**	**0.697**	**0.705**	**0.702**	0.701
0.929	**0.483**	**0.484**	**0.484**	**0.484**	**0.484**
1	**0.421**	**0.421**	**0.421**	**0.421**	**0.421**
$GranSize_{l_{i-1}} - GranSize_{l_i}$					
0	nil	0	0	0	0
0.071	nil	0.2	0	0	0
0.143	nil	1.6	0	0	0
0.214	nil	7.4	0	0	0
0.286	nil	19.2	0	0	0
0.357	nil	42.6	0	0	0
0.429	nil	102	0	0	0
0.5	nil	250.4	4	0	0
0.571	nil	600.8	25.6	2.4	0
0.643	nil	1363.4	111.6	15.2	2.8
0.714	nil	2766.4	331.2	52.8	12.4
0.786	nil	**4403.4**	**752.8**	**140.4**	**24.2**
0.857	nil	3380.2	681.4	118	16.4
0.929	nil	72.8	5.2	0	0
1	nil	0	0	0	0

4. For Pima Indians Diabetes data set, for radius of 0.25, we have *CombAGS* = 0.682, with accuracy error of 0.015; we have reduction in training object number in range of 74.74 %.

Table 6.36 CV-5; CombAGS—Combination of acc and GranSize; result of experiments for multiple concept-dependent granulation with use of kNN classifier; data set: **australian**; r_{gran} = Granulation radius

r_{gran}	Layer1	Layer2	Layer3	Layer4	Layer5
CombAGS					
0	0.882	0.882	0.882	0.882	0.882
0.071	0.884	0.884	0.884	0.884	0.884
0.143	0.846	0.885	0.885	0.885	0.885
0.214	0.885	0.886	0.886	0.886	0.886
0.286	0.894	0.897	0.897	0.897	0.897
0.357	0.892	0.888	0.888	0.888	0.888
0.429	0.892	0.849	0.833	0.833	0.833
0.5	0.859	0.851	0.678	0.678	0.678
0.571	**0.785**	0.836	0.802	0.775	0.774
0.643	**0.643**	0.698	0.696	0.686	0.676
0.714	**0.502**	**0.524**	**0.529**	**0.531**	**0.531**
0.786	**0.444**	**0.444**	**0.444**	**0.444**	**0.444**
0.857	**0.431**	**0.431**	**0.431**	**0.431**	**0.431**
0.929	**0.435**	**0.435**	**0.435**	**0.435**	**0.435**
1	**0.428**	**0.428**	**0.428**	**0.428**	**0.428**
$GranSize_{l_{i-1}} - GranSize_{l_i}$					
0	*nil*	0	0	0	0
0.071	*nil*	0	0	0	0
0.143	*nil*	0.6	0	0	0
0.214	*nil*	3.6	0	0	0
0.286	*nil*	4.8	0	0	0
0.357	*nil*	14.4	0	0	0
0.429	*nil*	26	1.4	0	0
0.5	*nil*	53.2	8	0	0
0.571	*nil*	**84.2**	**23.2**	**4.2**	0.4
0.643	*nil*	67.6	16.8	3.2	**1.4**
0.714	*nil*	23.4	4.8	1.4	0
0.786	*nil*	0.2	0	0	0
0.857	*nil*	0	0	0	0
0.929	*nil*	0	0	0	0
1	*nil*	0	0	0	0

5. For Fertility Diagnosis data set, for radius of 0.667, we have $CombAGS = 0.695$, with accuracy error of 0.03; we have reduction in training object number in range of 55.0 %.

Table 6.37 CV-5; CombAGS—Combination of acc and GranSize; result of experiments for multiple concept-dependent granulation with use of kNN classifier; data set: **car evaluation**; r_{gran} = Granulation radius

r_{gran}	Layer1	Layer2	Layer3	Layer4	Layer5
CombAGS					
0	0.656	0.656	0.656	0.656	0.656
0.167	0.694	0.647	0.647	0.647	0.647
0.333	0.736	0.673	0.661	0.661	0.661
0.5	0.818	0.733	0.684	0.645	0.642
0.667	0.868	0.844	0.805	0.791	0.786
0.833	0.799	0.848	0.863	0.867	0.868
1	**0.472**	**0.472**	**0.472**	**0.472**	**0.472**
$GranSize_{l_{i-1}} - GranSize_{l_i}$					
0	nil	0	0	0	0
0.167	nil	4.6	0	0	0
0.333	nil	10.2	1.6	0	0
0.5	nil	27.8	6.8	2.4	0.2
0.667	nil	55.4	17.6	4.4	0.6
0.833	nil	**170.2**	**52.6**	**10.2**	**2.2**
1	nil	0	0	0	0

6. For German Credit data set, for radius of 0.65, we have *CombAGS* = 0.661, with accuracy error of 0.006; we have reduction in training object number in range of 59.85 %.

7. For Heart Disease data set, for radius of 0.538, we have *CombAGS* = 0.753, with accuracy error of 0.007; we have reduction in training object number in range of 67.69 %.

8. For Hepatitis data set, for radius of 0.632, we have *CombAGS* = 0.742, with accuracy error of 0.006; we have reduction in training object number in range of 60.0 %.

9. For Congressional House Votes data set, for radius of 0.938, we have *CombAGS* = 0.631, with accuracy error of −0.005; we have reduction in training object number in range of 31.97 %.

10. For Mushroom data set, for radius of 0.955 we get *CombAGS* = 0.864, with accuracy error of 0.0; we have reduction in training object number in range of 72.88 %.

11. For Nursery data set, for radius of 0.875, we have *CombAGS* = 0.733, with accuracy error of −0.118; we have reduction in training object number in range of 77.09 %

Table 6.38 CV-5; CombAGS—Combination of acc and GranSize; result of experiments for multiple concept-dependent granulation with use of kNN classifier; data set: **diabetes**; r_{gran} = Granulation radius

r_{gran}	Layer1	Layer2	Layer3	Layer4	Layer5
0	0.796	0.796	0.796	0.796	0.796
0.125	0.777	0.804	0.804	0.804	0.804
0.25	0.682	0.748	0.72	0.715	0.715
0.375	0.509	**0.571**	0.581	0.576	0.576
0.5	**0.39**	**0.409**	**0.413**	**0.413**	**0.413**
0.625	**0.333**	**0.333**	**0.333**	**0.333**	**0.333**
0.75	**0.32**	**0.32**	**0.32**	**0.32**	**0.32**
0.875	**0.32**	**0.32**	**0.32**	**0.32**	**0.32**
1	**0.316**	**0.316**	**0.316**	**0.316**	**0.316**
$GranSize_{l_{i-1}} - GranSize_{l_i}$					
0	nil	0	0	0	0
0.125	nil	28.2	0	0	0
0.25	nil	**102.4**	16.8	0.2	0
0.375	nil	70.2	**18**	**4.2**	**0.6**
0.5	nil	21.4	2.8	0.2	0
0.625	nil	0.4	0	0	0
0.75	nil	0	0	0	0
0.875	nil	0	0	0	0
1	nil	0	0	0	0

12. For Soybean Large data set, for radius of 0.886, we have *CombAGS* = 0.774, with accuracy error of 0.057; we have reduction in training object number in range of 67.77 %.
13. For SPECT data set, for radius of 0.818 we get *CombAGS* = 0.70, with accuracy error of 0.0; we have reduction in training object number in range of 60.67 %.
14. For SPECTF data set, for radii of 0.114 and 0.136, we have *CombAGS* = 0.702, with accuracy error of −0.023; we have reduction in training object number in range of 60.3 %.

6.3.2 Comparison of Results Pointed by Double Granulation and Best CombAGS

There is the question about validity of optimal parameter estimation based on double granulation. In what regards the best value of CombAGS parameter, for data sets:

Table 6.39 CV-5; CombAGS—Combination of acc and GranSize; result of experiments for multiple concept-dependent granulation with use of kNN classifier; data set: **fertility diagnosis**; r_{gran} = Granulation radius

r_{gran}	Layer1	Layer2	Layer3	Layer4	Layer5
CombAGS					
0	0.703	0.703	0.703	0.703	0.703
0.111	0.703	0.703	0.703	0.703	0.703
0.222	0.731	0.728	0.728	0.728	0.728
0.333	0.77	0.701	0.701	0.701	0.701
0.444	0.778	0.789	0.724	0.724	0.724
0.556	0.816	0.758	0.69	0.684	0.684
0.667	0.695	0.75	0.746	0.734	0.734
0.778	0.568	0.599	0.606	0.608	0.608
0.889	**0.47**	**0.473**	**0.473**	**0.473**	**0.473**
1	**0.435**	**0.435**	**0.435**	**0.435**	**0.435**
$GranSize_{l_{i-1}} - GranSize_{l_i}$					
0	nil	0	0	0	0
0.111	nil	0	0	0	0
0.222	nil	1.8	0	0	0
0.333	nil	4.2	0	0	0
0.444	nil	6.6	0.8	0	0
0.556	nil	9	2	0.6	0
0.667	nil	**11.2**	**3.4**	**1.2**	0
0.778	nil	5	1.2	0.2	0
0.889	nil	0.4	0	0	0
1	nil	0	0	0	0

Adult, Australian Credit, Car, German Credit and Hepatitis, estimated radii are the same as those pointed by best CombAGS parameters.

In case of Pima Indians Diabetes the difference between accuracy estimated radius gives accuracy error equal 0.015, where reduction in training object number is 74.74 %, for best CombAGS we have accuracy error −0.028 and reduction 12.17 %.

In case of Fertility Diagnosis the difference between Acc estimated radius gives accuracy error equal 0.003, where reduction in training object number is 55 %, for best CombAGS we have accuracy error 0 and reduction 7 %.

In case of Heart Disease the difference between Acc estimated radius gives accuracy error equal 0.007, where reduction in training object number is 67.69 %, for best CombAGS we have accuracy error 0.004 and reduction 83.52 %.

In case of Congressional House Votes the difference between Acc estimated radius gives accuracy error equal −0.005, where reduction in training object number is 31.97 %, for best CombAGS we have accuracy error −0.002 and reduction 94.35 %.

Table 6.40 CV-5; CombAGS—Combination of acc and GranSize; result of experiments for multiple concept-dependent granulation with use of kNN classifier; data set: **german-credit**; r_{gran} = Granulation radius

r_{gran}	Layer1	Layer2	Layer3	Layer4	Layer5
CombAGS					
0	0.775	0.775	0.775	0.775	0.775
0.05	0.776	0.776	0.776	0.776	0.776
0.1	0.781	0.781	0.781	0.781	0.781
0.15	0.786	0.787	0.787	0.787	0.787
0.2	0.779	0.779	0.779	0.779	0.779
0.25	0.789	0.784	0.784	0.784	0.784
0.3	0.804	0.78	0.78	0.78	0.78
0.35	0.822	0.783	0.783	0.783	0.783
0.4	0.827	0.782	0.782	0.782	0.782
0.45	0.836	0.796	0.796	0.796	0.796
0.5	0.811	0.777	0.767	0.767	0.767
0.55	0.795	0.81	0.737	0.739	0.749
0.6	0.741	0.815	0.677	0.667	0.667
0.65	**0.661**	0.75	0.748	0.674	0.659
0.7	0.558	0.634	0.652	0.643	0.639
0.75	**0.465**	**0.492**	**0.501**	**0.503**	**0.504**
0.8	**0.41**	**0.412**	**0.412**	**0.412**	**0.412**
0.85	**0.375**	**0.375**	**0.375**	0.375	0.375
0.9	**0.373**	**0.374**	**0.374**	0.374	0.374
0.95	**0.364**	**0.364**	**0.364**	0.364	0.364
1	**0.365**	**0.365**	**0.365**	0.365	0.365
$GranSize_{l_{i-1}} - GranSize_{l_i}$					
0	nil	0	0	0	0
0.05	nil	0	0	0	0
0.1	nil	0.2	0	0	0
0.15	nil	0.8	0	0	0
0.2	nil	1	0	0	0
0.25	nil	1.8	0	0	0
0.3	nil	5	0	0	0
0.35	nil	7.2	0	0	0
0.4	nil	18.4	0	0	0
0.45	nil	29.6	0	0	0
0.5	nil	54.8	1.4	0	0
0.55	nil	80.4	10.8	1.2	0.2
0.6	nil	131.8	27.2	2.6	0
0.65	nil	**162.8**	**42.2**	9.4	2

(continued)

Table 6.40 (continued)

r_{gran}	Layer1	Layer2	Layer3	Layer4	Layer5
0.7	nil	124.2	**42.2**	**10.8**	**1.8**
0.75	nil	48.2	14.6	4.2	0.8
0.8	nil	10	0.8	0.2	0
0.85	nil	0	0	0	0
0.9	nil	0.2	0	0	0
0.95	nil	0	0	0	0
1	nil	0	0	0	0

In case of Mushroom the difference between Acc estimated radius gives accuracy error equal 0, where reduction in training object number is 72.88 %, for best CombAGS we have accuracy error 0.005 and reduction 99.26 %.

In case of Nursery the difference between Acc estimated radius gives accuracy error equal −0.118, where reduction in training object number is 77.09 %, for best CombAGS we have accuracy error −0.075 and reduction 94.69 %.

In case of Soybean Large the difference between Acc estimated radius gives accuracy error equal 0.057, where reduction in training object number is 67.77 %, for best CombAGS we have accuracy error 0.014 and reduction 21.05 %.

In case of SPECT the difference between Acc estimated radius gives accuracy error equal 0, where reduction in training object number is 60.67 %, for best CombAGS we have accuracy error −0.007 and reduction 87.65 %.

In case of SPECTF the difference between Acc estimated radius gives accuracy error equal −0.23, where reduction in training object number is 60.3 %, for best CombAGS we have accuracy error 0.004 and reduction 76.4 %.

Concluding, in the most cases the result of accuracy for the radii pointed by double granulation seems to be comparable with result pointed by CombAGS, which can be seen in Fig. 6.15. There are differences between the degree of approximation, for Pima, Fertility Diagnosis and Soybean Large, the approximation pointed by double granulation is better, for Heart Disease, Congressional House Votes, Mushroom, Nursery, SPECT and SPECTF the approximation for radius pointed by *CombAGS* is better. In case of Adult, Australian Credit, Car, German Credit and Hepatitis the result is the same, because the double granulation points to the same radius as best *CombAGS* does (Table 6.49).

Table 6.41 CV-5; CombAGS—Combination of acc and GranSize; result of experiments for multiple concept-dependent granulation with use of kNN classifier; data set: **heart disease**; r_{gran} = Granulation radius

r_{gran}	Layer1	Layer2	Layer3	Layer4	Layer5
CombAGS					
0	0.901	0.901	0.901	0.901	0.901
0.077	0.892	0.892	0.892	0.892	0.892
0.154	0.899	0.901	0.901	0.901	0.901
0.231	0.891	0.875	0.875	0.875	0.875
0.308	0.886	0.886	0.886	0.886	0.886
0.385	0.864	0.876	0.877	0.877	0.877
0.462	**0.834**	0.853	0.831	0.838	0.838
0.538	**0.753**	0.81	0.782	0.762	0.762
0.615	0.604	0.657	0.619	0.614	0.614
0.692	0.486	0.5	0.506	0.507	0.507
0.769	**0.427**	0.426	0.427	0.427	0.427
0.846	**0.415**	**0.415**	**0.415**	**0.415**	**0.415**
0.923	**0.417**	**0.417**	**0.417**	**0.417**	**0.417**
1	**0.419**	**0.419**	**0.419**	**0.419**	**0.419**
$GranSize_{l_{i-1}} - GranSize_{l_i}$					
0	*nil*	0	0	0	0
0.077	*nil*	0	0	0	0
0.154	*nil*	1	0	0	0
0.231	*nil*	1.2	0	0	0
0.308	*nil*	4.8	0	0	0
0.385	*nil*	14.8	0.2	0	0
0.462	*nil*	29	2.6	0.2	0
0.538	*nil*	**35.6**	**9.6**	**1.6**	0
0.615	*nil*	28.6	8.2	1.2	0
0.692	*nil*	7.6	1.6	0.2	0
0.769	*nil*	0.8	0.2	0	0
0.846	*nil*	0	0	0	0
0.923	*nil*	0	0	0	0
1	*nil*	0	0	0	0

Table 6.42 CV-5; CombAGS—Combination of acc and GranSize; result of experiments for multiple concept-dependent granulation with use of kNN classifier; data set: **hepatitis**; r_{gran} = Granulation radius

r_{gran}	Layer1	Layer2	Layer3	Layer4	Layer5
CombAGS					
0	0.892	0.892	0.892	0.892	0.892
0.053	0.895	0.895	0.895	0.895	0.895
0.105	0.898	0.898	0.898	0.898	0.898
0.158	0.905	0.905	0.905	0.905	0.905
0.211	0.905	0.905	0.905	0.905	0.905
0.263	0.894	0.898	0.898	0.898	0.898
0.316	0.892	0.895	0.895	0.895	0.895
0.368	0.88	0.901	0.901	0.901	0.901
0.421	0.889	0.895	0.895	0.895	0.895
0.474	0.886	0.891	0.892	0.892	0.892
0.526	0.857	0.878	0.84	0.84	0.84
0.579	0.803	0.786	0.796	0.796	0.796
0.632	**0.742**	0.794	0.761	0.757	0.757
0.684	**0.675**	0.712	0.72	0.712	0.712
0.737	**0.589**	0.62	0.622	0.623	0.623
0.789	**0.517**	**0.534**	**0.536**	**0.536**	**0.536**
0.842	**0.478**	**0.489**	**0.489**	**0.489**	**0.489**
0.895	**0.457**	**0.46**	**0.46**	**0.46**	**0.46**
0.947	**0.466**	**0.466**	**0.466**	**0.466**	**0.466**
1	**0.445**	**0.445**	**0.445**	**0.445**	**0.445**
$GranSize_{l_{i-1}} - GranSize_{l_i}$					
0	nil	0	0	0	0
0.053	nil	0	0	0	0
0.105	nil	0	0	0	0
0.158	nil	0	0	0	0
0.211	nil	0	0	0	0
0.263	nil	1	0	0	0
0.316	nil	0.8	0	0	0
0.368	nil	5.2	0	0	0
0.421	nil	4.8	0	0	0
0.474	nil	10.2	0.2	0	0

(continued)

Table 6.42 (continued)

r_{gran}	Layer1	Layer2	Layer3	Layer4	Layer5
0.526	nil	15.4	2	0	0
0.579	nil	21.6	3.2	0	0
0.632	nil	**22.6**	**4.6**	**0.6**	0
0.684	nil	12.4	2	0.4	0
0.737	nil	9.4	1.2	0.2	0
0.789	nil	4.2	0.6	0	0
0.842	nil	2.8	0	0	0
0.895	nil	0.8	0	0	0
0.947	nil	0	0	0	0
1	nil	0	0	0	0

Table 6.43 CV-5; CombAGS—Combination of acc and GranSize; result of experiments for multiple concept-dependent granulation with use of kNN classifier; Ddata set: **congressional voting records**; r_{gran} = Granulation radius

r_{gran}	Layer1	Layer2	Layer3	Layer4	Layer5
CombAGS					
0	0.946	0.946	0.946	0.946	0.946
0.063	0.946	0.946	0.946	0.946	0.946
0.125	0.945	0.946	0.946	0.946	0.946
0.188	0.945	0.946	0.946	0.946	0.946
0.25	0.94	0.942	0.942	0.942	0.942
0.313	0.943	0.946	0.946	0.946	0.946
0.375	0.941	0.944	0.944	0.944	0.944
0.438	0.941	0.945	0.945	0.945	0.945
0.5	0.936	**0.971**	**0.971**	**0.971**	**0.971**
0.563	0.938	0.953	0.953	0.953	0.953
0.625	**0.942**	**0.964**	**0.964**	**0.964**	**0.964**
0.688	**0.923**	0.949	0.947	0.947	0.947
0.75	**0.902**	0.881	0.933	0.927	0.927
0.813	**0.849**	0.91	0.905	0.911	0.911
0.875	**0.779**	0.825	0.775	0.775	0.775
0.938	**0.631**	**0.703**	0.704	0.706	0.706
1	**0.469**	**0.469**	**0.469**	**0.469**	**0.469**

(continued)

Table 6.43 (continued)

r_{gran}	Layer1	Layer2	Layer3	Layer4	Layer5
$GranSize_{l_{i-1}} - GranSize_{l_i}$					
0	nil	0	0	0	0
0.063	nil	0	0	0	0
0.125	nil	0.4	0	0	0
0.188	nil	0.2	0	0	0
0.25	nil	0.8	0	0	0
0.313	nil	1.4	0	0	0
0.375	nil	2	0	0	0
0.438	nil	4.8	0	0	0
0.5	nil	3.8	0	0	0
0.563	nil	5.8	0.6	0	0
0.625	nil	9	0	0	0
0.688	nil	14	0.6	0	0
0.75	nil	22.8	2.2	0.6	0
0.813	nil	31.8	4.4	0.8	0
0.875	nil	29.6	4.2	0.6	0
0.938	nil	**32.4**	**5**	**1**	0
1	nil	0	0	0	0

Table 6.44 CV-5; CombAGS—Combination of acc and GranSize; result of experiments for multiple concept-dependent granulation with use of kNN classifier; data set: **mushroom**; r_{gran} = Granulation radius

r_{gran}	Layer1	Layer2	Layer3	Layer4	Layer5
CombAGS					
0	0.944	0.944	0.944	0.944	0.944
0.045	0.943	0.943	0.943	0.943	0.943
0.091	0.943	0.943	0.943	0.943	0.943
0.136	0.943	0.943	0.943	0.943	0.943
0.182	0.944	0.944	0.944	0.944	0.944
0.227	0.943	0.943	0.943	0.943	0.943
0.273	0.944	0.944	0.944	0.944	0.944
0.318	0.944	0.944	0.944	0.944	0.944
0.364	0.943	0.943	0.943	0.943	0.943

(continued)

Table 6.44 (continued)

r_{gran}	Layer1	Layer2	Layer3	Layer4	Layer5
0.409	0.941	0.942	0.942	0.942	0.942
0.455	0.944	0.945	0.945	0.945	0.945
0.5	0.956	0.947	0.948	0.948	0.948
0.545	0.971	0.951	0.948	0.948	0.948
0.591	0.98	0.943	0.916	0.916	0.916
0.636	0.988	0.952	0.929	0.929	0.929
0.682	**0.994**	0.972	0.943	0.942	0.942
0.727	**0.994**	0.987	0.987	0.975	0.975
0.773	**0.994**	**0.994**	0.993	0.993	0.993
0.818	**0.993**	**0.995**	**0.995**	**0.994**	**0.994**
0.864	**0.985**	**0.994**	**0.995**	**0.995**	**0.995**
0.909	**0.961**	**0.983**	**0.989**	**0.99**	**0.99**
0.955	**0.864**	**0.927**	**0.948**	**0.953**	**0.953**
1	**0.5**	**0.5**	**0.5**	**0.5**	**0.5**
$GranSize_{l_{i-1}} - GranSize_{l_i}$					
0	nil	0	0	0	0
0.045	nil	0	0	0	0
0.091	nil	0	0	0	0
0.136	nil	0	0	0	0
0.182	nil	0	0	0	0
0.227	nil	0	0	0	0
0.273	nil	0.8	0	0	0
0.318	nil	2	0	0	0
0.364	nil	3.4	0	0	0
0.409	nil	7.4	0	0	0
0.455	nil	13.6	0	0	0
0.5	nil	17.4	0.6	0	0
0.545	nil	29	1.2	0	0
0.591	nil	32.2	4.2	0	0
0.636	nil	33	3.4	0.4	0
0.682	nil	28	3.6	0.6	0
0.727	nil	26.4	2.6	0.2	0
0.773	nil	31	5	0.2	0
0.818	nil	52.6	11.6	2.2	0
0.864	nil	117.6	31.8	6.2	0.2
0.909	nil	278.2	79.2	18.8	1.8
0.955	nil	**815.2**	**266.4**	**65.6**	**6.8**
1	nil	0	0	0	0

Table 6.45 CV-5; CombAGS—Combination of acc and GranSize; result of experiments for multiple concept-dependent granulation with use of kNN classifier; data set: **nursery**; r_{gran} = Granulation radius

r_{gran}	Layer1	Layer2	Layer3	Layer4	Layer5
CombAGS					
0	0.685	0.685	0.685	0.685	0.685
0.125	0.714	0.709	0.709	0.709	0.709
0.25	0.762	0.72	0.72	0.72	0.72
0.375	0.777	0.738	0.715	0.715	0.715
0.5	**0.785**	0.762	0.737	0.727	0.727
0.625	0.738	0.726	0.72	0.714	0.713
0.75	**0.8**	**0.808**	**0.795**	**0.79**	**0.788**
0.875	**0.733**	**0.79**	**0.796**	**0.796**	**0.796**
1	**0.289**	**0.289**	**0.289**	**0.289**	**0.289**
$GranSize_{l_{i-1}} - GranSize_{l_i}$					
0	nil	0	0	0	0
0.125	nil	2.4	0	0	0
0.25	nil	8.8	0	0	0
0.375	nil	17.2	2.8	0.2	0
0.5	nil	37.6	9.4	2	0
0.625	nil	97.8	26.6	7.6	1
0.75	nil	330.8	90	26.2	3.2
0.875	nil	**1151.4**	**369**	**94.4**	**14.4**
1	nil	0	0	0	0

Table 6.46 CV-5; CombAGS—Combination of acc and GranSize; result of experiments for multiple concept-dependent granulation with use of kNN classifier; data set: **soybean**; r_{gran} = Granulation radius

r_{gran}	Layer1	Layer2	Layer3	Layer4	Layer5
0	0.813	0.813	0.813	0.813	0.813
⋮	⋮	⋮	⋮	⋮	⋮
0.4	0.823	0.823	0.823	0.823	0.823
0.429	0.811	0.811	0.811	0.811	0.811
0.457	0.826	0.826	0.826	0.826	0.826
0.486	0.822	0.822	0.822	0.822	0.822
0.514	0.825	0.823	0.823	0.823	0.823
0.543	0.832	0.833	0.833	0.833	0.833
0.571	0.825	0.822	0.822	0.822	0.822
0.6	0.826	0.818	0.818	0.818	0.818

(continued)

Table 6.46 (continued)

r_{gran}	Layer1	Layer2	Layer3	Layer4	Layer5
0.629	0.828	0.826	0.826	0.826	0.826
0.657	0.825	0.828	0.828	0.828	0.828
0.686	0.832	0.822	0.82	0.82	0.82
0.714	0.833	0.824	0.823	0.823	0.823
0.743	0.852	0.839	0.824	0.824	0.824
0.771	0.85	0.85	0.84	0.84	0.84
0.8	0.85	0.86	0.855	0.855	0.855
0.829	0.853	0.88	0.872	0.872	0.872
0.857	0.815	0.858	0.852	0.852	0.852
0.886	0.774	0.814	0.824	0.824	0.824
0.914	0.726	0.759	0.763	0.764	0.764
0.943	0.655	0.693	0.694	0.694	0.694
0.971	0.562	0.584	0.587	0.587	0.587
1	**0.464**	**0.464**	**0.464**	**0.464**	**0.464**
$GranSize_{l_{i-1}} - GranSize_{l_i}$					
0	nil	0	0	0	0
⋮	⋮	⋮	⋮	⋮	⋮
0.4	nil	0	0	0	0
0.429	nil	0.4	0	0	0
0.457	nil	0.2	0	0	0
0.486	nil	0.2	0	0	0
0.514	nil	0.8	0	0	0
0.543	nil	0.6	0	0	0
0.571	nil	2.6	0	0	0
0.6	nil	4.8	0	0	0
0.629	nil	6.2	0.2	0	0
0.657	nil	11.6	0	0	0
0.686	nil	15.8	0.2	0	0
0.714	nil	19.6	1.2	0	0
0.743	nil	21.2	0.6	0.2	0
0.771	nil	28.8	2	0.2	0
0.8	nil	33.4	2.2	0	0
0.829	nil	35.8	2.4	0	0
0.857	nil	47	6.8	0	0
0.886	nil	**50.6**	**8.4**	0.2	0
0.914	nil	49.6	8	**0.6**	0
0.943	nil	45.8	7.2	0.4	0
0.971	nil	23	1.8	0	0
1	nil	0	0	0	0

Table 6.47 CV-5; CombAGS—Combination of acc and GranSize; result of experiments for multiple concept-dependent granulation with use of kNN classifier; data set: **SPECT**; r_{gran} = Granulation radius

r_{gran}	Layer1	Layer2	Layer3	Layer4	Layer5
CombAGS					
0	0.759	0.759	0.759	0.759	0.759
0.045	0.751	0.751	0.751	0.751	0.751
0.091	0.754	0.755	0.755	0.755	0.755
0.136	0.742	0.744	0.744	0.744	0.744
0.182	0.756	0.756	0.756	0.756	0.756
0.227	0.775	0.753	0.753	0.753	0.753
0.273	0.806	0.768	0.768	0.768	0.768
0.318	0.78	0.779	0.779	0.779	0.779
0.364	0.794	0.725	0.725	0.725	0.725
0.409	0.803	0.789	0.789	0.789	0.789
0.455	0.853	0.772	0.772	0.772	0.772
0.5	0.794	0.742	0.741	0.741	0.741
0.545	0.852	0.808	0.803	0.803	0.803
0.591	0.857	0.825	0.803	0.803	0.803
0.636	0.85	0.841	0.769	0.769	0.769
0.682	**0.839**	0.839	0.818	0.816	0.816
0.727	**0.817**	0.863	0.784	0.785	0.785
0.773	**0.767**	**0.836**	0.833	0.827	0.833
0.818	**0.7**	**0.779**	**0.796**	0.791	0.791
0.864	**0.621**	**0.698**	**0.707**	**0.714**	**0.714**
0.909	**0.542**	**0.589**	**0.596**	**0.596**	**0.596**
0.955	**0.456**	**0.474**	**0.476**	**0.476**	**0.476**
1	**0.397**	**0.397**	**0.397**	**0.397**	**0.397**
$GranSize_{l_{i-1}} - GranSize_{l_i}$					
0	*nil*	0	0	0	0
0.045	*nil*	0	0	0	0
0.091	*nil*	0.2	0	0	0
0.136	*nil*	0.4	0	0	0
0.182	*nil*	0	0	0	0
0.227	*nil*	1	0	0	0
0.273	*nil*	1.2	0	0	0
0.318	*nil*	1.2	0	0	0
0.364	*nil*	2.8	0	0	0
0.409	*nil*	2.4	0	0	0
0.455	*nil*	5.2	0	0	0
0.5	*nil*	3.6	0.2	0	0

(continued)

Table 6.47 (continued)

r_{gran}	Layer1	Layer2	Layer3	Layer4	Layer5
0.545	nil	6.6	0.2	0	0
0.591	nil	8.4	1	0	0
0.636	nil	10.2	2.2	0	0
0.682	nil	15.2	1.8	0.2	0
0.727	nil	22.4	4.4	0.2	0
0.773	nil	25.8	5.2	0.6	**0.2**
0.818	nil	**29**	**6.2**	1	0
0.864	nil	28.4	4.2	**1.2**	0
0.909	nil	17.4	2.4	0.2	0
0.955	nil	6.8	0.6	0	0
1	nil	0	0	0	0

Table 6.48 CV-5; CombAGS—Combination of acc and GranSize; result of experiments for multiple concept-dependent granulation with use of k-NN classifier; data set: **SPECTF**; r_{gran} = Granulation radius

r_{gran}	Layer1	Layer2	Layer3	Layer4	Layer5
CombAGS					
0	0.731	0.731	0.731	0.731	0.731
0.023	0.762	0.709	0.709	0.709	0.709
0.045	0.795	0.718	0.718	0.718	0.718
0.068	0.813	0.77	0.727	0.727	0.727
0.091	**0.77**	0.812	0.78	0.775	0.775
0.114	**0.702**	0.785	0.799	0.798	0.798
0.136	**0.595**	**0.711**	**0.753**	**0.761**	0.748
0.159	**0.508**	**0.561**	**0.574**	**0.575**	**0.576**
0.182	**0.447**	**0.453**	**0.454**	**0.454**	**0.454**
0.205	**0.399**	**0.404**	**0.404**	**0.404**	**0.404**
0.227	**0.394**	**0.394**	**0.394**	**0.394**	**0.394**
0.25	**0.399**	**0.399**	**0.399**	**0.399**	**0.399**
⋮	⋮	⋮	⋮	⋮	⋮
1	**0.39**	**0.39**	**0.39**	**0.39**	**0.39**
$GranSize_{l_{i-1}} - GranSize_{l_i}$					
0	nil	0	0	0	0
0.023	nil	4.4	0	0	0
0.045	nil	11.2	0	0	0
0.068	nil	22.4	1.8	0	0
0.091	nil	35.6	7.2	0.4	0

(continued)

Table 6.48 (continued)

r_{gran}	Layer1	Layer2	Layer3	Layer4	Layer5
0.114	nil	**46.2**	13.4	1.4	0
0.136	nil	**46.2**	**17.2**	**5.6**	**1.2**
0.159	nil	19	5.2	0.8	0.2
0.182	nil	3.2	0.4	0	0
0.205	nil	1.4	0	0	0
0.227	nil	0	0	0	0
0.25	nil	0	0	0	0
⋮	⋮	⋮	⋮	⋮	⋮
1	nil	0	0	0	0

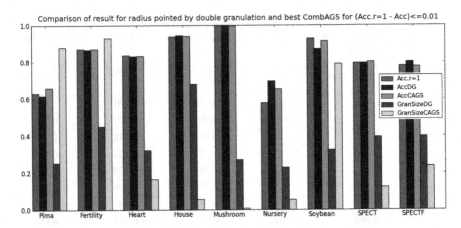

Fig. 6.15 The result comparison, for radius pointed by double granulation and best *CombAGS*. *AccDG* is the accuracy for radius pointed by double granulation, *AccCAGS* is accuracy for radius pointed by *CombAGS*, *GranSizeDG* is the percentage size of data set for radius pointed by double granulation, and *GranSizeCAGS* by best *CombAGS*

6.3.3 A Comparison for Accuracy Error $acc_{r=1} - acc \leq 0.01$ of CombAGS and $GranSize_{l_{i-1}} - GranSize_{l_i}$

We conclude the chapter with a summary table underlying our rule of thumb for the optimal radius of granulation.

Table 6.49 A summary of data concerning the heuristics for the optimal granulation radius

data set	$r(1,2)$	$\Delta(CombAGS)$	$\delta(r(1,2))$	$\Delta(1nbhd)$	$\Delta(2nbhd)$
adult	0.786	0	0	–	–
australian	0.571	0	0	–	–
car evaluation	0.833	0	0	–	–
diabetes	0.25	0.089	1	0	–
fertility diagnosis	0.6674	0	0	–	–
german credit	0.65	0.129	4	0.121	0.046
heart disease	0.538	0.115	3	0.048	0.009
hepatitis	0.579	0.145	4	0.061	0.009
congr. voting	0.813	0.060	5	0.044	0.024
mushroom	0.955	0.042	2	0.005	0
nursery	0.875	0	0	–	–
soybean	0.886	0.048	2	0.020	0
SPECT	0.818	0.042	1	0	–
SPECTF	0.114	0	0	–	–

In the table above, we show, for each data set, the radius of granulation which gives the maximal drop in size between the first and second granulation layers, within 5 % interval, $r(1, 2)$, the difference $\Delta(CombAGS)$ between the maximal value of $CombAGS$ after the 5th layer and the value of $CombAGS$ at $r(1, 2)$ after the 5th layer. We consider also the grid topology on the interval $[0, 1]$ with values of granulation radii as grid nodes, and $\delta(r(1, 2))$ denotes the grid distance from $r(1, 2)$ to the granulation radius which yields the maximal value of $CombAGS$. In the last two columns we give the minimal values of the difference between the maximal value of $CombAGS$ and values over the 1-neighborhood of $r(1, 2)$ and over the 2-neighborhood of $r(1, 2)$ denoted, respectively, as $\Delta(1nbhd)$, $\Delta(2nbhd)$ in the cases where the maximal value is not reached at $r(1, 2)$. Save for data sets **german-credit, heart disease, hepatitis, congressional voting records,** for which the grid distance from the maximal drop in size between the first and the second layers to the maximal value of $CombAGS$ is between 3 and 5, for the remaining 10 of data sets this distance is at most 2, and for 6 of them the heuristics points to exactly the optimal radius

References

1. Artiemjew, P.: Classifiers from granulated data sets: concept dependent and layered granulation. In: Proceedings RSKD'07. The Workshops at ECML/PKDD'07, pp. 1–9. Warsaw University Press, Warsaw (2007)
2. Artiemjew, P.: Rough mereological classifiers obtained from weak rough set inclusions. In: Proceedings of International Conference on Rough Set and Knowledge Technology RSKT'08, Chengdu China. Lecture Notes in Artificial Intelligence, vol. 5009, pp. 229–236. Springer, Berlin (2008)
3. Artiemjew, P.: On classification of data by means of rough mereological granules of objects and rules. In: Proceedings of International Conference on Rough Set and Knowledge Technology RSKT'08, Chengdu China. Lecture Notes in Artificial Intelligence, vol. 5009, pp. 221–228. Springer, Berlin (2008)
4. Artiemjew, P.: Natural versus granular computing. Classifiers from granular structures. In: Proceedings RSCTC 2008. Lecture Notes in Artificial Intelligence, vol. 5306, pp. 150–159. Springer, Berlin (2008)
5. Artiemjew, P.: On strategies for knowledge granulation and applications to decision systems. Ph.D. Dissertation, Polish Japanese Institute of Information Technology, Polkowski, L. supervisor, Warszawa (2009)
6. Pawlak, Z.: Rough sets. Int. J. Comput. Inf. Sci. **11**, 341–356 (1982)
7. Polkowski, L.: Formal granular calculi based on rough inclusions. In: Proceedings 2006 IEEE International Conference on Granular Computing GrC'06, pp. 57–62. IEEE Press (2006)
8. Polkowski, L.: The paradigm of granular rough computing. In: Proceedings ICCI'07. 6th IEEE International Conference on Cognitive Informatics, pp. 145–163. IEEE Computer Society, Los Alamitos (2007)
9. Polkowski, L.: A unified approach to granulation of knowledge and granular computing based on rough mereology: a survey. In: Pedrycz, W., Skowron, A., Kreinovich, V. (eds.) Handbook of Granular Computing, pp. 375–401. Wiley, New York (2008)
10. Polkowski, L., Artiemjew, P.: On classifying mappings induced by granular structures. Transactions on Rough Sets IX. A subseries of Lecture Notes in Computer Science, vol. 5390, pp. 264–286. Springer, Berlin (2008)
11. Polkowski, L., Artiemjew, P.: A study in granular computing: on classifiers induced from granular reflections of data. Transactions on Rough Sets IX. A subseries of Lecture Notes in Computer Science, vol. 5390, pp. 230–263. Springer, Berlin (2008)
12. Polkowski, L.: Granulation of knowledge: similarity based approach in information and decision systems. In: Meyers, R. A. (ed.) Encyclopedia of Complexity and System Sciences. Springer, Berlin, article 00788 (2009)
13. Quinlan, J.R.: C4.5: Programs for Machine Learning. Morgan Kaufmann Publishers, Burlington (1993)
14. Stone, P.: Layered learning in Multiagent Systems. A Winning Approach to Robotic Soccer. MIT Press, Cambridge (2000)
15. Stone, P., Veloso, M.: Layered learning. In: Proceedings ECML 2000. Lecture Notes in Artificial Intelligence, vol. 1810, pp. 369–381. Springer, Berlin (2000)
16. UCI Repository; http://www.ics.uci.edu/~mlearn/databases/. Accessed 6 Apr 2014

Chapter 7
Naive Bayes Classifier on Granular Reflections: The Case of Concept-Dependent Granulation

Given the number of times on which an unknown event hath hap-pende and failed: Required *the chance that the probability of its happening in a single trial lies somewhere between any two degrees of probability that can be named*
[Rev. Thomas Bayes. Philosophical Transactions RSL 53 (1763).]

In this chapter, we study the performance of Bayes classifier on granular structures. From results in Table 7.18, we infer that Bayes classifier is well suited for granular cases as for most tested data its accuracy in best cases is better in granulated cases then in non-granulated ones.

7.1 Naive Bayes Classifier

The Bayes classifier, for a general perspective, cf., Mitchell [7], Devroye et al. [4], Duda and Hart [5], or Bishop [2], for monographic expositions, and Langley et al. [6], and, Rish et al. [9] for analysis of classifier performance versus data structure, was introduced in Chap. 3. Its study in rough set framework was, given, e.g., in Pawlak [8], Al-Aidaroos et al. [1], Cheng et al. [3], Su et al. [10], Wang et al. [12, 13], Yao and Zhou [14], Zhang et al. [15].

Naive Bayes classifier owes its naivety epithet to the fact that one assumes the independence of attributes which condition in reality is not often met. Its working in the realm of decision systems can be described concisely as follows. For a given training decision system (U_{trn}, A, d) and a test system (U_{tst}, A, d), where $U = U_{trn} \cup U_{tst}$ is the set of objects, $A = \{a_1, a_2, \ldots, a_n\}$ is the conditional attribute set, and $d \in D = \{d_1, d_2, \ldots, d_k\}$ is the decision attribute.

The classification of a test object $v \in U_{tst}$ described by means of its information set $(a_1(v), a_2(v), \ldots, a_n(v))$ consists of computing for all decision classes the value of the parameter

$$P(d = d_i | b_1 = a_1(v), b_2 = a_2(v), \ldots, b_n = a_n(v))$$

and the decision on v is the decision value with the maximal value of the parameter.

© Springer International Publishing Switzerland 2015
L. Polkowski and P. Artiemjew, *Granular Computing in Decision Approximation*,
Intelligent Systems Reference Library 77, DOI 10.1007/978-3-319-12880-1_7

The Bayes theorem along with the frequency interpretation of probability allows to express this probability as

$$\frac{P(b_1 = a_1(v), b_2 = a_2(v), \ldots, b_n = a_n(v)|d = d_i) * P(d = d_i)}{P(b_1 = a_1(v), b_2 = a_2(v), \ldots, b_n = a_n(v))} \tag{7.1}$$

One usually dispenses with the denominator of Eq. (7.1), because it is constant for all decision classes. Assuming independence of attributes, the numerator of Eq. (7.1) can be computed as

$$P(d = d_i) * \prod_{m=1}^{n} P(b_m = a_m(v)|d = d_i).$$

In practice, we can use partial estimation

$$P(b_m = a_m(v)|d = d_i) = \frac{\text{number of test instances } b_m = a_m(v) \text{ in training class } d_i}{\text{cardinality of class } d_i}.$$

Each decision class is voting by submitting the value of the parameter

$$Param_{d=d_i} = P(d = d_i) * \prod_{m=1}^{n} P(b_m = a_m(v)|d = d_i). \tag{7.2}$$

In this approach, we could encounter a problem of zero frequency of a descriptor $b_m = a_m(v)$ in a class d_i, i.e., $P(b_m = a_m(v)|d = d_i) = 0$. One of the methods to avoid the problem of zero-valued decisive parameters, is to search among the remaining classes for the smallest non–zero value of $P(b_m = a_m(v)|d = d_j)$. The found value is additionally slightly lowered, and assigned instead of the zero value. In case of more than one class with the zero frequency of the descriptor $b_m = a_m(v)$, we could assign such reduced value to all of them. Another method to avoid this problem is to consider the remaining decision classes, which contain the value $b_m = a_m(v)$. In case of the zero frequency of the descriptor $b_m = a_m(v)$ in all training classes, this descriptor can be disregarded. In order to help ourselves with the task of computing with small numbers, we can use logarithms of probabilities. In practice, it is also acceptable to use sums instead of products in which case decision classes vote by the parameter

$$Param_{d=d_i} = P(d = d_i) * \sum_{m=1}^{n} P(b_m = a_m(v)|d = d_i). \tag{7.3}$$

This classifier is fit for symbolic attributes. In case of numerical data, assuming the normal distribution, the probability $P(b_m = a_m(v)|d = d_i)$ can be estimated based on the Gaussian function

$$f(x) = \frac{1}{\sqrt{(2 * \pi * \sigma_c^2)}} * e^{\frac{-(x-\mu_c)^2}{2*\sigma_c^2}}.$$

To compute this value, the estimates of mean values and variances in decision classes are necessary:

$$\mu_c = \frac{\sum_{i=1}^{cardinality\ of\ class\ c} a(v_i)}{cardinality\ of\ class\ c},$$

$$\sigma_c^2 = \frac{1}{cardinality\ of\ class\ c} * \sum_{i=1}^{cardinality\ of\ class\ c} (a(v_i) - \mu_c)^2.$$

7.1.1 An Example of Bayes Classification

In this section we show an exemplary classification on the lines of Eq. (7.2). The test decision system is given in Table 7.1 and the training system is given in Table 7.2.

We have $P(c = 2) = \frac{3}{6} = \frac{1}{2}, P(c = 4) = \frac{1}{2}$.

We start with classification of the test object x_1 whose information set is $(2, 4, 2, 1)$ and the decision $c = 4$.

According to the formula (7.2), we obtain
$P(a_1 = 2|c = 2) = \frac{1}{3}$.
$P(a_2 = 4|c = 2) = \frac{0}{3}$ we cannot handle it, there is no descriptor $a_2 = 4$ in all classes. Next,

$P(a_3 = 2|c = 2) = \frac{1}{3}$.
$P(a_4 = 1|c = 2) = \frac{3}{3}$.

Finally, $Param_{c=2} = \frac{1}{2} * (\frac{1}{3} + \frac{0}{3} + \frac{1}{3} + \frac{3}{3}) = \frac{5}{6}$.

Table 7.1 Test system (X, A, c)

	a_1	a_2	a_3	a_4	c
x_1	2	4	2	1	4
x_2	1	2	1	1	2
x_3	9	7	10	7	4
x_4	4	4	10	10	2

Table 7.2 Training system (Y, A, c)

	a_1	a_2	a_3	a_4	c
y_1	1	3	1	1	2
y_2	10	3	2	1	2
y_3	2	3	1	1	2
y_4	10	9	7	1	4
y_5	3	5	2	2	4
y_6	2	3	1	1	4

Continuing, we obtain
$P(a_1 = 2|c = 4) = \frac{1}{3}$.
$P(a_2 = 4|c = 4) = \frac{0}{3}$ we cannot handle it, there is no descriptor $a_2 = 4$ in all classes.
$P(a_3 = 2|c = 4) = \frac{1}{3}$.
$P(a_4 = 1|c = 4) = \frac{2}{3}$.
Finally, $Param_{c=4} = \frac{1}{2} * (\frac{1}{3} + \frac{0}{3} + \frac{1}{3} + \frac{2}{3}) = \frac{2}{3}$.
As $Param_{c=2} > Param_{c=4}$, the object x_1 is assigned the decision value of 2. This decision is inconsistent with the expert decision, this object is incorrectly classified.

For the second test object x_2, with the information set $(1, 2, 1, 1)$ and the decision value of 4, we obtain in the analogous manner:

$P(a_1 = 1|c = 2) = \frac{1}{3}$, we increase counter by 1 because $P(a_1|c = 4) = 0$ so finally
$P(a_1 = 1|c = 2) = \frac{2}{3}$.
$P(a_2 = 2|c = 2) = \frac{0}{3}$ we cannot handle it because the descriptor $a_2 = 2$ is missing in all classes.
$P(a_3 = 1|c = 2) = \frac{1}{3}$.
$P(a_4 = 1|c = 2) = \frac{3}{3}$,
so $Param_{c=2} = \frac{1}{2} * (\frac{2}{3} + \frac{0}{3} + \frac{1}{3} + \frac{3}{3}) = 1$.

$P(a_1 = 1|c = 4) = \frac{0}{3}$, in this case we have to increase counter of $P(a_1 = 1|c = 2)$ by one to account for the class, which contains at least one count of the descriptor $a_1 = 1$.

$P(a_2 = 2|c = 4) = \frac{0}{3}$, we cannot handle it, $a_2 = 2$ is missing in all classes.
$P(a_3 = 1|c = 4) = \frac{1}{3}$.
$P(a_4 = 1|c = 4) = \frac{2}{3}$, so, finally, $Param_{c=2} = \frac{1}{2} * (\frac{0}{3} + \frac{0}{3} + \frac{1}{3} + \frac{2}{3}) = \frac{1}{2}$.
As $Param_{c=2} > Param_{c=4}$, the object x_2 is assigned the decision value of 2; this decision is consistent with the expert decision so the object is correctly classified.

The next test object is x_3 with the information set $(9, 7, 10, 7, 4)$.
We have $Param_{c=2} = P(c = 2) * \sum_{i=1}^{4} P(a_i = v_i|c = 2)$, and,

$P(a_1 = 9|c = 2) = \frac{0}{3}$.
$P(a_2 = 7|c = 2) = \frac{0}{3}$.
$P(a_3 = 10|c = 2) = \frac{0}{3}$.
$P(a_4 = 7|c = 2) = \frac{0}{3}$,

so, finally, $Param_{c=2} = \frac{1}{2} * (\frac{0}{3} + \frac{0}{3} + \frac{0}{3} + \frac{0}{3}) = 0$.
Also, for $Param_{c=4} = P(c = 4) * \sum_{i=1}^{4} P(a_i = v_i|c = 4)$, we have
$P(a_1 = 9|c = 4) = \frac{0}{3}$.
$P(a_2 = 7|c = 4) = \frac{0}{3}$.

$P(a_3 = 10|c = 4) = \frac{0}{3}$.
$P(a_4 = 7|c = 4) = \frac{0}{3}$,

and, finally, $Param_{c=2} = \frac{1}{2} * (\frac{0}{3} + \frac{0}{3} + \frac{0}{3} + \frac{0}{3}) = 0$.
As $Param_{c=2} == Param_{c=4}$, the random decision $random(2, 4) = 4$ is assigned to x_3, so the object is correctly classified.

For the last test object, x_4 with the information set $(4, 4, 10, 10, 4)$, we compute

$Param_{c=2} = P(c = 2) * \sum_{i=1}^{4} P(a_i = v_i|c = 2)$:
$P(a_1 = 4|c = 2) = \frac{0}{3}$.
$P(a_2 = 4|c = 2) = \frac{0}{3}$.
$P(a_3 = 10|c = 2) = \frac{0}{3}$.
$P(a_4 = 10|c = 2) = \frac{0}{3}$,

and, finally, $Param_{c=2} = \frac{1}{2} * (\frac{0}{3} + \frac{0}{3} + \frac{0}{3} + \frac{0}{3}) = 0$.

For $Param_{c=4} = P(c = 4) * \sum_{i=1}^{4} P(a_i = v_i|c = 4)$, we need:
$P(a_1 = 4|c = 4) = \frac{0}{3}$.
$P(a_2 = 4|c = 4) = \frac{0}{3}$.
$P(a_3 = 10|c = 4) = \frac{0}{3}$.
$P(a_4 = 10|c = 4) = \frac{0}{3}$,

hence, $Param_{c=2} = \frac{1}{2} * (\frac{0}{3} + \frac{0}{3} + \frac{0}{3} + \frac{0}{3}) = 0$.

A random decision assignment $random(2, 4) = 4$ causes x_4 to be incorrectly classified.
We now compute parameters:

$$Global\_Accuracy = \frac{\text{number of test objects correctly classified in whole test system}}{\text{number of classified objects in whole test system}};$$

$$Balanced\_Accuracy = \frac{\sum_{i=1}^{number\ of\ classes} \frac{\text{number of test objects correctly classified in class } c_i}{\text{number of objects classified in class } c_i}}{number\ of\ classes}.$$

In our exemplary case, these values are (Table 7.3)

$$Global\_Accuracy = \frac{2}{4} = \frac{1}{2};$$

$$Balanced\_Accuracy = \frac{\frac{1}{2} + \frac{1}{2}}{2} = \frac{1}{2}.$$

Table 7.3 Consistency of
Bayes classifier on test data

Tst obj	Expert decision	Decision of our classifier
x_1	4	2
x_2	2	2
x_3	4	4
x_4	2	4

7.2 Results of an Experimental Session with Real Data

In this experimental session, we perform five times CV-5 validation, where for each
training part we apply concept-dependent granulation. In granulation, we apply random covering method.

7.2.1 Examined Variants of Bayes Classifier

We consider four variants of classification:

1. $Param^{V1}_{d=d_i} = \sum_{m=1}^{n} P(b_m = a_m(v)|d = d_i)$.
2. $Param^{V2}_{d=d_i} = P(d = d_i) * \sum_{m=1}^{n} P(b_m = a_m(v)|d = d_i)$.
3. $Param^{V3}_{d=d_i} = \prod_{m=1}^{n} P(b_m = a_m(v)|d = d_i)$.
4. $Param^{V4}_{d=d_i} = P(d = d_i) * \prod_{m=1}^{n} P(b_m = a_m(v)|d = d_i)$.

7.2.2 Evaluation of Results

In our experiments, we compute bias of accuracy on the basis of 5 times CV-5 result:

$$AccBias = \frac{\sum_{i=1}^{5}(max(acc_1^{CV5}, acc_2^{CV5}, \ldots, acc_5^{CV5}) - acc_i^{CV5})}{5}, \qquad (7.4)$$

where

$$Acc = \frac{\sum_{i=1}^{5} acc_i^{CV5}}{5}.$$

7.2.3 A Discussion of Results

A summary of best results for particular data sets is as follows.

For Adult data set variant V2 works best, for radius 0.429 we get $Acc = 0.788$,
$AccBias = 0.013$ with reduction in training object number in range of 99.73 %. In
this case, relative to non granulated data, we have gain in accuracy of 0.027.

For Australian Credit data set variant V1 works best, for radius 0.429 we get $Acc = 0.853, AccBias = 0.006$ with reduction in training object number in range of 94.05 %. In this case, relative to non granulated data, we have gain in accuracy of 0.01.

For Car Evaluation data set, variant V2 works best, for radius 0.333 we get $Acc = 0.725, AccBias = 0.013$ with reduction in training object number in range of 98.74 %. In this case, relative to non granulated data, we have gain in accuracy of 0.025.

For Pima Indians Diabetes data set, variant V2 works best, for radius 0.25 we get $Acc = 0.661, AccBias = 0.006$ with reduction in training object number in range of 74.86 %. In this case, relative to non granulated data, we have gain in accuracy of 0.003.

For Fertility Diagnosis data set, variant V2 works best, for radius 0.556 we get $Acc = 0.888, AccBias = 0$ with reduction in training object number in range of 76.3 %. In this case, relative to non granulated data, error in accuracy is equal 0.

For German Credit data set, variant V2 works best, for radius 0.5 we get $Acc = 0.703, AccBias = 0.006$ with reduction in training object number in range of 92.99 %. In this case, relative to non granulated data, we have gain in accuracy 0.003.

For Heart Disease data set, variant V1 works best, for radius 0.538 we get $Acc = 0.841, AccBias = 0.007$ with reduction in training object number in range of 67.85 %. In this case, relative to non granulated data, we have gain in accuracy 0.012.

For Hepatitis data set, variant V4 works best, for radius 0.526 we get $Acc = 0.885, AccBias = 0.018$ with reduction in training object number in range of 85.39 %. In this case, relative to non granulated data, we have gain in accuracy 0.009.

For Congressional House Votes data set, variant V4 works best, for radius 0.875 we get $Acc = 0.941, AccBias = 0.004$ with reduction in training object number in range of 57.7 %. In this case, relative to non granulated data, we have gain in accuracy 0.014.

For Mushroom data set, variant V4 works best, for radius 0.818 we get $Acc = 0.955, AccBias = 0.002$ with reduction in training object number in range of 98.58 %. In this case, relative to non granulated data, we have gain in accuracy of 0.037.

For Nursery data set, variant V2 works best, for radius 0.875 we get $Acc = 0.854, AccBias = 0.009$ with reduction in training object number in range of 77.15 %. In this case, relative to non granulated data, we have error in accuracy of 0.015.

For Soybean Large data set, variant V2 works best, for radius 0.4 we get $Acc = 0.68, AccBias = 0.011$ with reduction in training object number in range of 96.27 %. In this case, relative to non granulated data, we have gain in accuracy of 0.383.

For SPECT data set, variant V2 works best, for radius 0.318 we get $Acc = 0.794, AccBias = 0$ with reduction in training object number in range of 98.2 %. In this case, relative to non granulated data, we have error in accuracy of 0 (Tables 7.4, 7.5, 7.6, 7.7, 7.8, 7.9, 7.10, 7.11, 7.12, 7.13, 7.14, 7.15, 7.16, 7.17 and 7.18).

For SPECTF data set, variant V2 works best, for radius 0.91 we get $Acc = 0.794, AccBias = 0$ with reduction in training object number in range of 76.37 %. In this case, relative to non granulated data, we have error in accuracy of 0.

Summarizing, in almost all cases there is gain in accuracy in comparison with the non granulated case, and a significant reduction in object numbers in training data sets. The best variants of classification turn out to be $V2$ and $V4$, For Adult, Car, Pima

Table 7.4 $5 \times$ CV-5; The result of experiments for four variants of Bayes classifier; data set *adult*; Concept dependent granulation; $r_{gran} =$ Granulation radius; nil = result for data without missing values; $Acc =$ Accuracy of classification; $AccBias =$ Accuracy bias defined based on Eq. (7.4); $GranSize =$ The size of data set after granulation in the fixed r

r_{gran}	Acc				AccBias				GranSize			
	v1	v2	v3	v4	v1	v2	v3	v4	v1	v2	v3	v4
0	0.676	0.689	0.681	0.669	0.021	0.029	0.038	0.03	2	2	2	2
0.0714286	0.689	0.691	0.692	0.671	0.029	0.02	0.025	0.034	2.12	2.08	2.12	2.12
0.142857	0.687	0.574	0.678	0.647	0.026	0.06	0.024	0.03	3.6	3.8	4.28	3.68
0.214286	0.683	0.652	0.69	0.657	0.015	0.038	0.01	0.054	8.56	8.12	8.32	8.44
0.285714	0.694	0.7	0.506	0.59	0.003	0.079	0.094	0.089	19.32	19.68	19.04	19.8
0.357143	0.703	0.776	0.463	0.456	0.004	0.019	0.053	0.098	44.2	42.88	44.08	45.68
0.428571	0.679	0.788	0.643	0.651	0.005	0.013	0.018	0.019	109.64	105.72	106.24	103.96
0.5	0.677	0.769	0.673	0.686	0.003	0.004	0.003	0.01	263.52	264.24	264.28	265.6
0.571429	0.697	0.761	0.691	0.711	0.001	0	0.014	0.005	646.44	646.84	655.32	653.72
0.642857	0.709	0.761	0.706	0.742	0.001	0	0.01	0.005	1,599.28	1,614.92	1,620.04	1,610.4
0.714286	0.716	0.761	0.743	0.785	0.001	0	0.003	0.003	3,964.2	3,961.08	3,952.96	3,960.08
0.785714	0.72	0.761	0.76	0.802	0.001	0	0.001	0.001	9,651.08	9,666.44	9,673.8	9,659.12
0.857143	0.724	0.761	0.743	0.783	0	0	0.001	0.001	2,0881.2	2,0877.6	2,0889.5	2,0883.9
0.928571	0.722	0.761	0.727	0.767	0	0	0.001	0.001	3,4233.6	3,4234.2	3,4229.6	3,4239.7
1	0.721	0.761	0.723	0.764	0	0	0	0.001	3,9039.2	3,9039.5	3,9039	3,9039.3

Table 7.5 $5 \times$ CV-5; The result of experiments for four variants of Bayes classifier; data set *Australian credit*; Concept dependent granulation; r_{gran} = Granulation radius; nil = result for data without missing values; *Acc* = Accuracy of classification; *AccBias* = Accuracy bias defined based on Eq. (7.4); *GranSize* = The size of data set after granulation in the fixed r

r_{gran}	Acc				AccBias				GranSize			
	V1	V2	V3	V4	V1	V2	V3	V4	V1	V2	V3	V4
0	0.792	0.788	0.81	0.792	0.022	0.007	0.018	0.018	2	2	2	2
0.0714286	0.789	0.703	0.813	0.788	0.025	0.084	0.006	0.009	2.32	2.32	2.52	2.4
0.142857	0.788	0.682	0.812	0.76	0.01	0.096	0.013	0.039	3.4	3.84	3.52	3.76
0.214286	0.789	0.707	0.79	0.759	0.012	0.09	0.03	0.045	5.2	5.4	5.16	5.32
0.285714	0.806	0.738	0.656	0.628	0.013	0.058	0.021	0.037	8.8	9.08	8.56	9.36
0.357143	0.827	0.727	0.692	0.707	0.014	0.034	0.023	0.086	16.64	15.16	16.32	16.12
0.428571	0.853	0.772	0.717	0.745	0.006	0.044	0.031	0.038	32.84	30.72	32.28	31.28
0.5	0.85	0.814	0.749	0.732	0.002	0.019	0.019	0.012	71.56	70.76	71	69.68
0.571429	0.852	0.77	0.725	0.721	0.008	0.034	0.007	0.017	157	158.36	157.16	155.92
0.642857	0.857	0.764	0.734	0.732	0.009	0.005	0.008	0.013	319	320.4	317.8	318.08
0.714286	0.843	0.83	0.732	0.737	0.005	0.003	0.012	0.018	468.56	468.44	467.88	468.28
0.785714	0.843	0.813	0.732	0.739	0.006	0.006	0.013	0.018	536.28	536.24	536	536.04
0.857143	0.843	0.799	0.73	0.739	0.005	0.014	0.011	0.018	547.36	547.16	547.16	547.28
0.928571	0.843	0.8	0.73	0.739	0.004	0.013	0.01	0.018	548.92	548.76	548.72	548.8
1	0.843	0.799	0.729	0.739	0.005	0.012	0.011	0.018	552	552	552	552

Table 7.6 5 × CV-5; The result of experiments for four variants of Bayes classifier; data set *car evaluation*; Concept dependent granulation; r_{gran} = Granulation radius; nil = result for data without missing values; *Acc* = Accuracy of classification; *AccBias* = Accuracy bias defined based on Eq. (7.4); *GranSize* = The size of data set after granulation in the fixed *r*

r_{gran}	Acc				AccBias				GranSize			
	V1	V2	V3	V4	V1	V2	V3	V4	V1	V2	V3	V4
0	0.407	0.423	0.442	0.447	0.02	0.017	0.025	0.023	4	4	4	4
0.166667	0.315	0.653	0.092	0.369	0.022	0.052	0.037	0.071	8.12	8.48	7.72	8.52
0.333333	0.357	0.723	0.044	0.118	0.039	0.013	0.005	0.051	17.96	17.44	17.36	17.4
0.5	0.383	0.715	0.077	0.32	0.032	0.005	0.014	0.009	38.96	38.52	36.72	38.84
0.666667	0.403	0.7	0.108	0.382	0.015	0	0.011	0.034	105.28	106.12	106.84	107.32
0.833333	0.436	0.7	0.06	0.328	0.012	0	0.004	0.006	368.88	369.08	369.28	374.68
1	0.451	0.7	0.052	0.196	0.008	0	0.003	0.004	1382.4	1382.4	1382.4	1382.4

Table 7.7 5 × CV-5; The result of experiments for four variants of Bayes classifier; data set *diabetes*; Concept dependent granulation; r_{gran} = Granulation radius; nil = result for data without missing values; *Acc* = Accuracy of classification; *AccBias* = Accuracy bias defined based on Eq. (7.4); *GranSize* = The size of data set after granulation in the fixed r

r_{gran}	Acc				AccBias				GranSize			
	V1	V2	V3	V4	V1	V2	V3	V4	V1	V2	V3	V4
0	0.518	0.521	0.546	0.544	0.033	0.029	0.026	0.021	2	2	2	2
0.125	0.591	0.598	0.535	0.545	0.009	0.046	0.025	0.022	33.76	33.44	33.2	33.6
0.25	0.643	0.661	0.572	0.573	0.007	0.006	0.026	0.024	154.96	154.48	153.48	154.36
0.375	0.634	0.663	0.574	0.589	0.013	0.003	0.023	0.01	366.32	365.96	366.2	366.08
0.5	0.604	0.659	0.602	0.612	0.008	0.002	0.009	0.008	539.8	539.64	540.48	540.68
0.625	0.609	0.659	0.598	0.614	0.002	0.004	0.014	0.008	609.88	609.88	610	609.96
0.75	0.606	0.658	0.597	0.612	0.006	0.003	0.018	0.01	614.4	614.4	614.4	614.4
0.875	0.606	0.658	0.597	0.612	0.006	0.003	0.018	0.01	614.4	614.4	614.4	614.4
1	0.606	0.658	0.597	0.612	0.006	0.003	0.018	0.01	614.4	614.4	614.4	614.4

Table 7.8 $5 \times$ CV-5; The result of experiments for four variants of Bayes classifier; data set *fertility diagnosis*; Concept dependent granulation; r_{gran} = Granulation radius; nil = result for data without missing values; Acc = Accuracy of classification; $AccBias$ = Accuracy bias defined based on Eq. (7.4); $GranSize$ = The size of data set after granulation in the fixed r

r_{gran}	Acc				AccBias				GranSize			
	V1	V2	V3	V4	V1	V2	V3	V4	V1	V2	V3	V4
0	0.508	0.572	0.496	0.49	0.092	0.083	0.174	0.08	2	2	2	2
0.111111	0.508	0.648	0.446	0.526	0.092	0.102	0.034	0.104	2.4	2.44	2.24	2.2
0.222222	0.508	0.674	0.342	0.526	0.032	0.106	0.168	0.074	3.92	3.68	3.68	3.96
0.333333	0.588	0.826	0.262	0.414	0.042	0.054	0.018	0.086	6.36	5.96	5.6	6.2
0.444444	0.584	0.834	0.35	0.504	0.046	0.046	0.1	0.136	10.72	10.56	10.44	11
0.555556	0.594	0.88	0.55	0.63	0.036	0	0.05	0.03	19.12	18.96	18.08	19
0.666667	0.528	0.88	0.49	0.556	0.072	0	0.05	0.054	35.28	35.88	35.68	34.8
0.777778	0.49	0.88	0.438	0.55	0.04	0	0.032	0.04	58.76	57.96	57.56	58.28
0.888889	0.542	0.88	0.448	0.59	0.038	0	0.052	0.06	74.04	73.92	73.64	73.68
1	0.584	0.88	0.48	0.638	0.046	0	0.05	0.032	80	80	80	80

Table 7.9 $5 \times$ CV-5; The result of experiments for four variants of Bayes classifier; data set *german credit*; Concept dependent granulation; r_{gran} = Granulation radius; nil = result for data without missing values; Acc = Accuracy of classification; $AccBias$ = Accuracy bias defined based on Eq. (7.4); $GranSize$ = The size of data set after granulation in the fixed r

r_{gran}	Acc				AccBias				GranSize			
	V1	V2	V3	V4	V1	V2	V3	V4	V1	V2	V3	V4
0	0.627	0.619	0.625	0.625	0.024	0.038	0.024	0.021	2	2	2	2
0.05	0.627	0.616	0.625	0.625	0.024	0.041	0.024	0.021	2	2	2	2
0.1	0.624	0.613	0.625	0.635	0.027	0.044	0.025	0.039	2	2.16	2	2.08
0.15	0.605	0.583	0.612	0.624	0.005	0.08	0.026	0.038	2.52	2.44	2.56	2.44
0.2	0.621	0.588	0.613	0.616	0.029	0.074	0.033	0.038	3.64	3.32	3.72	3.52
0.25	0.61	0.554	0.574	0.598	0.015	0.094	0.063	0.05	4.92	4.72	4.84	5.24
0.3	0.626	0.614	0.469	0.538	0.007	0.058	0.04	0.11	7.44	6.76	7.16	7.4
0.35	0.641	0.646	0.468	0.458	0.013	0.054	0.096	0.058	11.16	11.08	11.32	11.28
0.4	0.635	0.684	0.488	0.514	0.016	0.032	0.057	0.019	18.76	19.36	19.64	18.2
0.45	0.646	0.69	0.56	0.554	0.01	0.012	0.046	0.031	33.88	32.72	32.84	32.52
0.5	0.649	0.703	0.56	0.588	0.02	0.006	0.046	0.034	59.32	56.12	58.4	58.12
0.55	0.686	0.701	0.586	0.594	0.008	0.001	0.02	0.039	105.32	104.52	102.76	105.72
0.6	0.698	0.7	0.609	0.625	0.005	0	0.016	0.013	187.28	187.28	188.32	186.84
0.65	0.706	0.7	0.636	0.667	0.022	0	0.012	0.021	318.72	321.6	317.96	319.28
0.7	0.69	0.7	0.652	0.687	0.008	0	0.019	0.018	486.28	486	485.6	487.28
0.75	0.677	0.7	0.666	0.7	0.007	0	0.016	0.01	650	647.92	648.96	650.72
0.8	0.669	0.7	0.67	0.699	0.011	0	0.005	0.012	751.28	750.92	751.32	751.12
0.85	0.679	0.7	0.67	0.703	0.005	0	0.017	0.006	789.56	789.68	789.8	789.56
0.9	0.678	0.7	0.67	0.704	0.006	0	0.014	0.01	796.48	796.44	796.64	796.44
0.95	0.679	0.7	0.671	0.705	0.005	0	0.014	0.006	798.68	798.72	798.72	798.76
1	0.677	0.7	0.671	0.704	0.005	0	0.015	0.009	800	800	800	800

Table 7.10 $5 \times$ CV-5: The result of experiments for four variants of Bayes classifier; data set *heart disease*; Concept dependent granulation; $r_{gram} =$ Granulation radius; nil = result for data without missing values; $Acc =$ Accuracy of classification; $AccBias =$ Accuracy bias defined based on Eq. (7.4); $GranSize =$ The size of data set after granulation in the fixed r

r_{gram}	Acc				AccBias				GranSize			
	V1	V2	V3	V4	V1	V2	V3	V4	V1	V2	V3	V4
0	0.8	0.787	0.793	0.794	0.011	0.013	0.01	0.013	2	2	2	2
0.0769231	0.801	0.774	0.785	0.793	0.01	0.026	0.019	0.015	2.04	2.2	2.12	2.16
0.153846	0.802	0.752	0.773	0.781	0.009	0.037	0.016	0.015	2.68	3.08	2.96	2.88
0.230769	0.807	0.736	0.731	0.758	0.022	0.057	0.036	0.057	4.56	4.96	4.72	4.56
0.307692	0.802	0.784	0.722	0.735	0.009	0.027	0.022	0.058	9.2	8.28	8.52	9
0.384615	0.824	0.806	0.79	0.79	0.013	0.035	0.024	0.018	16.6	16.04	16.48	16.72
0.461538	0.823	0.824	0.763	0.753	0.018	0.01	0.033	0.01	34.84	34.64	34.36	35.32
0.538462	0.841	0.814	0.722	0.709	0.007	0.016	0.033	0.017	69.44	70.2	69.44	70.32
0.615385	0.827	0.814	0.696	0.707	0.018	0.016	0.03	0.019	127.24	127.2	126.76	127.8
0.692308	0.83	0.821	0.73	0.727	0.007	0.008	0.026	0.017	181.36	181.28	181.28	180.28
0.769231	0.83	0.796	0.738	0.737	0.003	0.007	0.021	0.007	210.56	210.12	210.24	210.36
0.846154	0.829	0.776	0.739	0.739	0.004	0.01	0.017	0.005	216	216	216	216
0.923077	0.829	0.776	0.739	0.739	0.004	0.01	0.017	0.005	216	216	216	216
1	0.829	0.776	0.739	0.739	0.004	0.01	0.017	0.005	216	216	216	216

Table 7.11 5 × CV-5: The result of experiments for four variants of Bayes classifier; data set *hepatitis*; Concept dependent granulation; r_{gran} = Granulation radius; nil = result for data without missing values; Acc = Accuracy of classification; $AccBias$ = Accuracy bias defined based on Eq. (7.4); $GranSize$ = The size of data set after granulation in the fixed r

r_{gran}	Acc				AccBias				GranSize			
	V1	V2	V3	V4	V1	V2	V3	V4	V1	V2	V3	V4
0	0.836	0.834	0.846	0.821	0.015	0.012	0.018	0.025	2	2	2	2
0.0526316	0.839	0.828	0.846	0.821	0.013	0.017	0.018	0.025	2	2	2	2
0.105263	0.839	0.828	0.846	0.821	0.013	0.017	0.018	0.025	2	2	2	2
0.157895	0.827	0.831	0.846	0.821	0.018	0.021	0.018	0.025	2.2	2	2	2
0.210526	0.825	0.831	0.826	0.835	0.021	0.021	0.039	0.017	2.4	2.12	2.2	2.36
0.263158	0.826	0.841	0.791	0.844	0.019	0.017	0.048	0.014	2.6	2.68	2.68	2.76
0.315789	0.813	0.822	0.76	0.859	0.026	0.036	0.059	0.025	3.52	3.36	3.88	3.52
0.368421	0.822	0.836	0.693	0.855	0.01	0.035	0.075	0.054	5.32	5.08	4.96	4.68
0.421053	0.827	0.817	0.639	0.823	0.025	0.035	0.129	0.061	7.48	7.56	7.4	6.88
0.473684	0.868	0.827	0.761	0.84	0.035	0.025	0.026	0.057	11.64	12.16	11.44	11.72
0.526316	0.876	0.806	0.804	0.885	0.008	0.013	0.035	0.018	18.28	19.28	18.48	18.12
0.578947	0.871	0.796	0.8	0.863	0.026	0.004	0.013	0.021	31.36	30.84	29.8	30.68
0.631579	0.866	0.794	0.766	0.883	0.012	0	0.066	0.021	46.68	46.4	45.84	47.48
0.684211	0.857	0.794	0.804	0.871	0.014	0	0.015	0.026	70.28	70.04	69.4	70.2
0.736842	0.852	0.794	0.813	0.879	0.019	0	0.032	0.018	89.32	90.2	89.6	90.72
0.789474	0.855	0.794	0.83	0.886	0.009	0	0.028	0.023	109.28	110	109.88	110
0.842105	0.845	0.794	0.843	0.879	0.006	0	0.022	0.025	116.92	117.04	116.8	117.12
0.894737	0.845	0.794	0.844	0.876	0.006	0	0.021	0.021	121	121	121	121
0.947368	0.843	0.794	0.845	0.876	0.009	0	0.019	0.021	122	121.92	121.96	121.96
1	0.841	0.794	0.845	0.876	0.004	0	0.019	0.021	124	124	124	124

Table 7.12 $5 \times$ CV-5; The result of experiments for four variants of Bayes classifier; data set *congressional voting records*; Concept dependent granulation; r_{gran} = Granulation radius; nil = result for data without missing values; Acc = Accuracy of classification; $AccBias$ = Accuracy bias defined based on Eq. (7.4); $GranSize$ = The size of data set after granulation in the fixed r

r_{gran}	Acc				AccBias				GranSize			
	V1	V2	V3	V4	V1	V2	V3	V4	V1	V2	V3	V4
0	0.889	0.886	0.892	0.889	0.006	0.004	0.007	0.01	2	2	2	2
0.0625	0.887	0.887	0.892	0.889	0.007	0.005	0.007	0.01	2.04	2.08	2	2
0.125	0.883	0.888	0.876	0.889	0.009	0.009	0.016	0.01	2.24	2.24	2.2	2.04
0.1875	0.872	0.891	0.855	0.893	0.013	0.015	0.037	0.006	2.6	2.56	2.4	2.28
0.25	0.863	0.901	0.81	0.877	0.015	0.017	0.068	0.022	3.16	3.04	2.84	2.8
0.3125	0.867	0.904	0.782	0.853	0.011	0.008	0.08	0.036	3.6	3.6	3.48	3.32
0.375	0.874	0.881	0.767	0.781	0.02	0.02	0.095	0.06	4.56	4.44	4.32	4.48
0.4375	0.897	0.89	0.663	0.635	0.011	0.02	0.101	0.139	5.4	5.56	5.48	5.48
0.5	0.905	0.904	0.636	0.615	0.012	0.009	0.111	0.054	6.72	6.84	6.4	6.44
0.5625	0.91	0.885	0.629	0.642	0.007	0.028	0.091	0.061	8.72	9.08	8.88	8.68
0.625	0.913	0.886	0.738	0.794	0.007	0.026	0.076	0.024	11.92	13.8	12.56	12.52
0.6875	0.91	0.872	0.875	0.888	0.009	0.043	0.023	0.018	18.72	19.2	18.24	19.48
0.75	0.914	0.8	0.917	0.915	0.006	0.082	0.01	0.007	30.8	30.4	31.64	31.92
0.8125	0.926	0.697	0.924	0.934	0.002	0.053	0.014	0.013	52.56	52.44	53.44	54.96
0.875	0.921	0.623	0.942	0.941	0.003	0.014	0.014	0.004	91.08	88.92	90.64	90.52
0.9375	0.919	0.614	0.934	0.943	0	0	0.011	0.004	143.84	145.6	143	144.96
1	0.916	0.64	0.925	0.927	0.001	0.032	0.002	0.004	213.92	213.88	213.64	214

Table 7.13 5 × CV-5; The result of experiments for four variants of Bayes classifier; data set *mushroom*; Concept dependent granulation; r_{gran} = Granulation radius; nil = result for data without missing values; *Acc* = Accuracy of classification; *AccBias* = Accuracy bias defined based on Eq. (7.4); *GranSize* = The size of data set after granulation in the fixed *r*

r_{gran}	Acc				AccBias				GranSize			
	V1	V2	V3	V4	V1	V2	V3	V4	V1	V2	V3	V4
0	0.883	0.884	0.885	0.884	0.005	0.002	0.002	0.003	2	2	2	2
0.0454545	0.88	0.881	0.882	0.882	0.003	0.005	0.004	0.004	2	2	2	2
0.0909091	0.881	0.883	0.882	0.881	0.004	0.003	0.003	0.002	2	2	2	2
0.136364	0.882	0.882	0.882	0.88	0.005	0.003	0.001	0.006	2	2	2	2
0.181818	0.882	0.871	0.881	0.881	0.006	0.013	0.007	0.005	2.08	2.08	2.12	2.04
0.227273	0.882	0.847	0.876	0.876	0.002	0.035	0.014	0.007	2.04	2.24	2.12	2.12
0.272727	0.883	0.802	0.864	0.871	0.005	0.06	0.017	0.017	2.4	2.6	2.64	2.6
0.318182	0.883	0.742	0.861	0.86	0.005	0.069	0.004	0.026	3.72	3.96	3.76	3.84
0.363636	0.885	0.694	0.78	0.832	0.007	0.071	0.033	0.026	5.8	5.84	5.4	5.92
0.409091	0.887	0.613	0.598	0.734	0.002	0.026	0.039	0.048	9.44	9.72	10.68	9.24
0.454545	0.891	0.622	0.707	0.77	0.001	0.049	0.076	0.036	15.32	15.28	15.48	14.68
0.5	0.891	0.766	0.841	0.865	0.003	0.123	0.022	0.007	21	21.6	21.76	22.36
0.545455	0.9	0.676	0.863	0.879	0.008	0.124	0.02	0.012	29.28	29.48	27.4	28.56
0.590909	0.908	0.584	0.862	0.881	0.015	0.024	0.016	0.013	37.8	38.88	37.24	38.84
0.636364	0.91	0.538	0.842	0.844	0.002	0.081	0.017	0.016	45.8	47.16	43.84	46.24
0.681818	0.908	0.673	0.846	0.846	0.011	0.096	0.015	0.03	48.88	46.96	47.44	47.96
0.727273	0.903	0.652	0.904	0.914	0.002	0.099	0.012	0.016	47.4	47.72	46.64	48.12
0.772727	0.898	0.682	0.936	0.932	0.003	0.054	0.015	0.007	57.96	57.32	56.48	56.6
0.818182	0.908	0.818	0.954	0.955	0.002	0.024	0.006	0.002	92.44	91.96	93.72	92.4
0.863636	0.903	0.884	0.947	0.946	0.005	0.009	0.002	0.004	194.84	193.36	195.6	192.64
0.909091	0.895	0.894	0.938	0.938	0.001	0.003	0.002	0.001	516.36	514.4	514.16	518.68
0.954545	0.892	0.896	0.928	0.928	0	0.001	0.001	0	1753.72	1755.12	1752.6	1751.72
1	0.89	0.895	0.918	0.918	0	0	0	0	6499.2	6499.2	6499.2	6499.2

Table 7.14 $5 \times$ CV-5; The result of experiments for four variants of Bayes classifier; data set *nursery*; Concept dependent granulation; $r_{gran} =$ Granulation radius; nil = result for data without missing values; $Acc =$ Accuracy of classification; $AccBias =$ Accuracy bias defined based on Eq. (7.4); $GranSize =$ The size of data set after granulation in the fixed r

r_{gran}	Acc				AccBias				GranSize			
	V1	V2	V3	V4	V1	V2	V3	V4	V1	V2	V3	V4
0	0.398	0.38	0.387	0.418	0.045	0.047	0.035	0.04	4.92	5	4.92	4.92
0.125	0.316	0.487	0.237	0.419	0.029	0.004	0.007	0.031	7.72	7.76	7.76	7.36
0.25	0.375	0.542	0.036	0.206	0.012	0.024	0.015	0.033	14.12	13.96	13.76	13.68
0.375	0.435	0.568	0.016	0.162	0.033	0.03	0.015	0.043	26.28	26.96	26.44	26.24
0.5	0.464	0.62	0.031	0.361	0.017	0.043	0.05	0.052	57.76	59.12	58.56	58.12
0.625	0.48	0.681	0.038	0.514	0.03	0.089	0.062	0.009	158.96	160.96	161.44	160.8
0.75	0.491	0.819	0.04	0.566	0.027	0.019	0.063	0.024	546.56	545.88	546.88	545.52
0.875	0.497	0.854	0.032	0.57	0.024	0.009	0.05	0.025	2367.64	2369.36	2374	2364.36
1	0.522	0.869	0.018	0.541	0.026	0.001	0.028	0.025	10368	10368	10368	10368

Table 7.15 $5 \times$ CV-5; The result of experiments for four variants of Bayes classifier; data set *soybean large*; Concept dependent granulation; r_{gran} = Granulation radius; nil = result for data without missing values; Acc = Accuracy of classification; $AccBias$ = Accuracy bias defined based on Eq. (7.4); $GranSize$ = The size of data set after granulation in the fixed r

r_{gran}	Acc				AccBias				GranSize			
	V1	V2	V3	V4	V1	V2	V3	V4	V1	V2	V3	V4
0	0.672	0.678	0.672	0.672	0.03	0.01	0.01	0.019	19	19	19	19
...	...	...	...	...	...	...	...	...	...	...	...	...
0.4	0.674	0.68	0.672	0.667	0.026	0.011	0.009	0.016	19	19	19	19
0.428571	0.674	0.663	0.671	0.667	0.026	0.028	0.01	0.016	19	19.04	19	19
0.457143	0.669	0.622	0.667	0.664	0.012	0.05	0.01	0.02	19.12	19.16	19.12	19.04
0.485714	0.66	0.594	0.656	0.662	0.021	0.034	0.005	0.014	19.2	19.28	19.36	19.28
0.514286	0.651	0.556	0.647	0.659	0.03	0.064	0.014	0.018	19.52	19.48	19.56	19.6
0.542857	0.617	0.334	0.623	0.641	0.027	0.042	0.009	0.024	20.36	20.84	20.52	20.6
0.571429	0.59	0.221	0.609	0.627	0.058	0.032	0.016	0.013	21.8	22.48	21.8	21.88
0.6	0.498	0.201	0.488	0.605	0.028	0.049	0.052	0.01	24.12	24.4	23.96	23.88
0.628571	0.422	0.233	0.414	0.566	0.026	0.062	0.068	0.028	26.8	27.6	26.8	27.24
0.657143	0.382	0.239	0.361	0.493	0.036	0.033	0.045	0.018	30.28	31.52	30.16	31.16
0.685714	0.343	0.187	0.324	0.426	0.026	0.021	0.011	0.025	34.48	36.24	35.32	35.64
0.714286	0.286	0.193	0.278	0.35	0.023	0.044	0.041	0.014	40.04	40.48	40.72	40.76
0.742857	0.197	0.187	0.179	0.323	0.006	0.033	0.013	0.02	49.2	49.32	48.44	47.68
0.771429	0.185	0.181	0.155	0.302	0.01	0.027	0.01	0.015	57.96	58.64	58.16	57.8
0.8	0.159	0.195	0.144	0.268	0.008	0.018	0.005	0.009	72.44	73	72.28	73.12

(continued)

Table 7.15 (continued)

r_{gran}	Acc				AccBias				GranSize			
	V1	V2	V3	V4	V1	V2	V3	V4	V1	V2	V3	V4
0.828571	0.14	0.212	0.079	0.196	0.009	0.029	0.008	0.041	92.64	91.08	91.84	92.16
0.857143	0.116	0.245	0.056	0.142	0.001	0.014	0.023	0.013	119.44	119.36	117.88	119.44
0.885714	0.096	0.23	0.044	0.112	0.009	0.051	0	0.01	161.32	161.48	160.76	160.8
0.914286	0.081	0.212	0.044	0.113	0.006	0.02	0	0.008	223.8	221.08	222.48	222.44
0.942857	0.075	0.237	0.023	0.103	0.003	0.034	0	0.002	300.64	299.84	301.32	300.76
0.971429	0.417	0.222	0.043	0.144	0.031	0.028	0.006	0.011	405.6	404.72	404.64	404.72
1	0.69	0.297	0.037	0.141	0.006	0.017	0.011	0.003	509.36	509.68	509.76	509.8

Table 7.16 5 × CV-5; The result of experiments for four variants of Bayes classifier; data set *SPECT*; Concept dependent granulation; r_{gran} = Granulation radius; nil = result for data without missing values; *Acc* = Accuracy of classification; *AccBias* = Accuracy bias defined based on Eq. (7.4); *GranSize* = The size of data set after granulation in the fixed *r*

r_{gran}	Acc				AccBias				GranSize			
	V1	V2	V3	V4	V1	V2	V3	V4	V1	V2	V3	V4
0	0.594	0.593	0.567	0.578	0.031	0.036	0.058	0.052	2	2	2	2
0.0454545	0.594	0.602	0.567	0.578	0.031	0.031	0.058	0.052	2	2.04	2	2
0.0909091	0.594	0.625	0.564	0.593	0.031	0.053	0.062	0.037	2.08	2.16	2.04	2.04
0.136364	0.593	0.644	0.546	0.63	0.033	0.033	0.079	0.041	2.2	2.24	2.2	2.28
0.181818	0.593	0.686	0.54	0.655	0.066	0.077	0.048	0.046	2.28	2.4	2.48	2.56
0.227273	0.565	0.687	0.551	0.677	0.06	0.076	0.053	0.027	2.64	2.4	2.6	2.76
0.272727	0.59	0.788	0.525	0.674	0.036	0.006	0.048	0.056	2.96	3	2.96	3
0.318182	0.608	0.794	0.486	0.673	0.037	0	0.058	0.08	3.24	3.32	3.24	3.48
0.363636	0.628	0.794	0.4	0.638	0.019	0	0.058	0.013	3.72	3.76	3.68	4.04
0.409091	0.663	0.789	0.233	0.605	0.041	0.006	0.052	0.039	4.6	4.6	4.8	4.72
0.454545	0.64	0.794	0.23	0.435	0.015	0	0.08	0.068	5.76	5.68	5.52	5.76
0.5	0.631	0.794	0.206	0.334	0.024	0	0	0.088	6.88	6.56	6.52	7.32
0.545455	0.61	0.794	0.222	0.258	0.035	0	0.062	0.026	8.24	9.4	8.44	9.64
0.590909	0.574	0.794	0.266	0.293	0.028	0	0.069	0.055	11.4	12.32	11.52	12.24
0.636364	0.572	0.794	0.401	0.438	0.021	0	0.125	0.064	15.28	16.16	16.56	17.08
0.681818	0.498	0.794	0.428	0.503	0.023	0	0.11	0.048	22.56	23.16	23.2	23.2

(continued)

Table 7.16 (continued)

r_{gran}	Acc				AccBias				GranSize			
	V1	V2	V3	V4	V1	V2	V3	V4	V1	V2	V3	V4
0.727273	0.465	0.794	0.59	0.71	0.014	0	0.042	0.066	33.44	33.28	33.64	34.24
0.772727	0.438	0.794	0.7	0.757	0.008	0	0.022	0.018	50.2	50.6	51.2	49.6
0.818182	0.42	0.794	0.687	0.763	0.015	0	0.022	0.027	72.08	72.44	72.2	72.28
0.863636	0.404	0.794	0.646	0.742	0.019	0	0.017	0.034	100.84	100.48	101.8	100
0.909091	0.406	0.794	0.652	0.753	0.01	0	0.019	0.019	131.28	130.56	130.56	130.88
0.954545	0.427	0.794	0.664	0.767	0.007	0	0.006	0.02	162.84	162.72	163.04	162.64
1	0.438	0.794	0.682	0.772	0.008	0	0.007	0.007	184.84	184.72	184.84	184.68

Table 7.17 5 × CV-5; The result of experiments for four variants of Bayes classifier; data set *SPECTF*; Concept dependent granulation; r_{gran} = Granulation radius; nil = result for data without missing values; *Acc* = Accuracy of classification; *AccBias* = Accuracy bias defined based on Eq. (7.4); *GranSize* = The size of data set after granulation in the fixed *r*

r_{gran}	Acc				AccBias				GranSize			
	V1	V2	V3	V4	V1	V2	V3	V4	V1	V2	V3	V4
0	0.522	0.518	0.548	0.536	0.047	0.045	0.033	0.074	2	2	2	2
0.0227273	0.531	0.679	0.291	0.363	0.012	0.043	0.023	0.016	6.8	6.64	6.92	6.84
0.0454545	0.551	0.777	0.303	0.436	0.03	0.014	0.034	0.059	14.12	13.44	13.64	14.16
0.0681818	0.547	0.793	0.408	0.578	0.03	0.004	0.015	0.047	29.04	27.64	28	28.24
0.0909091	0.548	0.794	0.544	0.667	0.017	0	0.013	0.036	50.76	50.48	50.48	50.72
0.113636	0.493	0.794	0.659	0.715	0.058	0	0.045	0.011	85.72	84.92	84.68	84.88
0.136364	0.475	0.794	0.735	0.763	0.042	0	0.011	0.019	125.36	125.56	126.68	125.36
0.159091	0.441	0.794	0.779	0.797	0.021	0	0.011	0.009	167.28	166.56	167.12	167.36
0.181818	0.441	0.794	0.796	0.795	0.013	0	0.006	0.014	195.2	195.28	195.6	195.64
0.204545	0.43	0.794	0.792	0.792	0.027	0	0.006	0.013	209.4	209.32	209.44	209.4
0.227273	0.422	0.794	0.791	0.791	0.024	0	0.007	0.01	212.96	212.96	212.92	212.96
0.25	0.419	0.794	0.791	0.791	0.026	0	0.007	0.01	213.6	213.6	213.6	213.6
...	...	...	...	...	...	...	...	...	...	...	...	...
1	0.419	0.794	0.791	0.791	0.026	0	0.007	0.01	213.6	213.6	213.6	213.6

Table 7.18 We conclude Chap. 6 and this chapter with a comparison of the performance by either of classifiers on the best granulated case and on the non-granulated case, denoted *nil*

Name	k-NN (acc, red, r)	k-NN. nil(acc)	Bayes(acc, red, r, method)	Bayes.nil (acc, method)
(d1)*Adult*	0.832, 75.27, 0.786	0.841	0.788, 99.73, 0.429, V2	0.764, V4
(d2)*Australian-credit*	0.851, 71.86, 0.571	0.855	0.853, 94.05, 0.429, V1	0.843, V1
(d3)*Car Evaluation*	0.865, 73.23, 0.833	0.944	0.725, 98.74, 0.333, V2	0.7, V2
(d4)*Diabetes*	0.616, 74.74, 0.25	0.631	0.661, 74.86, 0.25, V2	0.658, V2
(d5)*Fertility_Diagnosis*	0.84, 55, 0.667	0.87	0.888, 76.3, 0.556, V2	0.88, V2
(d6)*German-credit*	0.724, 59.85, 0.65	0.73	0.703, 92.99, 0.5, V2	0.704, V4
(d7)*Heartdisease*	0.83, 67.69, 0.538	0.837	0.841, 67.85, 0.538, V1	0.829, V1
(d8)*Hepatitis*	0.884, 60, 0.632	0.89	0.885, 85.39, 0.526, V4	0.845, V3
(d9)*Congressional Voting Records*	0.943, 31.97, 0.938	0.938	0.941, 57.7, 0.875, V4	0.927, V4
(d10)*Mushroom*	1.0, 72.88, 0.955	1	0.955, 98.58, 0.818, V4	0.918, V3, V4
(d11)*Nursery*	0.696, 77.09, 0.875	0.578	0.854, 77.15, 0.875, V2	0.869, V2
(d12)*soybean-large*	0.871, 67.77, 0.886	0.928	0.68, 96.27, 0.4, V2	0.69, V1
(d13)*SPECT Heart*	0.794, 60.67, 0.818	0.794	0.794, 98.2, 0.318, V2	0.794, V2
(d14)*SPECTF Heart*	0.802, 60.3, 0.114	0.779	0.794, 76.37, 0.91, V2	0.794, V2

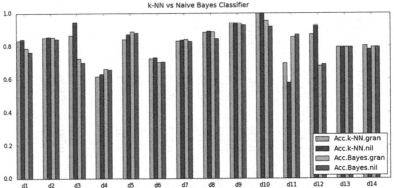

Results are taken for the maximal reduction in size of the training set for accuracy satisfactorily close to the best one; the results show no prevalence for one of them over the other and closeness of results by both as well. It is essential to notice the very large reduction in size of the training set, in some cases up to over 90 %, which reduction may be taken as the main feature of the granular preprocessing. Summary of results, k-NN vs Naive Bayes Classifier, granular and non–granular (nil) case, acc = accuracy of classification, red = percentage reduction in object number, r = granulation radius, *method* = variant of Naive Bayes classifier

Indians Diabetes, Fertility Diagnosis, German Credit, Nursery, Soybean, SPECT and SPECTF, the variant $V2$ wins. In case of Hepetitis, Congressional House Votes and Mushroom, the variant $V4$ wins. For Australian Credit and Heart Disease, the variant $V1$ wins, but the results for the variant $V2$ are also very good for these data sets.

References

1. Al-Aidaroos, K., Abu Bakar, A., Othman, Z.: Data classification using rough sets and naive Bayes. In: Proceedings of International Conference on Rough Sets and Knowledge Technology RSKT 2010. Lecture Notes in Computer Science, vol. 6401, pp. 134–142 (2010)
2. Bishop, Ch.: Pattern Recognition and Machine Learning. Springer, New York (2006)
3. Cheng, K., Luo, J., Zhang, C.: Rough set weighted naive Bayesian classifier in intrusion prevention system. In: Proceedings of the International Conference on Network Security, Wireless Communication and Trusted Computing NSWCTC 2009, pp. 25–28. IEEE Press, Wuhan, P.R. China (2009)
4. Devroye, L., Györfi, L., Lugosi, G.: A Probabilistic Theory of Pattern Recognition. Springer, New York (1996)
5. Duda, R., Hart, P.: Pattern Classification and Scene Analysis. Wiley, New York (1973)
6. Langley, P., Iba, W., Thompson, K.: An analysis of Bayesian classifiers. In: Proceedings of the 10th National Conference on Artificial Intelligence, pp. 399–406. AAAI Press, San Jose (1992)
7. Mitchell, T.: Machine Learning. McGraw-Hill, Englewood Cliffs (1997)
8. Pawlak, Z.: Bayes theorem: the rough set perspective. In: Inuiguchi, M., Tsumoto, S., Hirano, S. (eds.) Rough Set Theory and Granular Computing, pp. 1–12. Springer, Heidelberg (2003)
9. Rish, I., Hellerstein, J., Thathachar, J.: An analysis of data characteristics that affect naive Bayes performance. IBM Technology Report RC 21993 (2001)
10. Su, H., Zhang, Y., Zhao, F., Li, Q.: An ensemble deterministic model based on rough set and fuzzy set and Bayesian optimal classifier. Int. J. Innov. Comput. Inf. Control **3**(4), 977–986 (2007)
11. University of Irvine (UCI) Repository http://www.ics.uci.edu/~mlearn/databases/
12. Wang, Y., Wu, Z., Wu, R.: Spam filtering system based on rough set and Bayesian classifier. In: Proceedings of International IEEE Conference on Granular Computing GrC 2008, pp. 624–627. Hangzhou, P.R. China (2008)
13. Wang, Z., Webb, G.I., Zheng, F.: Selective augmented Bayesian network classifiers based on rough set theory. In: International Conference on Advances in Knowledge Discovery and Data Mining. Lecture Notes in Computer Science, vol. 3056, pp. 319–328 (2004)
14. Yao, Y., Zhou, B.: Naive Bayesian rough sets. In: Proceedings of the International Conference on Rough Sets and Knowledge Technology RSKT 2010. Lecture Notes in Computer Science, vol. 6401, pp. 719–726 (2010)
15. Zhang, H., Zhou, J., Miao, D., Gao, C.: Bayesian rough set model: a further investigation. Int. J. Approx. Reason. **53**, 541–557 (2012)

Chapter 8
Granular Computing in the Problem of Missing Values

What do You read, my lord? Words, words, words
[William Shakespeare. Hamlet.]

In this chapter we discuss methods for dealing with the problem of missing values already outlined in Chap. 4. There are four basic Strategies *A*, *B*, *C* and *D* which we examine on artificially damaged data sets.

8.1 Introduction

The problem of missing values is discussed and solved in theory of inductive reasoning, databases and data mining, see for a general discussion, e.g., Acuña and Rodrıguez [1], Allison [2], Batista and Monard [4], Bruha [5], Imieliński and Lipski [8], Little and Rubin [10], Quinlan [15]; an approach related to decision systems is discussed, e.g., in Grzymala-Busse [6], Grzymala-Busse and Hu [7], Li and Cercone [9], Nakata and Sakai [11], Sadiq et al. [16]. Initial experiments, in which we use selected by us strategies are available in Polkowski and Artiemjew [12, 13]. We begin with a reminder of details of chosen by us strategies.

8.1.1 A Survey of Strategies

We recall four strategies for our treatment of data with missing values,

1. *Strategy A: in building granules* $* = $ *don't care, in repairing values of* $*$, $* = $ *don't care.*
2. *Strategy B: in building granules* $* = $ *don't care, in repairing values of* $*$, $* = *$.
3. *Strategy C: in building granules* $* = *$, *in repairing values of* $*$, $* = don't care.*
4. *Strategy D: in building granules* $* = *$, *in repairing values of* $*$, $* = *$.

© Springer International Publishing Switzerland 2015
L. Polkowski and P. Artiemjew, *Granular Computing in Decision Approximation*,
Intelligent Systems Reference Library 77, DOI 10.1007/978-3-319-12880-1_8

In case of strategies A and B for building granules, stars are treated as all possible values, and have significant influence on the degree of approximation of granulated data set. In case of strategies C and D for building granules, stars are treated as new values which has a slight influence on the approximation level. We now recall the form of granules in each case.

8.1.1.1 The Granule Building Phase Where ∗ = *Don't Care*

Granules in case of A or B strategies centered at objects u in the training set TRN_i are of the form

$$g_{r_{gran}}^{cd,*=don't\ care}(u) = \{v \in TRN_i : \frac{|IND^{*=don't\ care}(u, v)|}{|A|}$$

$$\leq r_{gran}\ AND\ d(u) = d(v)\},$$

where

$$IND^{*=don't\ care}(u, v) = \{a \in A : a(u) = a(v)\ OR\ a(u) = *\ OR\ a(v) = *\}.$$

8.1.1.2 The Granule Building Phase Where ∗ = ∗

Granules used in C and D strategies have the form:

$$g_{r_{gran}}^{cd,*=*}(u) = \{v \in TRN_i : \frac{|IND^{*=*}(u, v)|}{|A|} \leq r_{gran}\ AND\ d(u) = d(v)\},$$

where
$$IND^{*=*}(u, v) = \{a \in A : a(u) = a(v)\}.$$

8.1.1.3 The Repairing Phase Where ∗ = *Don't Care*

In case of A and C strategies, in order to repair objects containing missing values after granulation, we immerse objects with stars on specific positions j into original disturbed training set. We fill the value for the star by means of majority voting on non missing values of the attribute j.

In case of the strategy A, the granule around the disturbed object $MV(g_{r_{gran}}^{cd,*=don't\ care}(u))$ can be defined as follows,

$$if\ a_j(MV(g_{r_{gran}}^{cd,*=don't\ care}(u))) = *,$$

then the missing value could be repaired by the granule,

$$g_{r_{gran},a_j}^{cd,*=don't\ care}(MV(g_{r_{gran}}^{cd,*=don't\ care}(u)))$$

$$= \{v \in TRN_i : \frac{|IND_{a_j}^{*=don't\ care}(MV(g_{r_{gran}}^{cd,*=don't\ care}(u)), v)|}{|A|}$$

$$\leq r_{gran}\ AND\ d(MV(g_{r_{gran}}^{cd,*=don't\ care}(u))) = d(v)\},$$

where

$$IND_{a_j}^{*=don't\ care}(MV(g_{r_{gran}}^{cd,*=don't\ care}(u)), v)$$

$$= \{a \in A : (a(MV(g_{r_{gran}}^{cd,*=don't\ care}(u)))$$

$$= a(v)\ OR\ a(MV(g_{r_{gran}}^{cd,*=don't\ care}(u))) = *\ OR\ a(v) = *)\ AND\ a_j(v)! = *\}.$$

In case of the strategy C, the granule around the disturbed object $MV(g_{r_{gran}}^{cd,*=*}(u))$ can be defined as follows,

$$if\ a_j(MV(g_{r_{gran}}^{cd,*=*}(u))) = *,$$

then the missing value could be repaired by the granule,

$$g_{r_{gran},a_j}^{cd,*=don't\ care}(MV(g_{r_{gran}}^{cd,*=*}(u)))$$

$$= \{v \in TRN_i : \frac{|IND_{a_j}^{*=don't\ care}(MV(g_{r_{gran}}^{cd,*=*}(u)), v)|}{|A|}$$

$$\leq r_{gran}\ AND\ d(MV(g_{r_{gran}}^{cd,*=*}(u))) = d(v)\},$$

where

$$IND_{a_j}^{*=don't\ care}(MV(g_{r_{gran}}^{cd,*=*}(u)), v)$$

$$= \{a \in A : (a(MV(g_{r_{gran}}^{cd,*=*}(u))) = a(v)\ OR\ a(MV(g_{r_{gran}}^{cd,*=*}(u)))$$

$$= *\ OR\ a(v) = *)\ AND\ a_j(v)! = *\}.$$

8.1.1.4 The Repairing Phase Where * = *

As above, also in case of B and D strategies, in order to repair objects containing missing values after granulation, we immerse objects with stars on specific positions j into original disturbed training data set. We fill the star based on majority voting from non missing values of attribute number j.

In case of the strategy B, the granule around the disturbed object MV $(g_{r_{gran}}^{cd,*=don't\,care}(u))$ can be defined as follows,

$$g_{r_{gran},a_j}^{cd,*=*}(MV(g_{r_{gran}}^{cd,*=don't\,care}(u)))$$

$$= \{v \in TRN_i : \frac{|IND_{a_j}^{*=*}(MV(g_{r_{gran}}^{cd,*=don't\,care}(u)), v)|}{|A|}$$

$$\leq r_{gran} \text{ AND } d(MV(g_{r_{gran}}^{cd,*=don't\,care}(u))) = d(v)\},$$

where

$$IND_{a_j}^{*=*}(MV(g_{r_{gran}}^{cd,*=don't\,care}(u)), v)$$

$$= \{a \in A : a(MV(g_{r_{gran}}^{cd,*=don't\,care}(u))) = a(v) \text{ AND } a_j(v)! = *\}.$$

In case of the strategy D, the granule around the disturbed object MV $(g_{r_{gran}}^{cd,*=*}(u))$ can be defined as follows,

$$g_{r_{gran},a_j}^{cd,*=*}(MV(g_{r_{gran}}^{cd,*=*}(u)))$$

$$= \{v \subset TRN_i : \frac{|IND_{a_j}^{*=*}(MV(g_{r_{gran}}^{cd,*=*}(u)), v)|}{|A|}$$

$$\leq r_{gran} \text{ AND } d(MV(g_{r_{gran}}^{cd,*=*}(u))) = d(v)\},$$

where

$$IND_{a_j}^{*=*}(MV(g_{r_{gran}}^{cd,*=*}(u)), v)$$

$$= \{a \in A : a(MV(g_{r_{gran}}^{cd,*=*}(u))) = a(v) \text{ AND } a_j(v)! = *\}.$$

8.1.2 Examples of Basic Strategies

In this section we show exemplary phases of our strategies for the radius $r_{gran} = 0.5$ and as a basic data set we use the Quinlan data set [13] used originally in the context of decision trees algorithms.

The data set from Table 8.1 contains 56 descriptors, we have disturbed this data set by randomly introducing 6 star values, as the 10% damage. The perturbed data set is shown in Table 8.2.

In case of strategies A and B, the first phase consists of steps,
$$g_{0.5}^{cd,*=don't\,care}(D_1) = \{D_1, D_2, D_8\}.$$
$$g_{0.5}^{cd,*=don't\,care}(D_2) = \{D_1, D_2, D_6, D_8, D_{14}\}.$$

Table 8.1 Exemplary complete decision system ($TRN_i^{complete}$, A, d) by Quinlan [14]

Day	Outlook	Temperature	Humidity	Wind	Play golf
D_1	Sunny	Hot	High	Weak	No
D_2	Sunny	Hot	High	Strong	No
D_3	Overcast	Hot	High	Weak	Yes
D_4	Rainy	Mild	High	Weak	Yes
D_5	Rainy	Cool	Normal	Weak	Yes
D_6	Rainy	Cool	Normal	Strong	No
D_7	Overcast	Cool	Normal	Strong	Yes
D_8	Sunny	Mild	High	Weak	No
D_9	Sunny	Cool	Normal	Weak	Yes
D_{10}	Rainy	Mild	Normal	Weak	Yes
D_{11}	Sunny	Mild	Normal	Strong	Yes
D_{12}	Overcast	Mild	High	Strong	Yes
D_{13}	Overcast	Hot	Normal	Weak	Yes
D_{14}	Rainy	Mild	High	Strong	No

Table 8.2 Exemplary incomplete decision system (TRN_i, A, d)

Day	Outlook	Temperature	Humidity	Wind	Play golf
D_1	Sunny	Hot	High	Weak	No
D_2	*	Hot	High	Strong	No
D_3	Overcast	Hot	*	Weak	Yes
D_4	Rainy	Mild	High	Weak	Yes
D_5	Rainy	Cool	*	Weak	Yes
D_6	Rainy	Cool	Normal	Strong	No
D_7	Overcast	*	Normal	Strong	Yes
D_8	Sunny	Mild	High	Weak	No
D_9	Sunny	Cool	*	Weak	Yes
D_{10}	Rainy	Mild	*	Weak	Yes
D_{11}	Sunny	Mild	Normal	Strong	Yes
D_{12}	Overcast	Mild	High	Strong	Yes
D_{13}	Overcast	Hot	Normal	Weak	Yes
D_{14}	Rainy	Mild	High	Strong	No

$$g_{0.5}^{cd,*=don't\ care}(D_3) = \{D_3, D_4, D_5, D_7, D_9, D_{10}, D_{12}, D_{13}\}.$$
$$g_{0.5}^{cd,*=don't\ care}(D_4) = \{D_3, D_4, D_5, D_9, D_{10}, D_{12}\}.$$
$$g_{0.5}^{cd,*=don't\ care}(D_5) = \{D_3, D_4, D_5, D_7, D_9, D_{10}, D_{13}\}.$$
$$g_{0.5}^{cd,*=don't\ care}(D_6) = \{D_2, D_6, D_{14}\}.$$
$$g_{0.5}^{cd,*=don't\ care}(D_7) = \{D_3, D_5, D_7, D_9, D_{10}, D_{11}, D_{12}, D_{13}\}.$$

$g_{0.5}^{cd,*=don't\ care}(D_8) = \{D_1, D_2, D_8, D_{14}\}$.

$g_{0.5}^{cd,*=don't\ care}(D_9) = \{D_3, D_4, D_5, D_7, D_9, D_{10}, D_{11}, D_{13}\}$.

$g_{0.5}^{cd,*=don't\ care}(D_{10}) = \{D_3, D_4, D_5, D_7, D_9, D_{10}, D_{11}, D_{12}, D_{13}\}$.

$g_{0.5}^{cd,*=don't\ care}(D_{11}) = \{D_7, D_9, D_{10}, D_{11}, D_{12}\}$.

$g_{0.5}^{cd,*=don't\ care}(D_{12}) = \{D_3, D_4, D_7, D_{10}, D_{11}, D_{12}\}$.

$g_{0.5}^{cd,*=don't\ care}(D_{13}) = \{D_3, D_5, D_7, D_9, D_{10}, D_{13}\}$.

$g_{0.5}^{cd,*=don't\ care}(D_{14}) = \{D_2, D_6, D_8, D_{14}\}$.

The covering phase yields,

$Cover_{TRN_i} \leftarrow \emptyset$.

$Cover_{TRN_i} \leftarrow g_{0.5}^{cd,*=don't\ care}(D_1)$, $Cover_{TRN_i} = \{D_1, D_2, D_8\}$.

$Cover_{TRN_i} \leftarrow g_{0.5}^{cd,*=don't\ care}(D_2)$, $Cover_{TRN_i} = \{D_1, D_2, D_6, D_8, D_{14}\}$.

$Cover_{TRN_i} \leftarrow g_{0.5}^{cd,*=don't\ care}(D_3)$.

$Cover_{TRN_i} = \{D_1, D_2, D_3, D_4, D_5, D_6, D_7, D_8, D_9, D_{10}, D_{12}, D_{13}, D_{14}\}$.

$Cover_{TRN_i} \not\leftarrow g_{0.5}^{cd,*=don't\ care}(D_4)$, without any change.

$Cover_{TRN_i} \not\leftarrow g_{0.5}^{cd,*=don't\ care}(D_5)$, without any change.

$Cover_{TRN_i} \not\leftarrow g_{0.5}^{cd,*=don't\ care}(D_6)$, without any change.

$Cover_{TRN_i} \leftarrow g_{0.5}^{cd,*=don't\ care}(D_7)$, $Cover_{TRN_i} = TRN_i$, the training decision system has been covered.

Thus,

$$Cover_{TRN_i} = \{g_{0.5}^{cd,*=don't\ care}(D_1), g_{0.5}^{cd,*=don't\ care}(D_2), g_{0.5}^{cd,*=don't\ care}(D_3),$$

$$g_{0.5}^{cd,*=don't\ care}(D_7)\}.$$

Majority voting gives granules,

$MV(g_{0.5}^{cd,*=don't\ care}(D_1))$: Sunny Hot High Weak No.

$MV(g_{0.5}^{cd,*=don't\ care}(D_2))$: Sunny Hot High Strong No.

$MV(g_{0.5}^{cd,*=don't\ care}(D_3))$: Overcast Mild $*$ Weak Yes.

$MV(g_{0.5}^{cd,*=don't\ care}(D_7))$: Overcast Mild $*$ Weak Yes.

The repairing phase for the strategy A follows,

$g_{0.5,a_3}^{cd,*=don't\ care}(MV(g^{cd,*=don't\ care}(D_3))) = \{D_4, D_7, D_{11}, D_{12}, D_{13}\}$,
$*$ is replaced by Normal, we get Overcast Mild Normal Weak Yes.

$g_{0.5,a_3}^{cd,*=don't\ care}(MV(g^{cd,*=don't\ care}(D_7))) = \{D_4, D_7, D_{11}, D_{12}, D_{13}\}$,
$*$ is replaced by Normal, we get Overcast Mild Normal Weak Yes.

The repairing phase for the strategy B follows,

$g_{0.5,a_3}^{cd,*=*}(MV(g^{cd,*=don't\ care}(D_3))) = D_4, D_{12}, D_{13}$,

$*$ is replaced by *High*, we get *Overcast Mild High Weak Yes*.

$g_{0.5,a_3}^{cd,*=*}(MV(g^{cd,*=don't\ care}(D_7))) = D_4, D_{12}, D_{13}$,

$*$ is replaced by *High*, we get *Overcast Mild High Weak Yes*.

In case of strategies C and D, the first phase consists of the following steps,

$g_{0.5}^{cd,*=don't\ care}(D_1) = \{D_1, D_2, D_8\}$.

$g_{0.5}^{cd,*=don't\ care}(D_2) = \{D_1, D_2, D_{14}\}$.

$g_{0.5}^{cd,*=don't\ care}(D_3) = \{D_3, D_5, D_9, D_{10}, D_{13}\}$.

$g_{0.5}^{cd,*=don't\ care}(D_4) = \{D_4, D_5, D_{10}, D_{12}\}$.

$g_{0.5}^{cd,*=don't\ care}(D_5) = \{D_3, D_5, D_9, D_{10}\}$.

$g_{0.5}^{cd,*=don't\ care}(D_6) = \{D_6, D_{14}\}$.

$g_{0.5}^{cd,*=don't\ care}(D_7) = \{D_7, D_{11}, D_{12}, D_{13}\}$.

$g_{0.5}^{cd,*=don't\ care}(D_8) = \{D_1, D_8, D_{14}\}$.

$g_{0.5}^{cd,*=don't\ care}(D_9) = \{D_3, D_5, D_9, D_{10}\}$.

$g_{0.5}^{cd,*=don't\ care}(D_{10}) = \{D_3, D_4, D_5, D_9, D_{10}\}$.

$g_{0.5}^{cd,*=don't\ care}(D_{11}) = \{D_7, D_{11}, D_{12}\}$.

$g_{0.5}^{cd,*=don't\ care}(D_{12}) = \{D_4, D_7, D_{11}, D_{12}\}$.

$g_{0.5}^{cd,*=don't\ care}(D_{13}) = \{D_3, D_7, D_{13}\}$.

$g_{0.5}^{cd,*=don't\ care}(D_{14}) = \{D_2, D_6, D_8, D_{14}\}$.

The covering phase yields,

$Cover_{TRN_i} \leftarrow \emptyset$.

$Cover_{TRN_i} \leftarrow g_{0.5}^{cd,*=*}(D_1)$, $Cover_{TRN_i} = \{D_1, D_2, D_8\}$.

$Cover_{TRN_i} \leftarrow g_{0.5}^{cd,*=*}(D_2)$, $Cover_{TRN_i} = \{D_1, D_2, D_8, D_{14}\}$.

$Cover_{TRN_i} \leftarrow g_{0.5}^{cd,*=*}(D_3)$, $Cover_{TRN_i} = \{D_1, D_2, D_3, D_5, D_8, D_9, D_{10}, D_{13}, D_{14}\}$.

$Cover_{TRN_i} \leftarrow g_{0.5}^{cd,*=*}(D_4)$, $Cover_{TRN_i} = \{D_1, D_2, D_3, D_4, D_5, D_8, D_9, D_{10}, D_{12},$
$D_{13}, D_{14}\}$.

$Cover_{TRN_i} \nleftarrow g_{0.5}^{cd,*=*}(D_5)$, with no change.

$Cover_{TRN_i} \leftarrow g_{0.5}^{cd,*=*}(D_6)$, $Cover_{TRN_i} = \{D_1, D_2, D_3, D_4, D_5, D_6, D_8, D_9, D_{10},$
$D_{12}, D_{13}, D_{14}\}$.

$Cover_{TRN_i} \leftarrow g_{0.5}^{cd,*=*}(D_7)$, $Cover_{TRN_i} = TRN_i$, the training decision system has
been covered.

Thus

$$Cover_{TRN_i} = \{g_{0.5}^{cd,*=*}(D_1), g_{0.5}^{cd,*=*}(D_2), \ldots, g_{0.5}^{cd,*=*}(D_4),$$
$$g_{0.5}^{cd,*=*}(D_6), g_{0.5}^{cd,*=*}(D_7)\}.$$

Majority voting follows,

$MV(g_{0.5}^{cd,*=*}(D_1))$: *Sunny Hot High Weak No.*

$MV(g_{0.5}^{cd,*=*}(D_2))$: *Sunny Hot High Strong No.*

$MV(g_{0.5}^{cd,*=*}(D_3))$: *Rainy Cool * Weak Yes.*

$MV(g_{0.5}^{cd,*=*}(D_4))$: *Rainy Mild * Weak Yes.*

$MV(g_{0.5}^{cd,*=*}(D_6))$: *Rainy Cool Normal Strong No.*

$MV(g_{0.5}^{cd,*=*}(D_7))$: *Overcast Mild Normal Strong Yes.*

The repairing phase for the strategy C follows,

$g_{0.5,a_3}^{cd,*=don't\ care}(MV(g^{cd,*=*}(D_3))) = \{D_4, D_7, D_{13}\}$,

$*$ is replaced by *Normal*, we get *Rainy Cool Normal Weak Yes.*

$g_{0.5,a_3}^{cd,*=don't\ care}(MV(g^{cd,*=*}(D_4))) = \{D_4, D_7, D_{11}, D_{12}, D_{13}\}$,

$*$ is replaced by *Normal*, we get *Rainy Mild Normal Weak Yes.*

The repairing phase for the strategy D follows,

$g_{0.5,a_3}^{cd,*=don't\ care}(MV(g^{cd,*=*}(D_3))) = \{D_4\}$,

$*$ is replaced by *High*, we get *Rainy Cool High Weak Yes.*

$g_{0.5,a_3}^{cd,*=don't\ care}(MV(g^{cd,*=*}(D_4))) = \{D_4\}$,

$*$ is replaced by *High*, we get *Rainy Mild High Weak Yes.*

8.2 The Experimental Session

In this section we show basic procedure for experiments, the way of result validation and results of experiments with data selected from UCI Repository [12].

8.2.1 The Methodology of the Experiment

General procedure:

(1) The data set has been input.
(2) The data set has been split five times into five parts, for multiple Cross Validation.
(3) The training systems $TRN_i^{complete}$ have been granulated in decision concepts.
(4) The TST_i systems have been classified by proper complete granular reflections of $TRN_i^{complete}$ in radius r_{gran} using kNN classifier (the nil result).
(5) We have filled $TRN_i^{complete}$ with a fixed percent of randomly located stars (in our case it is 5 or 10 %).
(6) We have applied into TRN_i the strategy of missing values handling A, B, C or D.
(7) All TST_i systems have been classified by proper, repaired granular reflections of TRN_i for the radius r_{gran} by kNN classifier.
(8) The result has been evaluated by average value od classification from all Cross Validation tests.

8.2.2 Evaluation of Results

In our experiments we have computed bias of accuracy from 5 times Cross Validation results, by the already familiar formula,

$$AccBias = \frac{\sum_{i=1}^{5}(max(acc_1^{CV5}, acc_2^{CV5}, ..., acc_5^{CV5}) - acc_i^{CV5})}{5}, \tag{8.1}$$

where

$$Acc = \frac{\sum_{i=1}^{5} acc_i^{CV5}}{5}.$$

8.2.3 The Results of Experiments for Data Sets Damaged in 5 and 10%

To check the effectiveness of our methods we have carried out a series of experiments with data from UCI Repository [12]. We have performed five times Cross Validation five method. The result is evaluated with use of accuracy bias—see Eq. (8.1). As a reference classifier we use the same classification method as in Chaps. 5 and 6, viz., kNN in decision classes, where a class is winning if the summary distance of k-nearest objects from the class is the smallest. Parameter k is estimated on the sample of data based on Cross Validation five method.

We have two series of experiments, first one from Tables 8.3, 8.4, 8.5, 8.6, 8.7, 8.8, 8.9, 8.10, 8.11, 8.12, 8.13, 8.14 and 8.15 is with training data damaged in 5 %. In Tables 8.16, 8.17, 8.18, 8.19, 8.20, 8.21, 8.22, 8.23, 8.24, 8.25, 8.26, 8.27 and 8.28 data disturbed by stars in 10 % are examined (Table 8.29).

The result show effectiveness of our approach, for large enough enough radii of granulation we obtain results fully comparable to those for data with no missing values with additional reduction in object numbers, even up to 80 % of original training data size.

The effectiveness of methods and their behavior depend strictly on the type of data set. For instance for typical data sets with enough diversity of attribute values like, Australian Credit, Pima Indians Diabetes, Fertility Diagnosis, German Credit, Heart Disease, Hepatitis, Mushroom and SPECTF, the result is predictable, for strategies A and B we have a faster approximation for lower values of granulation radii, because in case of granulation with $* = don't\ care$ the granules contain more objects and thus the approximation is faster. In case of $* = *$, the approximation is similar to the *nil* result, but is slightly slower, because the stars could increase diversity of data, and the granules could contain a smaller number of objects which in consequence gives a larger number of granules inn coverings.

In case of data with less diversity of attribute domains, like Car Evaluation, Congressional Voting Records, Nursery, Soybean Large, SPECT, strategies work

Table 8.3 $5 \times$ CV-5; The result of experiments for A, B, C, D strategies of missing values handling versus complete data; data set *australian credit*; Concept dependent granulation; 5% *of missing values*; r_{gran} = Granulation radius; nil = result for data without missing values; *Acc* = Accuracy of classification; *AccBias* = Accuracy bias defined based on equation efAccCBiasEquation; *GranSize* = The size of data set after granulation in the fixed r

r_{gran}	Acc					AccBias					GranSize				
	nil	A	B	C	D	nil	A	B	C	D	nil	A	B	C	D
0	0.774	0.772	0.772	0.772	0.772	0.002	0.002	0.002	0.002	0.002	2	2	2	2	2
0.0714286	0.774	0.771	0.771	0.771	0.771	0.003	0.003	0.003	0.003	0.003	2.44	2.24	2.24	2.6	2.44
0.142857	0.765	0.772	0.772	0.772	0.773	0.01	0.002	0.002	0.024	0.023	3.32	2.6	2.56	4.12	4.04
0.214286	0.784	0.775	0.775	0.788	0.786	0.006	0.011	0.011	0.011	0.012	5.2	3.76	3.76	6.72	6.72
0.285714	0.794	0.776	0.776	0.799	0.798	0.013	0.009	0.021	0.012	0.015	8.56	5.6	5.76	12.56	12.2
0.357143	0.814	0.806	0.803	0.818	0.821	0.01	0.017	0.014	0.01	0.016	15.04	8.92	8.8	25.36	25.52
0.428571	0.841	0.801	0.803	0.837	0.839	0.007	0.021	0.013	0.008	0.011	31.52	15.84	15.88	50.8	50.56
0.5	0.842	0.824	0.824	0.842	0.846	0.007	0.012	0.012	0.016	0.006	69.24	32.52	32.28	112.72	112.56
0.571429	0.84	0.838	0.833	0.846	0.843	0.008	0.01	0.01	0.005	0.008	156.16	69.04	67.24	231.36	231.6
0.642857	0.846	0.835	0.839	0.845	0.843	0.003	0.009	0.018	0.016	0.012	317	149.6	154	390.4	390.52
0.714286	0.845	0.84	0.839	0.848	0.849	0.012	0.007	0.01	0.011	0.005	467.28	308.16	308.2	504.32	504.84
0.785714	0.852	0.844	0.843	0.849	0.848	0.016	0.009	0.009	0.019	0.006	536.4	462.52	462.6	544.36	544.44
0.857143	0.856	0.849	0.85	0.847	0.848	0.012	0.005	0.007	0.007	0.008	547.32	532.44	532.56	549.32	549.36
0.928571	0.858	0.852	0.851	0.85	0.851	0.013	0.008	0.008	0.011	0.01	548.88	547.76	547.76	550.96	550.96
1	0.857	0.85	0.85	0.849	0.849	0.014	0.011	0.011	0.01	0.01	552	551.64	551.64	552	552

Table 8.4 5 × CV-5; The result of experiments for A, B, C, D strategies of missing values handling versus complete data; data set *car evaluation*; Concept dependent granulation; 5% *of missing values*; r_{gran} = Granulation radius; nil = result for data without missing values; *Acc* = Accuracy of classification; *AccBias* = Accuracy bias defined based on equation efAcccBiasEquation; *GranSize* = The size of data set after granulation in the fixed r

r_{gran}	Acc					AccBias					GranSize				
	nil	A	B	C	D	nil	A	B	C	D	nil	A	B	C	D
0	0.324	0.322	0.322	0.322	0.322	0.008	0.011	0.011	0.011	0.011	4	4	4	4	4
0.166667	0.393	0.371	0.367	0.412	0.421	0.012	0.021	0.027	0.019	0.027	8.16	6.88	7.16	9.96	10.08
0.333333	0.535	0.463	0.473	0.561	0.559	0.011	0.049	0.063	0.016	0.012	16.52	13.08	13.44	21.16	21.52
0.5	0.679	0.599	0.577	0.734	0.727	0.027	0.019	0.03	0.015	0.022	38.32	28.4	26.76	53.44	53.36
0.666667	0.805	0.758	0.768	0.825	0.826	0.022	0.029	0.022	0.005	0.007	106.4	72.64	74.88	163.76	163.76
0.833333	0.869	0.845	0.848	0.868	0.868	0.011	0.008	0.012	0.008	0.006	372.24	240.44	241.16	537.08	537.72
1	0.942	0.91	0.908	0.925	0.925	0.008	0.011	0.006	0.006	0.006	1382.4	971.72	973.8	1371.6	1371.6

Table 8.5 5 × CV-5; The result of experiments for A, B, C, D strategies of missing values handling versus complete data; data set *diabetes*; Concept dependent granulation; 5% *of missing values*; r_{gran} = Granulation radius; nil = result for data without missing values; *Acc* = Accuracy of classification; *AccBias* = Accuracy bias defined based on equation efAcccBiasEquation; *GranSize* = The size of data set after granulation in the fixed *r*

r_{gran}	Acc					AccBias					GranSize				
	nil	A	B	C	D	*nil*	A	B	C	D	*nil*	A	B	C	D
0	0.596	0.601	0.601	0.601	0.601	0.003	0.011	0.011	0.011	0.011	2	2	2	2	2
0.125	0.621	0.592	0.592	0.61	0.609	0.004	0.029	0.007	0.017	0.018	33.68	6.16	5.52	34.64	34.76
0.25	0.629	0.592	0.565	0.624	0.62	0.019	0.019	0.024	0.023	0.007	154.04	20.48	18.96	159.04	159.2
0.375	0.641	0.601	0.601	0.636	0.642	0.014	0.016	0.022	0.023	0.017	366.84	84.68	79.52	380.52	379.84
0.5	0.649	0.624	0.629	0.649	0.651	0.013	0.018	0.015	0.011	0.005	539.76	257.32	255.36	551	550.72
0.625	0.655	0.642	0.644	0.648	0.652	0.009	0.015	0.025	0.008	0.01	609.92	478.04	482.36	610.6	610.56
0.75	0.654	0.649	0.651	0.651	0.649	0.01	0.012	0.013	0.014	0.011	614.4	592.16	593.64	614.4	614.4
0.875	0.654	0.65	0.649	0.649	0.649	0.01	0.01	0.009	0.011	0.011	614.4	612.92	612.96	614.4	614.4
1	0.654	0.649	0.649	0.649	0.649	0.01	0.011	0.011	0.011	0.011	614.4	614.36	614.36	614.4	614.4

Table 8.6 5 × CV-5: The result of experiments for A, B, C, D strategies of missing values handling versus complete data; data set *fertility diagnosis*; Concept dependent granulation; 5% *of missing values*; r_{gran} = Granulation radius; nil = result for data without missing values; *Acc* = Accuracy of classification; *AccBias* = Accuracy bias defined based on equation efAcccBiasEquation; *GranSize* = The size of data set after granulation in the fixed *r*

r_{gran}	Acc					AccBias					GranSize				
	nil	A	B	C	D	nil	A	B	C	D	nil	A	B	C	D
0	0.438	0.446	0.446	0.446	0.446	0.032	0.024	0.024	0.024	0.024	2	2	2	2	2
0.111111	0.438	0.45	0.45	0.476	0.476	0.032	0.02	0.02	0.044	0.044	2.24	2.04	2.04	2.64	2.64
0.222222	0.498	0.464	0.464	0.522	0.522	0.012	0.036	0.036	0.038	0.038	3.76	3.12	3.12	4.16	4.16
0.333333	0.586	0.564	0.57	0.602	0.604	0.044	0.046	0.04	0.058	0.066	5.84	4.84	4.84	7.24	7.24
0.444444	0.718	0.648	0.648	0.778	0.78	0.052	0.052	0.052	0.042	0.05	10.44	7.72	7.72	13.72	13.72
0.555556	0.838	0.804	0.806	0.844	0.844	0.012	0.026	0.024	0.016	0.016	18.88	13.88	13.88	24.56	24.56
0.666667	0.866	0.85	0.848	0.868	0.864	0.034	0.03	0.032	0.012	0.016	35.36	26.88	26.84	44.36	44.44
0.777778	0.874	0.868	0.868	0.88	0.874	0.016	0.012	0.012	0.01	0.016	58.04	46.96	46.96	65.4	65.4
0.888889	0.876	0.874	0.876	0.876	0.876	0.004	0.016	0.024	0.014	0.014	74.12	68.12	68.28	76.76	76.8
1	0.876	0.876	0.876	0.874	0.874	0.004	0.014	0.014	0.006	0.006	80	79.28	79.28	80	80

Table 8.7 $5 \times$ CV-5; The result of experiments for A, B, C, D strategies of missing values handling versus complete data; data set *german credit*; Concept dependent granulation; 5% *of missing values*; $r_{gran} =$ Granulation radius; nil = result for data without missing values; $Acc =$ Accuracy of classification; $AccBias =$ Accuracy bias defined based on equation efAcccBiasEquation; $GranSize =$ The size of data set after granulation in the fixed r

r_{gran}	Acc					AccBias					GranSize				
	nil	A	B	C	D	nil	A	B	C	D	nil	A	B	C	D
0	0.563	0.564	0.564	0.564	0.564	0.013	0.01	0.01	0.01	0.01	2	2	2	2	2
0.05	0.563	0.564	0.564	0.564	0.564	0.013	0.01	0.01	0.01	0.01	2	2	2	2	2
0.1	0.563	0.564	0.564	0.565	0.565	0.013	0.01	0.01	0.012	0.012	2	2	2	2.24	2.28
0.15	0.564	0.564	0.564	0.565	0.565	0.012	0.01	0.01	0.009	0.009	2.6	2.24	2.16	3.08	3.16
0.2	0.572	0.567	0.57	0.58	0.578	0.015	0.018	0.015	0.009	0.004	3.4	2.6	2.52	4.36	4.4
0.25	0.594	0.572	0.575	0.59	0.588	0.016	0.015	0.017	0.012	0.023	4.92	3.6	3.4	6.52	6.4
0.3	0.603	0.581	0.582	0.629	0.627	0.023	0.007	0.012	0.007	0.007	7	4.8	4.96	10.76	10.24
0.35	0.632	0.606	0.611	0.655	0.665	0.01	0.015	0.024	0.014	0.02	11.6	7.2	7.28	16.84	16.4
0.4	0.654	0.635	0.636	0.672	0.68	0.017	0.025	0.007	0.021	0.017	18.84	10.96	10.84	29.92	30.56
0.45	0.687	0.65	0.656	0.686	0.691	0.009	0.011	0.014	0.01	0.025	32.92	16.6	17.6	54.84	54.48
0.5	0.689	0.67	0.66	0.699	0.704	0.016	0.025	0.011	0.004	0.012	57.4	29.56	28.68	97.64	97.04
0.55	0.706	0.67	0.676	0.71	0.709	0.017	0.037	0.032	0.008	0.004	105.96	51.32	52.52	176.52	178
0.6	0.708	0.695	0.701	0.721	0.709	0.009	0.012	0.018	0.009	0.014	187.52	93.28	91.4	301.2	301.08
0.65	0.717	0.707	0.706	0.719	0.719	0.013	0.012	0.007	0.003	0.01	322.16	166.56	162.68	463.88	463.48
0.7	0.721	0.714	0.714	0.715	0.72	0.005	0.01	0.005	0.008	0.012	485	292.04	290.4	624.64	621.56
0.75	0.728	0.711	0.719	0.722	0.726	0.007	0.006	0.027	0.006	0.012	648.52	468.08	468.56	735.24	735.56
0.8	0.733	0.72	0.725	0.721	0.724	0.008	0.014	0.003	0.008	0.007	751.04	644.36	643.92	783.32	783.24
0.85	0.729	0.724	0.724	0.724	0.723	0.01	0.006	0.008	0.007	0.004	789.56	751.12	751.44	796.44	796.32
0.9	0.729	0.724	0.725	0.721	0.723	0.007	0.005	0.006	0.007	0.005	796.52	789.32	789.28	798.92	798.92
0.95	0.729	0.723	0.723	0.722	0.723	0.006	0.009	0.008	0.008	0.007	798.64	797.8	797.72	799.8	799.8
1	0.729	0.723	0.723	0.723	0.723	0.006	0.007	0.007	0.007	0.007	800	799.84	799.84	800	800

Table 8.8 $5 \times$ CV-5; The result of experiments for A, B, C, D strategies of missing values handling versus complete data; *heart disease*; Concept dependent granulation; 5% *of missing values*; r_{gran} = Granulation radius; nil = result for data without missing values; *Acc* = Accuracy of classification; *AccBias* = Accuracy bias defined based on equation efAcccBiasEquation; *GranSize* = The size of data set after granulation in the fixed r

r_{gran}	Acc					AccBias					GranSize				
	nil	A	B	C	D	*nil*	A	B	C	D	*nil*	A	B	C	D
0	0.788	0.784	0.784	0.784	0.784	C.023	0.027	0.027	0.027	0.027	2	2	2	2	2
0.0769231	0.788	0.784	0.784	0.784	0.784	C.023	0.027	0.027	0.027	0.027	2.04	2	2	2.16	2.16
0.153846	0.787	0.784	0.784	0.786	0.786	C.024	0.027	0.027	0.025	0.025	2.96	2.32	2.32	3.76	3.76
0.230769	0.791	0.786	0.784	0.789	0.79	C.035	0.033	0.034	0.022	0.021	4.72	3.32	3.4	6.12	6.2
0.307692	0.803	0.792	0.793	0.804	0.808	C.019	0.027	0.026	0.007	0.014	8.48	6.04	6.24	11.56	11.44
0.384615	0.827	0.803	0.803	0.81	0.811	C.014	0.016	0.019	0.013	0.015	16.8	10.4	10.4	23.68	23.6
0.461538	0.819	0.824	0.826	0.821	0.822	C.021	0.024	0.022	0.035	0.03	35.6	18.88	18.88	49.32	49.08
0.538462	0.82	0.819	0.817	0.816	0.812	C.024	0.018	0.024	0.014	0.007	70.28	35.44	34.84	94.4	94.6
0.615385	0.814	0.821	0.821	0.808	0.811	C.004	0.023	0.016	0.014	0.011	127.76	71.92	71.92	151.6	151.72
0.692308	0.821	0.82	0.819	0.821	0.819	C.009	0.01	0.011	0.008	0.014	180.64	127	126.76	195.24	195.08
0.769231	0.82	0.821	0.823	0.823	0.821	C.006	0.009	0.007	0.01	0.013	210	182.16	181.96	213.24	213.24
0.846154	0.821	0.823	0.823	0.82	0.821	C.008	0.01	0.01	0.013	0.013	216	210.08	210.08	215.96	215.96
0.923077	0.821	0.821	0.821	0.821	0.821	C.008	0.016	0.016	0.013	0.013	216	215.64	215.64	216	216
1	0.821	0.821	0.821	0.821	0.821	0.008	0.013	0.013	0.013	0.013	216	216	216	216	216

Table 8.9 5 × CV-5; The result of experiments for A, B, C, D strategies of missing values handling versus complete data; data set *hepatitis*; Concept dependent granulation; *5% of missing values*; r_{gran} = Granulation radius; nil = result for data without missing values; *Acc* = Accuracy of classification; *AccBias* = Accuracy bias defined based on equation efAcccBiasEquation; *GranSize* = The size of data set after granulation in the fixed r

r_{gran}	Acc					AccBias					GranSize				
	nil	A	B	C	D	nil	A	B	C	D	nil	A	B	C	D
0	0.815	0.817	0.817	0.817	0.817	0.017	0.015	0.015	0.015	0.015	2	2	2	2	2
0.0526316	0.815	0.817	0.817	0.817	0.817	0.017	0.015	0.015	0.015	0.015	2	2	2	2	2
0.105263	0.815	0.817	0.817	0.817	0.817	0.017	0.015	0.015	0.015	0.015	2	2	2	2	2
0.157895	0.815	0.817	0.817	0.817	0.817	0.017	0.015	0.015	0.015	0.015	2.04	2.04	2.04	2.24	2.24
0.210526	0.815	0.817	0.817	0.817	0.817	0.017	0.015	0.015	0.015	0.015	2.24	2.12	2.12	2.52	2.52
0.263158	0.818	0.817	0.817	0.821	0.821	0.014	0.015	0.015	0.012	0.012	2.84	2.24	2.24	3.28	3.2
0.315789	0.817	0.817	0.817	0.836	0.836	0.015	0.015	0.015	0.028	0.028	3.84	2.92	2.92	4.6	4.6
0.368421	0.817	0.819	0.819	0.844	0.843	0.022	0.013	0.013	0.021	0.022	5.12	3.68	3.6	6.8	6.84
0.421053	0.839	0.83	0.83	0.848	0.85	0.013	0.022	0.022	0.017	0.021	7.44	5.08	5.08	10.84	10.88
0.473684	0.855	0.839	0.839	0.88	0.877	0.022	0.013	0.013	0.03	0.032	10.8	7.44	7.44	17.48	17.28
0.526316	0.886	0.852	0.852	0.872	0.874	0.017	0.006	0.006	0.025	0.023	18.32	11.16	11.2	29.28	29.28
0.578947	0.883	0.874	0.872	0.885	0.894	0.014	0.036	0.037	0.018	0.015	30.48	17.6	17.6	44.48	44.6
0.631579	0.884	0.88	0.883	0.879	0.877	0.019	0.03	0.027	0.018	0.019	46.88	29.68	29.64	65.6	65.96
0.684211	0.883	0.862	0.858	0.893	0.886	0.008	0.028	0.039	0.01	0.01	69.84	44.68	44.8	86.88	86.88
0.736842	0.876	0.885	0.889	0.884	0.888	0.027	0.012	0.008	0.032	0.009	89.84	67.72	68.2	104.96	104.8
0.789474	0.88	0.886	0.889	0.886	0.886	0.017	0.017	0.014	0.017	0.01	109.76	90.88	91.32	115.92	115.92
0.842105	0.883	0.894	0.894	0.895	0.886	0.021	0.022	0.015	0.021	0.017	116.92	108.6	109	120.68	120.6
0.894737	0.883	0.893	0.894	0.893	0.889	0.021	0.017	0.015	0.017	0.021	121	118.08	118.08	122.84	122.84
0.947368	0.883	0.889	0.889	0.889	0.889	0.021	0.014	0.014	0.014	0.014	121.92	121.2	121.24	123.68	123.68
1	0.883	0.888	0.888	0.888	0.888	0.021	0.015	0.015	0.015	0.015	124	123.68	123.68	124	124

Table 8.10 $5 \times$ CV-5; The result of experiments for A, B, C, D strategies of missing values handling versus complete data; data set *congressional voting records*; Concept dependent granulation; 5% *of missing values*; r_{gran} = Granulation radius; nil = result for data without missing values; *Acc* = Accuracy of classification; *AccBias* = Accuracy bias defined based on equation efAcccBiasEquation; *GranSize* = The size of data set after granulation in the fixed r

r_{gran}	Acc					AccBias					GranSize				
	nil	A	B	C	D	nil	A	B	C	D	nil	A	B	C	D
0	0.901	0.9	0.9	0.9	0.9	0	0.001	0.001	0.001	0.001	2	2	2	2	2
0.0625	0.901	0.9	0.9	0.9	0.9	0	0.001	0.001	0.001	0.001	2	2	2	2	2
0.125	0.901	0.9	0.9	0.9	0.9	0	0.001	0.001	0.001	0.001	2	2	2	2.16	2.16
0.1875	0.901	0.9	0.9	0.9	0.9	0	0.001	0.001	0.001	0.001	2.28	2.04	2.04	2.8	2.8
0.25	0.901	0.9	0.9	0.901	0.901	0	0.001	0.001	0	0	3.04	2.24	2.2	3.12	3.12
0.3125	0.901	0.9	0.899	0.901	0.901	0.002	0.001	0.001	0.007	0.007	3.44	2.8	2.8	4.2	4.2
0.375	0.898	0.899	0.899	0.899	0.899	0.003	0.002	0.002	0.004	0.004	4.68	3.84	3.84	5.4	5.28
0.4375	0.902	0.898	0.9	0.905	0.906	0.011	0.005	0.003	0.01	0.009	5.68	4.88	4.92	7.12	7.2
0.5	0.905	0.904	0.904	0.916	0.916	0.012	0.011	0.011	0.008	0.008	6.92	6	5.92	9.72	9.72
0.5625	0.919	0.913	0.912	0.931	0.929	0.006	0.005	0.005	0.021	0.027	9	7.88	8	13.12	13.64
0.625	0.921	0.913	0.913	0.947	0.948	0.024	0.009	0.009	0.007	0.006	12.48	10.44	10.44	21.4	21
0.6875	0.943	0.939	0.938	0.95	0.952	0.005	0.008	0.005	0.007	0.006	19.44	15.44	15.88	34.76	34.84
0.75	0.954	0.949	0.95	0.949	0.949	0.007	0.011	0.011	0.01	0.014	31.52	24.44	24.28	58.88	58.64
0.8125	0.952	0.949	0.947	0.952	0.951	0.014	0.01	0.005	0.004	0.003	54.16	41.68	41.84	97.4	97.36
0.875	0.95	0.94	0.942	0.947	0.949	0.011	0.005	0.012	0.007	0.008	91.28	74.68	74.68	154.24	153.8
0.9375	0.948	0.943	0.947	0.948	0.946	0.006	0.007	0.005	0.006	0.003	144.96	124.4	124.44	227.52	227.56
1	0.949	0.949	0.949	0.948	0.948	0.002	0.01	0.01	0.009	0.009	214.12	198.32	198.48	299.24	299.24

Table 8.11 5 × CV-5; The result of experiments for A, B, C, D strategies of missing values handling versus complete data; data set *mushroom*; Concept dependent granulation; 5% *of missing values*; r_{gran} = Granulation radius; nil = result for data without missing values; *Acc* = Accuracy of classification; *AccBias* = Accuracy bias defined based on equation efAccBiasEquation; *GranSize* = The size of data set after granulation in the fixed *r*

r_{gran}	Acc					AccBias					GranSize				
	nil	A	B	C	D	nil	A	B	C	D	nil	A	B	C	D
0	0.887	0.887	0.887	0.887	0.887	0.001	0.001	0.001	0.001	0.001	2	2	2	2	2
0.0454545	0.887	0.887	0.887	0.887	0.887	0.001	0.001	0.001	0.001	0.001	2	2	2	2	2
0.0909091	0.887	0.887	0.887	0.887	0.887	0.001	0.001	0.001	0.001	0.001	2	2	2	2.04	2.08
0.136364	0.887	0.887	0.887	0.887	0.887	0.001	0.001	0.001	0.001	0.001	2	2	2	2.12	2.12
0.181818	0.887	0.887	0.887	0.887	0.887	0.001	0.001	0.001	0.001	0.001	2.08	2	2	2.52	2.72
0.227273	0.887	0.887	0.887	0.887	0.887	0.001	0.001	0.001	0.001	0.001	2.16	2.16	2.08	3.72	3.48
0.272727	0.887	0.887	0.887	0.888	0.887	0.001	0.001	0.001	0.002	0.001	2.4	2.28	2.4	5.56	5.72
0.318182	0.887	0.887	0.887	0.885	0.886	0	0.001	0.001	0.004	0.008	3.92	2.64	2.88	8.76	9.72
0.363636	0.887	0.887	0.888	0.89	0.886	0.003	0.001	0.001	0.003	0.004	5.92	4.2	3.84	14.68	14.52
0.409091	0.883	0.885	0.888	0.896	0.895	0.003	0.005	0.003	0.006	0.003	9.56	6.76	6.8	21.92	20.96
0.454545	0.893	0.887	0.887	0.914	0.912	0.003	0.002	0.005	0.005	0.009	15.88	10.56	10.28	34.04	33.4
0.5	0.917	0.894	0.893	0.946	0.942	0.014	0.007	0.012	0.005	0.011	21.56	16.88	15.16	43.24	40.52
0.545455	0.944	0.911	0.909	0.974	0.971	0.007	0.006	0.014	0.002	0.007	27.4	21.36	23.6	54.04	54
0.590909	0.967	0.936	0.932	0.983	0.988	0.003	0.004	0.008	0.004	0.002	40.28	29.88	29.48	62.6	62.32
0.636364	0.984	0.964	0.966	0.992	0.992	0.002	0.005	0.006	0.003	0.002	45.08	43.08	41.52	75.04	75.36
0.681818	0.991	0.978	0.981	0.997	0.997	0.002	0.005	0.004	0.001	0.001	48.64	49.68	50.72	102	104.12
0.727273	0.994	0.991	0.989	0.999	0.999	0.001	0.001	0.002	0	0.001	48.76	52.84	55	175.52	171.8

(continued)

Table 8.11 (continued)

r_{gran}	Acc					AccBias					GranSize				
0.772727	0.998	0.995	0.996	1	1	0.001	0.002	0.001	0	0	57.4	61.08	61.84	327.68	332.64
0.818182	0.999	0.998	0.999	1	1	0	0.001	0.001	0	0	92.4	80.44	79.16	687.16	687.56
0.863636	1	1	1	1	1	0	0	0	0	0	193.64	148.4	149.36	1472.44	1467.12
0.909091	1	1	1	1	1	0	0	0	0	0	515.92	369.72	361.88	2940	2936.04
0.954545	1	1	1	1	1	0	0	0	0	0	1752.48	1159.84	1155.32	5029.6	5023.52
1	1	1	1	1	1	0	0	0	0	0	6499.2	4463.8	4455.84	6490.48	6490.48

Table 8.12 5 × CV-5: The result of experiments for A, B, C, D strategies of missing values handling versus complete data; *Nursery*; Concept dependent granulation; 5% *of missing values*; r_{gran} = Granulation radius; nil = result for data without missing values; *Acc* = Accuracy of classification; *AccBias* = Accuracy bias defined based on equation efAccBiasEquation; *GranSize* = The size of data set after granulation in the fixed r

r_{gran}	Acc					AccBias					GranSize				
	nil	A	B	C	D	nil	A	B	C	D	nil	A	B	C	D
0	0.373	0.364	0.364	0.364	0.364	0.041	0.047	0.047	0.047	0.047	4.92	4.92	4.92	4.92	4.92
0.125	0.4	0.379	0.385	0.435	0.43	0.038	0.039	0.041	0.05	0.057	7.28	6.72	6.68	10.96	10.72
0.25	0.477	0.44	0.439	0.517	0.513	0.054	0.055	0.052	0.053	0.051	13.96	10.48	11	19.88	20.36
0.375	0.516	0.493	0.484	0.545	0.54	0.054	0.052	0.053	0.055	0.052	26.84	19.32	19.68	44.08	44.6
0.5	0.533	0.526	0.524	0.568	0.561	0.049	0.051	0.056	0.048	0.047	59.56	41.92	43.8	114.48	114.72
0.625	0.562	0.551	0.542	0.588	0.589	0.053	0.054	0.065	0.056	0.053	162.88	105	105.52	350.2	348.76
0.75	0.596	0.584	0.579	0.64	0.64	0.055	0.055	0.056	0.066	0.067	550.36	326.32	327.24	1209.52	1208.04
0.875	0.643	0.637	0.639	0.679	0.683	0.059	0.067	0.06	0.092	0.096	2369.72	1324.12	1326.68	4146.4	4141.92
1	0.556	0.684	0.684	0.596	0.596	0.028	0.047	0.043	0.069	0.069	10368	6443.28	6443.92	10290.6	10290.6

Table 8.13 5 × CV-5; The result of experiments for A, B, C, D strategies of missing values handling versus complete data; data set *soybean large*; Concept dependent granulation; 5% *of missing values*; r_{gran} = Granulation radius; nil = result for data without missing values; *Acc* = Accuracy of classification; *AccBias* = Accuracy bias defined based on equation efAcccBiasEquation; *GranSize* = The size of data set after granulation in the fixed r

r_{gran}	Acc					AccBias					GranSize				
	nil	A	B	C	D	nil	A	B	C	D	nil	A	B	C	D
0	0.681	0.686	0.686	0.686	0.686	0.007	0.016	0.016	0.016	0.016	19	19	19	19	19
...	...	...	...	...	...	...	...	...	...	...	...	...	...	...	...
0.371429	0.681	0.686	0.686	0.687	0.687	0.007	0.016	0.016	0.016	0.016	19	19	19	19.16	19.16
0.4	0.681	0.686	0.686	0.688	0.688	0.007	0.016	0.016	0.015	0.015	19	19	19	19.12	19.12
0.428571	0.681	0.686	0.686	0.689	0.689	0.007	0.016	0.016	0.017	0.017	19	19	19	19.44	19.48
0.457143	0.681	0.686	0.686	0.689	0.689	0.007	0.016	0.016	0.016	0.016	19.12	19	19	19.96	19.92
0.485714	0.682	0.688	0.688	0.692	0.693	0.009	0.02	0.02	0.017	0.016	19.36	19.16	19.16	21.16	21.2
0.514286	0.684	0.687	0.687	0.699	0.7	0.009	0.02	0.02	0.014	0.013	19.68	19.28	19.24	22.8	22.56
0.542857	0.686	0.687	0.687	0.706	0.706	0.005	0.016	0.016	0.022	0.022	20.48	19.6	19.6	26.04	26.04
0.571429	0.693	0.693	0.693	0.717	0.716	0.014	0.019	0.019	0.014	0.015	21.96	20.16	20.16	30.04	29.84
0.6	0.7	0.695	0.693	0.736	0.736	0.012	0.018	0.02	0.017	0.016	23.96	21.56	21.36	34.8	34.92
0.628571	0.71	0.701	0.702	0.769	0.771	0.014	0.016	0.017	0.014	0.013	27.24	23.56	23.6	39.88	40.08
0.657143	0.735	0.716	0.716	0.787	0.788	0.011	0.014	0.013	0.031	0.032	31.04	26.32	26.28	48.28	48.28
0.685714	0.751	0.725	0.724	0.816	0.814	0.031	0.006	0.007	0.012	0.006	34.6	30.56	30.36	58.72	58.6
0.714286	0.768	0.74	0.741	0.846	0.843	0.013	0.006	0.008	0.009	0.013	42.08	36.24	36	72.2	72.08
0.742857	0.801	0.76	0.764	0.85	0.848	0.008	0.011	0.013	0.025	0.028	48.24	41.64	41.4	93.8	93.8
0.771429	0.831	0.799	0.802	0.864	0.862	0.017	0.02	0.022	0.012	0.023	58.48	50.36	50.48	123.6	123.96
0.8	0.845	0.826	0.826	0.87	0.873	0.008	0.022	0.024	0.015	0.018	72.8	61.32	61.04	163.48	163.4

(continued)

Table 8.13 (continued)

r_{gran}	Acc					AccBias					GranSize				
0.828571	0.864	0.852	0.851	0.893	0.889	0.01	0.011	0.01	0.02	0.01	91.96	77.84	77.2	218.4	217.92
0.857143	0.867	0.854	0.852	0.896	0.896	0.013	0.006	0.008	0.011	0.008	118.12	101.52	101.2	285.68	285.16
0.885714	0.878	0.874	0.876	0.907	0.907	0.016	0.013	0.012	0.011	0.014	161.44	138.24	138.48	361.32	360.84
0.914286	0.886	0.876	0.877	0.91	0.906	0.003	0.013	0.016	0.005	0.009	221.68	193.36	193.8	437.4	437.32
0.942857	0.889	0.881	0.888	0.91	0.905	0.004	0.009	0.009	0.006	0.004	300.08	267.52	267.8	497.8	498.24
0.971429	0.915	0.905	0.907	0.91	0.909	0.007	0.004	0.009	0.005	0.006	404.8	372.96	372.52	533.28	533.12
1	0.915	0.904	0.904	0.904	0.904	0.005	0.004	0.005	0.006	0.006	509.72	491.36	491.52	544.96	544.96

Table 8.14 5 × CV-5; The result of experiments for A, B, C, D strategies of missing values handling versus complete data; data set *SPECT*; Concept dependent granulation; 5% *of missing values*; r_{gran} = Granulation radius; nil = result for data without missing values; *Acc* = Accuracy of classification; *AccBias* = Accuracy bias defined based on equation efAccBiasEquation; *GranSize* = The size of data set after granulation in the fixed *r*

r_{gran}	Acc					AccBias					GranSize				
	nil	A	B	C	D	nil	A	B	C	D	nil	A	B	C	D
0	0.557	0.539	0.539	0.539	0.539	0.053	0.041	0.041	0.041	0.041	2	2	2	2	2
0.0454545	0.557	0.539	0.539	0.539	0.539	0.053	0.041	0.041	0.041	0.041	2	2	2	2	2
0.0909091	0.557	0.539	0.539	0.539	0.539	0.053	0.041	0.041	0.041	0.041	2	2	2	2.04	2.04
0.136364	0.557	0.539	0.539	0.539	0.539	0.053	0.041	0.041	0.041	0.041	2.12	2.12	2.12	2.28	2.28
0.181818	0.566	0.539	0.539	0.541	0.541	0.045	0.041	0.041	0.047	0.047	2.48	2.2	2.2	2.56	2.56
0.227273	0.572	0.539	0.539	0.59	0.59	0.065	0.041	0.041	0.046	0.046	2.6	2.36	2.36	2.68	2.68
0.272727	0.576	0.551	0.561	0.611	0.611	0.06	0.067	0.057	0.029	0.029	2.88	2.48	2.52	3.08	3.08
0.318182	0.609	0.582	0.582	0.617	0.62	0.028	0.051	0.051	0.023	0.02	3.28	2.92	2.92	3.6	3.6
0.363636	0.622	0.612	0.61	0.638	0.641	0.029	0.021	0.023	0.04	0.044	3.8	3.24	3.2	4.24	4.2
0.409091	0.625	0.599	0.598	0.689	0.689	0.046	0.027	0.027	0.056	0.056	4.4	3.68	3.64	5.68	5.68
0.454545	0.691	0.641	0.641	0.719	0.718	0.017	0.055	0.055	0.019	0.016	5.8	4.56	4.56	7.6	7.52
0.5	0.714	0.665	0.665	0.749	0.75	0.028	0.021	0.021	0.007	0.006	7.04	5.52	5.52	9.84	9.92
0.545455	0.721	0.709	0.709	0.769	0.765	0.016	0.033	0.033	0.018	0.018	9.2	7	6.96	12.76	12.72
0.590909	0.766	0.739	0.738	0.772	0.772	0.036	0.033	0.034	0.03	0.03	11.76	8.64	8.52	17.32	17.32
0.636364	0.775	0.765	0.765	0.784	0.784	0.018	0.021	0.021	0.021	0.021	15.12	11.28	11.28	25.04	25.16
0.681818	0.785	0.771	0.776	0.79	0.786	0.017	0.019	0.014	0.015	0.02	22.08	16.04	16.12	38.92	39.24

(continued)

Table 8.14 (continued)

r_{gran}	Acc						AccBias					GranSize				
0.727273	0.799	0.786	0.786	0.798	0.798	0.798	0.013	0.023	0.023	0.015	0.014	33.32	24.48	24.48	58.36	58.32
0.772727	0.797	0.798	0.798	0.793	0.793	0.793	0.008	0.014	0.014	0.001	0.001	49.72	36.96	36.92	82.96	82.92
0.818182	0.793	0.795	0.794	0.794	0.794	0.794	0.001	0.014	0.015	0	0	71.8	56.04	56.52	112.4	112.4
0.863636	0.794	0.792	0.795	0.793	0.793	0.796	0	0.002	0.006	0.001	0.009	101.04	82.92	82.8	146.16	146.08
0.909091	0.794	0.794	0.794	0.807	0.807	0.801	0	0	0	0.024	0.019	131.16	117.04	117.04	173.6	173.6
0.954545	0.794	0.798	0.798	0.813	0.813	0.81	0	0.015	0.015	0.03	0.017	162.56	152.84	152.36	194.6	194.48
1	0.797	0.803	0.803	0.805	0.805	0.805	0.008	0.017	0.017	0.015	0.015	184.8	180.84	180.84	207.84	207.84

Table 8.15 5 × CV-5; The result of experiments for A, B, C, D strategies of missing values handling versus complete data; data set *SPECTF*; Concept dependent granulation; 5% *of missing values*; r_{gran} = Granulation radius; nil = result for data without missing values; *Acc* = Accuracy of classification; *AccBias* = Accuracy bias defined based on equation efAcccBiasEquation; *GranSize* = The size of data set after granulation in the fixed *r*

r_{gran}	Acc					AccBias					GranSize				
	nil	A	B	C	D	*nil*	A	B	C	D	*nil*	A	B	C	D
0	0.426	0.437	0.437	0.437	0.437	0.027	0.046	0.046	0.046	0.046	2	2	2	2	2
0.0227273	0.565	0.439	0.438	0.566	0.571	0.064	0.044	0.049	0.03	0.024	6.72	2.12	2.12	6.88	6.88
0.0454545	0.678	0.443	0.448	0.666	0.683	0.019	0.033	0.039	0.046	0.04	13.24	2.64	2.64	14.12	13.76
0.0681818	0.764	0.481	0.481	0.748	0.75	0.034	0.058	0.058	0.012	0.011	28.52	3.88	3.96	28.24	28.64
0.0909091	0.777	0.493	0.508	0.781	0.788	0.013	0.058	0.054	0.01	0.006	50.6	5.4	5.4	52.68	52.68
0.113636	0.786	0.557	0.566	0.785	0.79	0.008	0.035	0.038	0.013	0.008	85.08	7.96	7.84	90.84	90.48
0.136364	0.776	0.583	0.584	0.782	0.783	0.007	0.039	0.06	0.008	0.008	125.56	11.36	11.8	132.08	132.28
0.159091	0.796	0.637	0.653	0.787	0.79	0.009	0.019	0.029	0.023	0.005	166.44	21.2	19.44	174.16	174
0.181818	0.798	0.715	0.695	0.788	0.789	0.004	0.041	0.05	0.006	0.005	195.68	33.64	34.12	200.28	200.08
0.204545	0.79	0.745	0.737	0.791	0.788	0.008	0.035	0.024	0.007	0.01	209.32	53.44	53.04	210.36	210.4
0.227273	0.788	0.718	0.721	0.788	0.79	0.017	0.061	0.073	0.006	0.012	213	74.68	73.64	212.96	212.96
0.25	0.788	0.764	0.755	0.785	0.788	0.017	0.031	0.031	0.006	0.01	213.6	114.44	115.6	213.6	213.6
0.272727	0.788	0.776	0.779	0.789	0.788	0.017	0.023	0.026	0.013	0.01	213.6	152.4	152.68	213.6	213.6
0.295455	0.788	0.784	0.79	0.789	0.788	0.017	0.021	0.015	0.009	0.01	213.6	179.4	180.84	213.6	213.6

(continued)

Table 8.15 (continued)

r_{gran}	Acc						AccBias						GranSize		
0.318182	0.788	0.785	0.787	0.788	0.788	0.017	0.013	0.019	0.006	0.01	213.6	196.2	196.12	213.6	213.6
0.340909	0.788	0.79	0.788	0.788	0.788	0.017	0.012	0.01	0.01	0.01	213.6	206.36	206.56	213.6	213.6
0.363636	0.788	0.789	0.791	0.788	0.788	0.017	0.009	0.01	0.01	0.01	213.6	210.8	210.76	213.6	213.6
0.386364	0.788	0.785	0.787	0.789	0.788	0.017	0.009	0.007	0.009	0.01	213.6	212.72	212.68	213.6	213.6
0.409091	0.788	0.787	0.787	0.788	0.788	0.017	0.01	0.01	0.01	0.01	213.6	213.36	213.28	213.6	213.6
0.431818	0.788	0.788	0.788	0.788	0.788	0.017	0.01	0.01	0.01	0.01	213.6	213.52	213.52	213.6	213.6
...	...	...	...	...	...	...	...	...	...	...	...	...	...	...	...
1	0.788	0.788	0.788	0.788	0.788	0.017	0.01	0.01	0.01	0.01	213.6	213.6	213.6	213.6	213.6

Table 8.16 5 × CV-5; The result of experiments for A, B, C, D strategies of missing values handling versus complete data; data set *australian credit*; Concept dependent granulation; 10 % *of missing values*; r_{gran} = Granulation radius; nil = result for data without missing values; *Acc* = Accuracy of classification; *AccBias* = Accuracy bias defined based on equation efAcccBiasEquation; *GranSize* = The size of data set after granulation in the fixed r

r_{gran}	Acc					AccBias					GranSize				
	nil	A	B	C	D	nil	A	B	C	D	nil	A	B	C	D
0	0.772	0.77	0.77	0.77	0.77	0.009	0.006	0.006	0.006	0.006	2	2	2	2	2
0.0714286	0.772	0.77	0.77	0.772	0.772	0.01	0.006	0.006	0.008	0.008	2.32	2	2	2	2
0.142857	0.77	0.77	0.771	0.773	0.773	0.006	0.006	0.007	0.011	0.011	3.24	2.16	2.16	3	2.96
0.214286	0.781	0.766	0.767	0.786	0.785	0.008	0.01	0.012	0.02	0.018	5.16	2.52	2.52	4.64	4.68
0.285714	0.799	0.775	0.777	0.811	0.81	0.014	0.012	0.007	0.015	0.009	8.4	4.04	3.84	8.68	8.4
0.357143	0.82	0.786	0.786	0.826	0.832	0.01	0.014	0.014	0.015	0.004	16.08	7.12	6.96	16.2	16.32
0.428571	0.841	0.806	0.8	0.838	0.838	0.007	0.032	0.012	0.009	0.002	32	10.08	9.76	32.44	31.92
0.5	0.838	0.817	0.818	0.84	0.847	0.005	0.012	0.012	0.008	0.004	70.8	18.28	18	72.04	72.24
0.571429	0.839	0.828	0.826	0.847	0.844	0.006	0.019	0.021	0.007	0.01	156.6	34.6	34.72	150.04	149.6
0.642857	0.848	0.832	0.826	0.847	0.839	0.007	0.007	0.017	0.007	0.008	318.12	73.44	73.32	286.24	284.8
0.714286	0.853	0.833	0.841	0.844	0.843	0.009	0.019	0.007	0.011	0.012	467.6	164.2	164.44	438.08	438.28
0.785714	0.857	0.843	0.843	0.847	0.843	0.007	0.01	0.012	0.008	0.014	536.12	325.92	328.04	524.64	525.08
0.857143	0.86	0.838	0.838	0.845	0.844	0.007	0.01	0.014	0.01	0.008	547.16	476.76	476.76	547	546.96
0.928571	0.862	0.842	0.841	0.844	0.843	0.005	0.005	0.014	0.014	0.013	548.84	537.8	537.36	551.28	551.28
1	0.861	0.843	0.843	0.843	0.843	0.004	0.014	0.013	0.014	0.014	552	550.84	550.8	551.88	552

Table 8.17 5 × CV-5; The result of experiments for A, B, C, D strategies of missing values handling versus complete data; data set *car evaluation*; Concept dependent granulation; 10 % *of missing values*; r_{gran} = Granulation radius; nil = result for data without missing values; *Acc* = Accuracy of classification; *AccBias* = Accuracy bias defined based on equation efAccCBiasEquation; *GranSize* = The size of data set after granulation in the fixed *r*

r_{gran}	Acc					AccBias					GranSize				
	nil	A	B	C	D	nil	A	B	C	D	nil	A	B	C	D
0	0.329	0.318	0.318	0.318	0.318	0.009	0.01	0.01	0.01	0.01	4	4	4	4	4
0.166667	0.405	0.336	0.34	0.438	0.434	0.005	0.012	0.023	0.033	0.018	8.28	6.2	6.28	10.48	10.76
0.333333	0.557	0.418	0.401	0.597	0.59	0.019	0.055	0.026	0.027	0.024	17.28	10.56	10.12	25.08	25.24
0.5	0.666	0.544	0.563	0.757	0.759	0.009	0.045	0.038	0.027	0.025	38.08	21.96	21.84	63.44	63.72
0.666667	0.805	0.704	0.685	0.84	0.841	0.008	0.025	0.033	0.007	0.008	106.68	51.64	50.56	201.4	201.12
0.833333	0.871	0.819	0.815	0.855	0.856	0.017	0.013	0.016	0.009	0.01	369.72	159.92	158.08	633.4	633.36
1	0.945	0.864	0.861	0.908	0.908	0.005	0.011	0.006	0.003	0.003	1382.4	643.72	653.48	1357.24	1357.24

Table 8.18 5 × CV-5; The result of experiments for A, B, C, D strategies of missing values handling versus complete data; data set *diabetes*; Concept dependent granulation; 10 % of missing values; r_{gran} = Granulation radius; nil = result for data without missing values; Acc = Accuracy of classification; AccBias = Accuracy bias defined based on equation efAcccBiasEquation; GranSize = The size of data set after granulation in the fixed r

r_{gran}	Acc					AccBias					GranSize				
	nil	A	B	C	D	nil	A	B	C	D	nil	A	B	C	D
0	0.605	0.609	0.609	0.609	0.609	0.009	0.012	0.012	0.012	0.012	2	2	2	2	2
0.125	0.608	0.615	0.61	0.609	0.617	0.006	0.009	0.027	0.011	0.019	35.2	3.16	3.2	33.16	31.68
0.25	0.632	0.624	0.61	0.634	0.62	0.013	0.013	0.015	0.018	0.024	155.88	8.96	8.8	145.96	145.44
0.375	0.639	0.6	0.602	0.636	0.641	0.009	0.018	0.017	0.02	0.015	365.52	29.04	26.72	364.84	363.6
0.5	0.649	0.602	0.618	0.647	0.648	0.017	0.02	0.021	0.018	0.02	540.28	87	84.24	546.72	546.48
0.625	0.647	0.614	0.61	0.645	0.646	0.009	0.013	0.026	0.019	0.019	609.72	282.04	282	609.24	609.16
0.75	0.648	0.637	0.639	0.647	0.647	0.009	0.012	0.013	0.029	0.023	614.4	491.2	488.04	614.24	614.24
0.875	0.648	0.639	0.645	0.65	0.647	0.009	0.015	0.017	0.021	0.023	614.4	593.64	593.6	614.4	614.4
1	0.648	0.647	0.647	0.647	0.647	0.009	0.023	0.023	0.023	0.023	614.4	613.64	613.6	614.4	614.4

Table 8.19 5 × CV-5; The result of experiments for A, B, C, D strategies of missing values handling versus complete data; data set *fertility diagnosis*; Concept dependent granulation; 10 % *of missing values*; r_{gran} = Granulation radius; nil = result for data without missing values; Acc = Accuracy of classification; AccBias = Accuracy bias defined based on equation efAcccBiasEquation; GranSize = The size of data set after granulation in the fixed r

r_{gran}	Acc					AccBias					GranSize				
	nil	A	B	C	D	nil	A	B	C	D	nil	A	B	C	D
0	0.394	0.38	0.38	0.38	0.38	0.066	0.06	0.06	0.06	0.06	2	2	2	2	2
0.111111	0.404	0.38	0.38	0.428	0.428	0.056	0.06	0.06	0.042	0.042	2.32	2	2	2.88	2.88
0.222222	0.492	0.39	0.38	0.504	0.496	0.048	0.05	0.06	0.066	0.054	3.76	2.32	2.32	4.4	4.4
0.333333	0.618	0.45	0.446	0.644	0.652	0.062	0.11	0.114	0.036	0.048	5.6	3.64	3.64	8.12	8.12
0.444444	0.694	0.55	0.558	0.802	0.826	0.056	0.04	0.062	0.018	0.014	10.4	5.64	5.72	15.48	15.68
0.555556	0.842	0.75	0.75	0.85	0.854	0.018	0.07	0.07	0.02	0.026	18.12	9.88	10.04	29.76	29.76
0.666667	0.848	0.832	0.838	0.868	0.868	0.012	0.058	0.042	0.012	0.012	35.44	19.12	19.12	50.12	50.12
0.777778	0.876	0.86	0.856	0.87	0.88	0.014	0.02	0.024	0.01	0.01	57.88	36.12	36.2	69.04	69.04
0.888889	0.882	0.872	0.874	0.88	0.876	0.008	0.008	0.006	0.01	0.014	73.84	59.36	59.36	78.48	78.48
1	0.88	0.874	0.874	0.874	0.874	0	0.016	0.016	0.016	0.016	80	76	76	79.96	79.96

Table 8.20 $5 \times$ CV-5; The result of experiments for A, B, C, D strategies of missing values handling versus complete data; data set *german credit*; Concept dependent granulation; 10% *of missing values*; r_{gran} = Granulation radius; nil = result for data without missing values; *Acc* = Accuracy of classification; *AccBias* = Accuracy bias defined based on equation efAcccBiasEquation; *GranSize* = The size of data set after granulation in the fixed r

r_{gran}	Acc					AccBias					GranSize				
	nil	A	B	C	D	nil	A	B	C	D	nil	A	B	C	D
0	0.563	0.567	0.567	0.567	0.567	0.01	0.009	0.009	0.009	0.009	2	2	2	2	2
0.05	0.563	0.567	0.567	0.567	0.567	0.01	0.009	0.009	0.009	0.009	2	2	2	2.04	2.04
0.1	0.563	0.567	0.567	0.567	0.567	0.01	0.009	0.009	0.009	0.009	2.04	2	2	2.52	2.6
0.15	0.564	0.567	0.567	0.567	0.567	0.009	0.009	0.009	0.008	0.008	2.44	2	2	3.64	3.64
0.2	0.564	0.567	0.567	0.577	0.579	0.009	0.009	0.009	0.007	0.005	3.12	2.2	2.2	5.72	5.6
0.25	0.584	0.567	0.566	0.618	0.627	0.007	0.009	0.012	0.013	0.013	4.96	2.44	2.24	8.56	8.48
0.3	0.614	0.571	0.567	0.648	0.644	0.02	0.024	0.025	0.012	0.036	7.08	3.44	3.32	13.28	12.88
0.35	0.644	0.577	0.578	0.671	0.666	0.016	0.011	0.014	0.016	0.014	11.04	4.48	4.56	23.4	24.24
0.4	0.677	0.611	0.603	0.678	0.689	0.017	0.028	0.024	0.021	0.012	18.96	6.12	6.36	43	44.44
0.45	0.685	0.625	0.623	0.7	0.69	0.01	0.032	0.005	0.008	0.015	34	10.96	10.28	80.12	79.24
0.5	0.684	0.639	0.641	0.702	0.71	0.017	0.019	0.048	0.011	0.019	59	16.2	16.64	142.76	143.56
0.55	0.711	0.655	0.648	0.718	0.712	0.022	0.036	0.017	0.023	0.013	104.08	27.8	26.4	249.76	248.2
0.6	0.718	0.666	0.67	0.723	0.712	0.01	0.028	0.017	0.017	0.011	188.52	47.16	46.08	400.16	401.6
0.65	0.72	0.696	0.695	0.723	0.72	0.011	0.013	0.012	0.014	0.018	319.12	87.36	84.4	569.08	567.68
0.7	0.731	0.71	0.708	0.722	0.72	0.012	0.009	0.007	0.014	0.005	486	157.52	152.44	702.88	702.08
0.75	0.729	0.715	0.719	0.718	0.724	0.005	0.011	0.01	0.011	0.007	647.36	282.08	278.16	771.2	771.36
0.8	0.736	0.719	0.72	0.719	0.723	0.005	0.011	0.008	0.011	0.009	750.48	462.92	465.68	793.72	793.56
0.85	0.733	0.722	0.722	0.722	0.723	0.014	0.012	0.014	0.006	0.009	789.68	649.8	649.56	798.6	798.56
0.9	0.731	0.722	0.721	0.721	0.721	0.014	0.008	0.008	0.009	0.005	796.6	756.28	756.56	799.68	799.68
0.95	0.731	0.722	0.723	0.721	0.722	0.014	0.007	0.006	0.005	0.004	798.72	793.04	793.08	799.92	799.92
1	0.731	0.722	0.722	0.722	0.722	0.013	0.004	0.004	0.005	0.005	800	799.56	799.48	800	800

Table 8.21 5 × CV-5; The result of experiments for A, B, C, D strategies of missing values handling versus complete data; data set *heart disease*; Concept dependent granulation; 10 % *of missing values*; r_{gran} = Granulation radius; nil = result for data without missing values; Acc = Accuracy of classification; AccBias = Accuracy bias defined based on equation efAcccBiasEquation; GranSize = The size of data set after granulation in the fixed r

r_{gran}	Acc					AccBias					GranSize				
	nil	A	B	C	D	nil	A	B	C	D	nil	A	B	C	D
0	0.799	0.789	0.789	0.789	0.789	0.008	0.011	0.011	0.011	0.011	2	2	2	2	2
0.0769231	0.799	0.789	0.789	0.79	0.79	0.009	0.011	0.011	0.01	0.01	2.2	2	2	2.76	2.8
0.153846	0.799	0.789	0.79	0.788	0.79	0.012	0.011	0.014	0.012	0.014	3.16	2.2	2.2	4.4	4.44
0.230769	0.801	0.791	0.792	0.799	0.801	0.01	0.009	0.012	0.023	0.021	4.88	2.76	2.76	8.08	8.12
0.307692	0.804	0.79	0.792	0.809	0.807	0.007	0.01	0.012	0.021	0.03	8.72	4.2	4.16	15.8	15.64
0.384615	0.814	0.793	0.796	0.821	0.827	0.016	0.033	0.019	0.008	0.01	16.88	7.04	7.08	29.64	29.84
0.461538	0.82	0.807	0.804	0.827	0.831	0.006	0.026	0.029	0.006	0.013	35.32	11.56	11.44	59.92	59.32
0.538462	0.824	0.802	0.801	0.827	0.826	0.016	0.016	0.007	0.01	0.007	68.88	20.28	19.44	110.32	111
0.615385	0.811	0.804	0.801	0.829	0.83	0.007	0.014	0.01	0.016	0.007	127.12	38.08	37.56	167.92	168
0.692308	0.823	0.806	0.808	0.828	0.822	0.007	0.031	0.029	0.005	0.007	180.96	78.64	77.32	204.04	204.08
0.769231	0.823	0.825	0.828	0.83	0.829	0.007	0.012	0.013	0.01	0.012	210.4	138.52	138.4	204.04	214.44
0.846154	0.824	0.826	0.829	0.827	0.828	0.005	0.007	0.008	0.014	0.013	216	190.52	191.2	214.44	216
0.923077	0.824	0.829	0.828	0.828	0.828	0.005	0.016	0.013	0.013	0.013	216	212.36	212.44	216	216
1	0.824	0.828	0.828	0.828	0.828	0.005	0.013	0.013	0.013	0.013	216	215.92	215.92	216	216

Table 8.22 5 × CV-5; The result of experiments for A, B, C, D strategies of missing values handling versus complete data; data set *hepatitis*; Concept dependent granulation; 10 % of missing values; r_{gran} = Granulation radius; nil = result for data without missing values; Acc = Accuracy of classification; AccBias = Accuracy bias defined based on equation efAcccBiasEquation; GranSize = The size of data set after granulation in the fixed r

r_{gran}	Acc					AccBias					GranSize				
	nil	A	B	C	D	nil	A	B	C	D	nil	A	B	C	D
0	0.808	0.818	0.818	0.818	0.818	0.005	0.014	0.014	0.014	0.014	2	2	2	2	2
0.0526316	0.808	0.818	0.818	0.818	0.818	0.005	0.014	0.014	0.014	0.014	2	2	2	2	2
0.105263	0.808	0.818	0.818	0.818	0.818	0.005	0.014	0.014	0.014	0.014	2	2	2	2	2
0.157895	0.808	0.818	0.818	0.818	0.818	0.005	0.014	0.014	0.014	0.014	2	2	2	2	2
0.210526	0.808	0.818	0.818	0.817	0.817	0.005	0.014	0.014	0.014	0.014	2.16	2	2	2.08	2.08
0.263158	0.808	0.818	0.818	0.823	0.823	0.005	0.014	0.014	0.009	0.009	2.52	2.08	2.08	2.8	2.76
0.315789	0.815	0.815	0.815	0.831	0.831	0.01	0.017	0.017	0.021	0.021	3.4	2.24	2.24	4	4.08
0.368421	0.822	0.815	0.813	0.837	0.836	0.01	0.017	0.019	0.014	0.015	4.88	2.64	2.64	6.04	6.04
0.421053	0.841	0.821	0.818	0.88	0.889	0.004	0.025	0.027	0.023	0.027	6.84	3.52	3.52	9.12	9.04
0.473684	0.87	0.831	0.825	0.877	0.874	0.021	0.014	0.021	0.032	0.03	10.84	5.28	5.28	15.4	15.56
0.526316	0.865	0.84	0.843	0.865	0.879	0.026	0.012	0.009	0.026	0.018	18.6	7.8	7.8	25.12	25.08
0.578947	0.879	0.861	0.858	0.883	0.883	0.037	0.017	0.013	0.021	0.027	30.52	12.36	12.36	39.36	39.44
0.631579	0.885	0.875	0.871	0.889	0.876	0.018	0.015	0.013	0.027	0.027	46.28	18.84	18.96	58.28	58.2
0.684211	0.898	0.888	0.88	0.888	0.881	0.025	0.022	0.023	0.022	0.048	69.76	32.16	32.16	80.76	80.6
0.736842	0.897	0.875	0.871	0.875	0.879	0.019	0.028	0.019	0.035	0.05	90.32	48.68	48.68	100.04	100.08
0.789474	0.899	0.867	0.868	0.889	0.885	0.017	0.03	0.028	0.034	0.025	110.12	71.56	71.96	113.4	113.28
0.842105	0.898	0.886	0.889	0.884	0.889	0.018	0.023	0.034	0.034	0.034	116.84	95.04	94.6	120.12	120
0.894737	0.899	0.898	0.886	0.89	0.889	0.017	0.037	0.03	0.058	0.034	121	112.08	112.08	122.4	122.4
0.947368	0.899	0.883	0.886	0.88	0.881	0.017	0.027	0.023	0.03	0.028	122	120.04	120.04	123.52	123.96
1	0.899	0.881	0.881	0.881	0.881	0.017	0.028	0.028	0.028	0.028	124	123.48	123.48	124	124

Table 8.23 5 × CV-5; The result of experiments for A, B, C, D strategies of missing values handling versus complete data; data set *congressional voting records*; Concept dependent granulation; 10 % *of missing values*; r_{gran} = Granulation radius; nil = result for data without missing values; Acc = Accuracy of classification; AccBias = Accuracy bias defined based on equation efAcccBiasEquation; GranSize = The size of data set after granulation in the fixed r

r_{gran}	Acc					AccBias					GranSize				
	nil	A	B	C	D	nil	A	B	C	D	nil	A	B	C	D
0	0.901	0.901	0.901	0.901	0.901	0	0	0	0	0	2	2	2	2	2
0.0625	0.901	0.901	0.901	0.901	0.901	0	0	0	0	0	2.04	2.04	2.04	2.12	2.12
0.125	0.901	0.901	0.901	0.901	0.901	0	0	0	0	0	2.2	2.04	2.04	2.44	2.44
0.1875	0.9	0.901	0.901	0.9	0.9	0.001	0	0	0.001	0.001	2.56	2.08	2.08	2.88	2.92
0.25	0.901	0.901	0.901	0.901	0.901	0.003	0	0	0.005	0.005	2.8	2.28	2.28	3.84	3.84
0.3125	0.902	0.901	0.901	0.9	0.901	0.004	0	0	0.001	0.003	3.24	2.56	2.6	4.92	5.24
0.375	0.898	0.901	0.901	0.898	0.898	0.005	0	0	0.005	0.005	4.8	2.96	3	7.28	7.28
0.4375	0.9	0.9	0.9	0.904	0.904	0.006	0.004	0.004	0.006	0.006	5.64	3.84	3.84	9.72	9.56
0.5	0.903	0.899	0.899	0.916	0.915	0.01	0.005	0.005	0.012	0.013	7.12	5.4	5.28	13.08	13.16
0.5625	0.916	0.898	0.898	0.941	0.942	0.008	0.005	0.007	0.008	0.008	8.96	7.2	7.08	20.56	20.56
0.625	0.933	0.903	0.902	0.946	0.945	0.018	0.01	0.011	0.004	0.007	12.6	8.64	8.4	33.84	33.96
0.6875	0.945	0.915	0.914	0.947	0.946	0.007	0.005	0.005	0.01	0.006	19.24	12.08	12.16	54.96	55.44
0.75	0.949	0.94	0.937	0.949	0.95	0.006	0.012	0.006	0.003	0.009	31.84	17.56	17.52	92.6	92.48
0.8125	0.947	0.947	0.949	0.948	0.943	0.007	0.009	0.007	0.004	0.009	52.04	29.92	29.72	147.4	147.92
0.875	0.948	0.951	0.949	0.949	0.943	0.009	0.006	0.005	0.005	0.011	90.92	56.84	57.08	214.76	214.64
0.9375	0.944	0.951	0.948	0.946	0.944	0.01	0.003	0.004	0.006	0.005	143.8	105.08	104.28	283.68	283.28
1	0.947	0.939	0.939	0.939	0.939	0.007	0.006	0.006	0.003	0.003	213.92	180.56	181	330.88	330.88

Table 8.24 5 × CV-5; The result of experiments for A, B, C, D strategies of missing values handling versus complete data; data set *mushroom*; Concept dependent granulation; 10 % *of missing values*; r_{gran} = Granulation radius; nil = result for data without missing values; *Acc* = Accuracy of classification; *AccBias* = Accuracy bias defined based on equation efAcccBiasEquation; *GranSize* = The size of data set after granulation in the fixed r

r_{gran}	Acc					AccBias					GranSize				
	nil	A	B	C	D	nil	A	B	C	D	nil	A	B	C	D
0	0.887	0.887	0.887	0.887	0.887	0	0.001	0.001	0.001	0.001	2	2	2	2	2
0.0454545	0.887	0.887	0.887	0.887	0.887	0	0.001	0.001	0.001	0.001	2	2	2	2	2.04
0.0909091	0.887	0.887	0.887	0.888	0.887	0	0.001	0.001	0.001	0.001	2	2	2	2.08	2.2
0.136364	0.887	0.887	0.887	0.887	0.887	0	0.001	0.001	0.001	0.001	2	2	2	2.64	2.56
0.181818	0.887	0.887	0.887	0.887	0.888	0	0.001	0.001	0.001	0.002	2	2	2	3.6	3.52
0.227273	0.887	0.887	0.887	0.888	0.887	0	0.001	0.001	0.002	0.002	2.24	2.04	2	5.68	5.24
0.272727	0.888	0.887	0.887	0.885	0.888	0.001	0.001	0.001	0.006	0.004	2.48	2.12	2.04	8.56	8.16
0.318182	0.887	0.887	0.887	0.887	0.887	0.001	0.001	0.001	0.003	0.006	3.68	2.48	2.56	13.04	13.52
0.363636	0.885	0.888	0.888	0.895	0.89	0.002	0.001	0.001	0.006	0.003	6.52	3.12	2.88	23.04	22.6
0.409091	0.885	0.887	0.888	0.909	0.908	0.004	0.002	0.002	0.011	0.012	9.4	4.4	4.28	35.08	32.08
0.454545	0.891	0.886	0.888	0.935	0.934	0.005	0.005	0.004	0.006	0.011	15.92	6.12	6.8	45.32	47.24
0.5	0.911	0.888	0.886	0.966	0.965	0.009	0.001	0.002	0.004	0.008	21.12	9.44	10.2	59.64	60
0.545455	0.941	0.895	0.889	0.985	0.984	0.006	0.003	0.006	0.002	0.004	27.64	15.24	14.64	79.68	75.88
0.590909	0.969	0.905	0.908	0.992	0.992	0.005	0.005	0.009	0.002	0.002	40.44	24.64	22.36	101.48	102.84
0.636364	0.983	0.93	0.931	0.997	0.996	0.002	0.006	0.009	0.001	0	43.8	31.64	33.04	152.92	149.48

(continued)

Table 8.24 (continued)

r_{gran}	Acc					AccBias					GranSize				
0.681818	0.992	0.965	0.96	0.999	0.999	0.001	0.009	0.009	0	0.001	47	44.36	41.52	245.96	251.04
0.727273	0.995	0.977	0.979	1	1	0.001	0.004	0.007	0	0	47.96	51.4	52.6	462.72	467.2
0.772727	0.998	0.99	0.99	1	1	0.001	0.002	0.001	0	0	54.68	63.84	64.24	888.28	894.36
0.818182	0.999	0.996	0.997	1	1	0.001	0.001	0	0	0	94.36	78.4	77.72	1708.36	1711.84
0.863636	1	0.999	0.999	1	1	0	0	0	0	0	192.6	120.68	120.48	3075.8	3065.56
0.909091	1	1	1	1	1	0	0	0	0	0	511.76	263.64	269.28	4781.84	4781.92
0.954545	1	1	1	1	1	0	0	0	0	0	1742.84	782.16	789.08	6113.96	6114.12
1	1	1	1	1	1	0	0	0	0	0	6499.2	3017.8	3026.48	6494.08	6494.08

Table 8.25 $5 \times$ CV-5; The result of experiments for A, B, C, D strategies of missing values handling versus complete data; data set *nursery*; Concept dependent granulation; 10 % *of missing values*; r_{gran} = Granulation radius; nil = result for data without missing values; *Acc* = Accuracy of classification; *AccBias* = Accuracy bias defined based on equation efAcccBiasEquation; *GranSize* = The size of data set after granulation in the fixed r

r_{gran}	Acc					AccBias					GranSize				
	nil	A	B	C	D	*nil*	A	B	C	D	*nil*	A	B	C	D
0	0.391	0.399	0.399	0.399	0.399	0.018	0.031	0.031	0.031	0.031	4.96	4.96	4.96	4.96	4.96
0.125	0.408	0.41	0.409	0.471	0.467	0.024	0.033	0.031	0.025	0.035	7.32	5.84	6.16	11.92	12.08
0.25	0.49	0.443	0.442	0.544	0.542	0.042	0.031	0.037	0.032	0.03	13.48	8.96	9.2	24.28	23.52
0.375	0.538	0.481	0.494	0.569	0.573	0.027	0.032	0.028	0.039	0.032	27.12	14.76	14.6	54.36	53.84
0.5	0.553	0.53	0.528	0.601	0.603	0.033	0.027	0.046	0.03	0.033	59.12	31.08	30.44	144.12	148.32
0.625	0.583	0.563	0.563	0.629	0.628	0.026	0.032	0.034	0.041	0.048	159.28	70.56	71.16	459.84	460.36
0.75	0.623	0.595	0.599	0.696	0.687	0.027	0.041	0.033	0.031	0.035	547.28	201.16	206.08	1609.52	1598.28
0.875	0.672	0.628	0.631	0.735	0.749	0.03	0.035	0.029	0.021	0.026	2371.24	748.2	757.08	5142.36	5147.2
1	0.568	0.746	0.744	0.686	0.686	0.012	0.03	0.027	0.052	0.052	10368	3760.96	3724.6	10206	10206

Table 8.26 5 × CV-5; The result of experiments for A, B, C, D strategies of missing values handling versus complete data; data set *soybean large*; Concept dependent granulation; 10 % *of missing values*; r_{gran} = Granulation radius; nil = result for data without missing values; *Acc* = Accuracy of classification; *AccBias* = Accuracy bias defined based on equation efAcccBiasEquation; *GranSize* = The size of data set after granulation in the fixed *r*

r_{gran}	Acc					AccBias					GranSize				
	nil	*A*	*B*	*C*	*D*	*nil*	*A*	*B*	*C*	*D*	*nil*	*A*	*B*	*C*	*D*
0	0.691	0.694	0.694	0.694	0.694	0.028	0.013	0.013	0.013	0.013	19	19	19	19	19
...	...	...	...	...	...	...	...	...	...	...	...	...	...	...	...
0.4	0.691	0.694	0.694	0.695	0.695	0.028	0.013	0.013	0.014	0.013	19	19	19	20.32	20.4
0.428571	0.691	0.694	0.694	0.696	0.696	0.028	0.013	0.013	0.014	0.014	19	19	19	21.6	21.68
0.457143	0.692	0.694	0.694	0.701	0.701	0.027	0.013	0.013	0.01	0.01	19.04	19	19	23.52	23.52
0.485714	0.693	0.694	0.694	0.705	0.705	0.026	0.013	0.013	0.009	0.009	19.12	19	19	26.6	26.6
0.514286	0.693	0.694	0.694	0.713	0.712	0.026	0.013	0.013	0.013	0.014	19.52	19	19.04	30.32	30.12
0.542857	0.697	0.694	0.694	0.748	0.747	0.029	0.013	0.013	0.009	0.01	20.12	19.16	19.16	35.16	35.16
0.571429	0.7	0.694	0.694	0.772	0.77	0.029	0.013	0.013	0.017	0.022	21.64	19.36	19.36	42.04	42.72
0.6	0.708	0.697	0.697	0.792	0.79	0.03	0.016	0.016	0.026	0.02	24.16	20.24	20.24	52.36	52.16
0.628571	0.719	0.705	0.706	0.819	0.82	0.032	0.015	0.014	0.016	0.012	27.12	21.96	21.96	65.8	65.48
0.657143	0.725	0.71	0.71	0.843	0.843	0.025	0.019	0.019	0.006	0.004	29.8	23.88	23.96	82.84	83.28
0.685714	0.738	0.712	0.713	0.86	0.859	0.013	0.023	0.022	0.01	0.011	35.44	26.16	25.92	107.64	106.8

(continued)

Table 8.26 (continued)

r_{gran}	Acc					AccBias					GranSize				
0.714286	0.758	0.725	0.729	0.87	0.864	0.006	0.026	0.019	0.014	0.015	41.72	30.68	30.4	142.4	141.88
0.742857	0.794	0.738	0.739	0.886	0.881	0.013	0.015	0.013	0.016	0.014	49.2	34.92	34.92	184.72	184.2
0.771429	0.831	0.77	0.772	0.894	0.894	0.019	0.007	0.009	0.012	0.018	57.72	41.8	41.44	242.84	242.76
0.8	0.854	0.813	0.813	0.909	0.9	0.002	0.01	0.011	0.012	0.012	72.44	51.56	51.32	310.44	309.08
0.828571	0.863	0.835	0.834	0.905	0.9	0.009	0.006	0.008	0.012	0.012	90.44	63.72	63.48	377.8	376.68
0.857143	0.864	0.851	0.848	0.905	0.901	0.01	0.019	0.008	0.011	0.008	118.88	84.88	85.52	445.04	445.36
0.885714	0.88	0.865	0.863	0.907	0.902	0.013	0.006	0.007	0.008	0.006	162.52	117.24	117.6	497.72	497.52
0.914286	0.887	0.873	0.869	0.905	0.902	0.007	0.004	0.006	0.007	0.007	220.96	165.44	165.32	528.12	528.08
0.942857	0.889	0.883	0.867	0.902	0.893	0.007	0.009	0.01	0.004	0.008	301.28	235.4	233.96	542.48	542.48
0.971429	0.918	0.9	0.889	0.902	0.893	0.007	0.004	0.013	0.007	0.009	405.2	340.16	339.8	546.04	546.04
1	0.918	0.894	0.894	0.893	0.893	0.004	0.008	0.009	0.009	0.009	509.88	470.76	470.28	546.4	546.4

Table 8.27 $5 \times$ CV-5; The result of experiments for A, B, C, D strategies of missing values handling versus complete data; data set *SPECT*; Concept dependent granulation; 10% of missing values; r_{gran} = Granulation radius; nil = result for data without missing values; *Acc* = Accuracy of classification; *AccBias* = Accuracy bias defined based on equation efAcccBiasEquation; *GranSize* = The size of data set after granulation in the fixed r

r_{gran}	Acc					AccBias					GranSize				
	nil	A	B	C	D	nil	A	B	C	D	nil	A	B	C	D
0	0.526	0.526	0.526	0.526	0.526	0.029	0.029	0.029	0.029	0.029	2	2	2	2	2
0.0454545	0.526	0.526	0.526	0.526	0.526	0.029	0.029	0.029	0.029	0.029	2	2	2	2	2
0.0909091	0.526	0.526	0.526	0.526	0.526	0.029	0.029	0.029	0.029	0.029	2	2	2	2.04	2.04
0.136364	0.526	0.526	0.526	0.543	0.543	0.029	0.029	0.029	0.06	0.06	2.08	2	2	2.36	2.36
0.181818	0.538	0.526	0.526	0.542	0.542	0.05	0.029	0.029	0.064	0.064	2.36	2	2	2.68	2.68
0.227273	0.549	0.526	0.526	0.587	0.587	0.058	0.029	0.029	0.038	0.038	2.48	2.08	2.08	3.2	3.2
0.272727	0.559	0.529	0.529	0.616	0.606	0.044	0.028	0.028	0.024	0.035	2.96	2.16	2.16	3.92	3.88
0.318182	0.585	0.541	0.541	0.642	0.641	0.048	0.074	0.074	0.059	0.06	3.36	2.44	2.44	4.52	4.36
0.363636	0.623	0.549	0.549	0.684	0.688	0.044	0.024	0.024	0.043	0.039	3.8	2.56	2.56	5.8	5.8
0.409091	0.651	0.579	0.579	0.684	0.684	0.038	0.065	0.065	0.054	0.054	4.56	3	3	7.28	7.36
0.454545	0.679	0.61	0.61	0.733	0.733	0.047	0.024	0.024	0.02	0.02	5.72	3.8	3.8	9.96	10.16
0.5	0.696	0.628	0.628	0.76	0.757	0.031	0.039	0.039	0.011	0.015	6.92	4.36	4.36	13.4	13.44
0.545455	0.73	0.662	0.665	0.761	0.763	0.035	0.028	0.039	0.011	0.009	8.84	5.36	5.28	19.24	19.16
0.590909	0.751	0.7	0.7	0.784	0.784	0.024	0.023	0.022	0.025	0.022	11.16	6.96	6.76	28.8	28.76

(continued)

Table 8.27 (continued)

r_{gran}	Acc					AccBias					GranSize				
0.636364	0.776	0.734	0.734	0.793	0.794	0.033	0.023	0.023	0.027	0.026	15.84	9.48	9.52	42.36	42.4
0.681818	0.785	0.765	0.765	0.796	0.8	0.024	0.021	0.021	0.002	0.017	23.16	13.32	13.32	63.32	63.16
0.727273	0.775	0.787	0.79	0.795	0.794	0.023	0.015	0.019	0.003	0	33.76	18.92	18.88	89.48	89.32
0.772727	0.788	0.778	0.778	0.794	0.794	0.006	0.034	0.035	0	0	49.6	28.72	28.72	120.28	120.28
0.818182	0.793	0.796	0.795	0.805	0.797	0.001	0.017	0.018	0.008	0.005	70.8	41.84	41.64	152.52	152.48
0.863636	0.794	0.796	0.796	0.822	0.822	0	0.005	0.005	0.013	0.02	100.4	65.68	66	176.68	176.68
0.909091	0.794	0.796	0.796	0.813	0.818	0	0.009	0.009	0.018	0.01	130.16	99.24	99.24	195.68	195.68
0.954545	0.794	0.794	0.794	0.818	0.819	0	0	0	0.01	0.013	162.56	139.32	139.28	207.52	207.6
1	0.796	0.802	0.805	0.819	0.819	0.009	0.015	0.027	0.013	0.013	185	175.2	175.24	212.44	212.44

Table 8.28 $5 \times$ CV-5; The result of experiments for A, B, C, D strategies of missing values handling versus complete data; data set *SPECTF*; Concept dependent granulation; 10 % *of missing values*; r_{gran} = Granulation radius; nil = result for data without missing values; *Acc* = Accuracy of classification; *AccBias* = Accuracy bias defined based on equation efAcccBiasEquation; *GranSize* = The size of data set after granulation in the fixed r

r_{gran}	Acc					AccBias					GranSize				
	nil	A	B	C	D	nil	A	B	C	D	nil	A	B	C	D
0	0.437	0.457	0.457	0.457	0.457	0.032	0.019	0.019	0.019	0.019	2	2	2	2	2
0.0227273	0.586	0.455	0.457	0.514	0.533	0.035	0.021	0.022	0.033	0.017	6.68	2	2	5.68	5.8
0.0454545	0.691	0.456	0.455	0.608	0.634	0.028	0.019	0.039	0.025	0.021	14.12	2.04	2	11.4	11.64
0.0681818	0.749	0.456	0.446	0.703	0.714	0.034	0.027	0.037	0.008	0.024	27.88	2.08	2.08	24.24	24.16
0.0909091	0.792	0.443	0.457	0.759	0.766	0.005	0.028	0.045	0.023	0.013	50.16	2.56	2.6	46.48	46.24
0.113636	0.791	0.452	0.478	0.773	0.78	0.022	0.032	0.047	0.017	0.017	85.16	2.8	2.84	77.92	78.6
0.136364	0.785	0.475	0.48	0.774	0.784	0.024	0.027	0.026	0.013	0.022	125.08	3.24	3.12	121.56	122.08
0.159091	0.792	0.512	0.531	0.778	0.784	0.01	0.072	0.09	0.02	0.006	166.64	5.04	5.04	165.84	165.52
0.181818	0.789	0.514	0.483	0.784	0.79	0.009	0.045	0.115	0.017	0.004	195.12	5.68	5.8	197	197.04
0.204545	0.791	0.555	0.498	0.781	0.794	0.022	0.066	0.09	0.032	0.019	209.72	8.92	8.72	209.32	209.24
0.227273	0.791	0.537	0.53	0.784	0.789	0.026	0.035	0.038	0.025	0.017	212.92	15.12	14.96	212.88	212.88
0.25	0.793	0.602	0.563	0.782	0.789	0.028	0.075	0.06	0.012	0.017	213.6	21.16	19.84	213.48	213.48
0.272727	0.793	0.638	0.615	0.785	0.789	0.028	0.075	0.097	0.013	0.017	213.6	27.2	26.64	213.6	213.6
0.295455	0.793	0.691	0.654	0.787	0.789	0.028	0.088	0.11	0.014	0.017	213.6	47.96	47.68	213.6	213.6
0.318182	0.793	0.68	0.683	0.786	0.789	0.028	0.08	0.1	0.012	0.017	213.6	68.84	68.64	213.6	213.6
0.340909	0.793	0.739	0.724	0.792	0.789	0.028	0.033	0.044	0.021	0.017	213.6	99.16	100.44	213.6	213.6
0.363636	0.793	0.751	0.742	0.784	0.789	0.028	0.04	0.044	0.021	0.017	213.6	130.96	127.44	213.6	213.6

(continued)

Table 8.28 (continued)

r_{gran}	Acc					AccBias					GranSize				
0.386364	0.793	0.772	0.772	0.786	0.789	0.028	0.022	0.029	0.02	0.017	213.6	161.68	161.72	213.6	213.6
0.409091	0.793	0.775	0.773	0.789	0.789	0.028	0.011	0.013	0.017	0.017	213.6	183.2	182.32	213.6	213.6
0.431818	0.793	0.786	0.783	0.789	0.789	0.028	0.012	0.026	0.017	0.017	213.6	199.6	198.32	213.6	213.6
0.454545	0.793	0.788	0.794	0.789	0.789	0.028	0.017	0.012	0.017	0.017	213.6	207.52	207.68	213.6	213.6
0.477273	0.793	0.788	0.79	0.789	0.789	0.028	0.014	0.016	0.017	0.017	213.6	211.08	210.4	213.6	213.6
0.5	0.793	0.79	0.79	0.789	0.789	0.028	0.016	0.016	0.017	0.017	213.6	212.72	212.72	213.6	213.6
0.522727	0.793	0.789	0.789	0.789	0.789	0.028	0.017	0.017	0.017	0.017	213.6	213.32	213.4	213.6	213.6
0.545455	0.793	0.789	0.789	0.789	0.789	0.028	0.017	0.017	0.017	0.017	213.6	213.6	213.6	213.6	213.6
...	...	...	...	...	...	...	...	...	...	...	...	...	...	...	...
1	0.793	0.789	0.789	0.789	0.789	0.028	0.017	0.017	0.017	0.017	213.6	213.6	213.6	213.6	213.6

Table 8.29 An extract of results on missing values recovery

data set	r_{gran}	$nil-5$	5granA	5granB	5granC	5granD	$nil-10$	10granA	10granB	10granC	10granD
australian	0.928	0.858	0.852	0.851	0.850	0.851	0.862	0.842	0.841	0.844	0.843
car	0.833	0.869	0.845	0.848	0.868	0.868	0.871	0.819	0.815	0.855	0.856
diabetes	0.875	0.654	0.650	0.649	0.649	0.649	0.648	0.639	0.645	0.650	0.647
fertility	0.888	0.876	0.876	0.876	0.876	0.876	0.882	0.872	0.874	0.880	0.876
german	0.9	0.729	0.723	0.723	0.722	0.723	0.731	0.722	0.721	0.721	0.721
heart	0.769	0.820	0.821	0.823	0.823	0.821	0.823	0.825	0.828	0.830	0.823
hepatitis	0.842	0.883	0.894	0.894	0.895	0.886	0.898	0.886	0.889	0.884	0.880
congr. voting	0.750	0.954	0.949	0.950	0.949	0.949	0.949	0.940	0.937	0.949	0.950
mushroom	0.863	1.0	1.0	1.0	1.0	1.0	1.0	0.999	0.999	0.999	0.999
nursery	0.875	0.643	0.637	0.639	0.679	0.683	0.672	0.628	0.631	0.735	0.749
soybean	0.971	0.915	0.905	0.907	0.910	0.909	0.918	0.900	0.889	0.902	0.893
SPECT	0.727	0.799	0.786	0.786	0.798	0.793	0.775	0.787	0.790	0.795	0.794
SPECTF	0.431	0.788	0.788	0.788	0.788	0.788	0.793	0.788	0.794	0.789	0.789

For data sets examined, we give the best granulation radius r_{gran}, and for this radius in sequence, the accuracy without granulation as $nil-5$ in the 5 % perturbation case followed by accuracies in this case for strategies A, B, C, D as 5granA, 5granB, 5granC, 5granD, and then, the accuracy without granulation $nil-10$ in the 10 % perturbation case, followed by accuracies in this case for strategies A, B, C, D as 10granA, 10granB, 10granC, 10granD. It follows that those strategies for recovery of missing values provide very good results and recovered systems approximate closely the unperturbed ones in a characteristic way. Having small diversity in case of granulation with strategy $* = don't\ care$ it is easy to create equal objects and the approximation even for radius 1 is significant. What could be interesting in case of C and D strategies, for instance for Congressional House Votes, Soybean Large and SPECT, which contain similar objects, and even copies of the same objects, the number of objects is larger than in case of *nil* because the stars increase diversity in data and the number of equal objects is decreasing. The detailed information about the diversity of data is presented in the Appendix I.

Summarizing, strategies A, B reduce diversity in data sets, whereas strategies C, D increase diversity in data, due to respective strategies for treatment of missing values either $* = don't\ care$ or $* = *$ during the granulation process. Most stars are repaired during granulation process thus strategies of repairing applied after granulation have slight influence on the quality of classification. Overall, the methods work in stable ways and results are fully comparable with *nil* results.

References

1. Acuña, E., Rodriguez, C.: The treatment of missing values and its effect in the classifier accuracy. In: Banks, D., House, L., McMorris, F.R., Arabie, P., Gaul, W. (eds.) Classification, Clustering and Data Mining Applications, pp. 639–646. Springer, Heidelberg (2004)
2. Allison, P.D.: Missing Data. Sage Publications, Thousand Oaks (2002)
3. Artiemjew, P.: Classifiers from granulated data sets: concept dependent and layered granulation. In: Proceedings RSKD'07. The Workshops at ECML/PKDD'07, pp. 1–9. Warsaw University Press, Warsaw (2007)
4. Batista, G.E., Monard, M.C.: An analysis of four missing data treatment methods for supervised learning. Appl. Artif. Intell. **17**(5/6), 519–533 (2003)
5. Bruha, I.: Meta-learner for unknown attribute value processing: dealing with inconsistency of meta-databases. J. Intell. Inf. Syst. **22**, 71–87 (2004)
6. Grzymala-Busse, J.W.: Data with missing attribute values: generalization of indiscernibility relation and rule induction. Transactions on Rough Sets I. Lecture Notes in Computer Science subseries, pp. 78–95. Springer, Berlin (2004)
7. Grzymala-Busse, J.W., Hu, M.: A comparison of several approaches to missing attribute values in data mining. In: Proceedings of the International Conference on Rough Sets and Current Trends in Computing RSCTC 2000. Lecture Notes in Computer Science, vol. 2005, pp. 378–385. Springer, Berlin (2001)
8. Imieliński, T., Lipski Jr, W.: Incomplete information in relational databases. J. ACM **31**, 761–791 (1984)
9. Li, J., Cercone, N.: Comparisons on different approaches to assign missing attribute values. Technical Report CS-2006-04, SCS, University of Waterloo (2006)
10. Little, R.J.A., Rubin, D.B.: Statistical Analysis with Missing Data. Wiley, New York (2002)
11. Nakata, M., Sakai, H.: Applying rough sets to information tables containing missing values. In: Proceedings of the International Symposium on Multiple-Valued Logic ISMVL 2009, pp. 286–291. Naha, Okinawa (2009)
12. Polkowski, L., Artiemjew, P.: On granular rough computing with missing values. In: Proceedings of the International Conference on Rough Sets and Intelligent Systems Paradigms RSEiSP'07. Lecture Notes in Computer Science, vol. 4585, pp. 271–279. Springer, Berlin (2007)

13. Polkowski, L., Artiemjew, P.: Granular computing: granular classifiers and missing values. In: Proceedings of the 6th IEEE International Conference on Cognitive Informatics ICCI'07, pp. 186–194 (2007)
14. Quinlan, J.R.: C4.5, Programs for Machine Learning. Morgan Kaufmann Publishers, San Francisco (1993)
15. Quinlan, J.R.: Unknown attribute values in induction. In: Proceedings of the 6th International Workshop on Machine Learning, pp. 64–168. Ithaca, New York (1989)
16. Sadiq, A.T., Duaimi, M.G., Shaker, S.A.: Data mining solutions using rough set theory and swarm intelligence. Int. J. Adv. Comput. Sci. Inf. Technol. (IJACSIT) 2(3), 1–16 (2013)
17. University of Irvine (UCI) Repository. archive.ics.uci.edu/ml/, Accessed 11.11.2014

Chapter 9
Granular Classifiers Based on Weak Rough Inclusions

'*ei*' (ε) is an indispensable word in logic for the construction of a syllogism ('*ei*' = '*if*')

[Comments on 'E' in Plutarch: De E apud Delphos. www.perseus.tufts.edu/hopper]

In this chapter we discuss the problem of classification with use of descriptor indiscernibility ratio (ε) and object indiscernibility ratio (r_{catch}). Theoretical background and description of those classifiers were introduced in Sect. 4.6, see [1–5].

9.1 Introduction

The main reason for this discussion is to show the level of classifiers effectiveness depending on the values of parameters. In order to investigate these problem, we have carried out a series of experiments with use of five times cross validation five method. We have selected for experimentation best classifiers shown initially in Table 4.8.

As reference data sets, we have chosen mixed attribute value Australian Credit, Pima Indians Diabetes and Heart Disease data sets. Additionally we have performed classification on the Wisconsin Diagnostic Breast Cancer data set, which was used for metrics evaluation in k-NN example in Chap. 4, cf. Table 4.2.

9.2 Results of Experiments with Classifiers 5_v1, 6_v1, 7_v1, 8_v1–8_v5 Based on the Parameter ε

In Figs. 9.1, 9.2, 9.3, and 9.4 we have results for epsilon classifiers 8_v1 to 8_v5 in comparison with the simplest one 1_v1 and classifiers based on weighted voting by t-norms 5_v1, 6_v1, 7_v1 methods, cf. Sect. 4.6. It turns out that 8_v1.4 and 8_v1.5 algorithms work best for most data sets. In case of Australian Credit, 8_v1.5 wins

© Springer International Publishing Switzerland 2015
L. Polkowski and P. Artiemjew, *Granular Computing in Decision Approximation*,
Intelligent Systems Reference Library 77, DOI 10.1007/978-3-319-12880-1_9

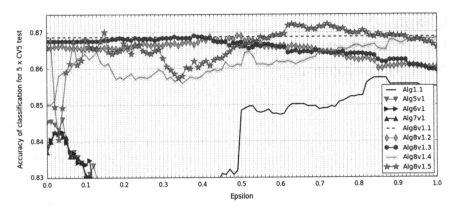

Fig. 9.1 5xCV5 test; Comparison of epsilon strategies; Data set: Australian Credit

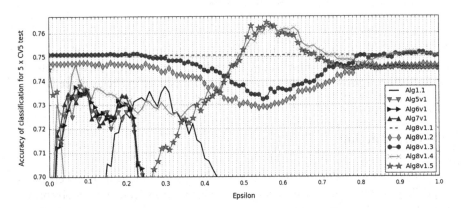

Fig. 9.2 5xCV5 test; Comparison of epsilon strategies; Data set: Pima Indians Diabetes

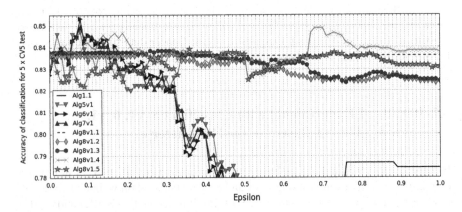

Fig. 9.3 5xCV5 test; Comparison of epsilon strategies; Data set: Heart Disease

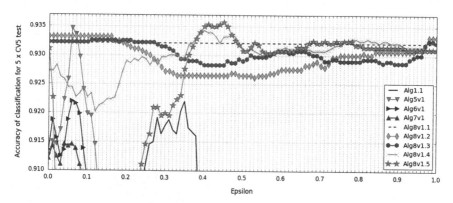

Fig. 9.4 5xCV5 test; Comparison of epsilon strategies; Data set: Wisconsin Diagnostic Breast Cancer

for epsilon from interval [0.6, 0.9], in case of Pima Indians Diabetes, 8_v1.4 and 8_v1.5 win at least for epsilon from the interval [0.5, 0.7]. For Heart Disease data set 8_v1.4 wins for epsilon in [0.1, 0.25] and [0.65, 1.0], additionally weighted voting classifiers based on t-norms work well for epsilon in [0.07, 0.17]. For the last data set Wisconsin Diagnostic Breast Cancer in the interval for epsilon [0.38, 0.5], classifiers 8_v1.4 and 8_v1.5 win.

In Figs. 9.5, 9.6, 9.7, 9.8, 9.9, 9.10, 9.11 and 9.12 we have result for 1v2 classifier. The result shown the wide spectrum of parameters which gives optimal accuracy of classification.

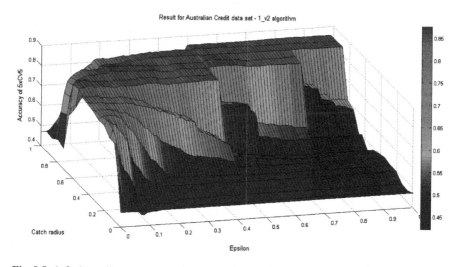

Fig. 9.5 1v2. Australian

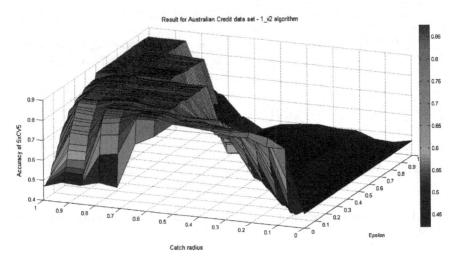

Fig. 9.6 1v2. Australian v2

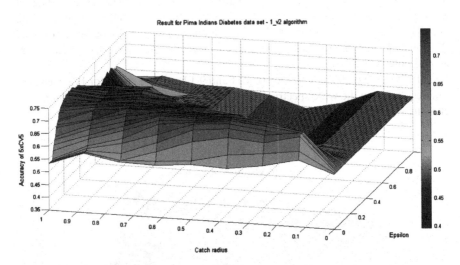

Fig. 9.7 1v2. Diabetes

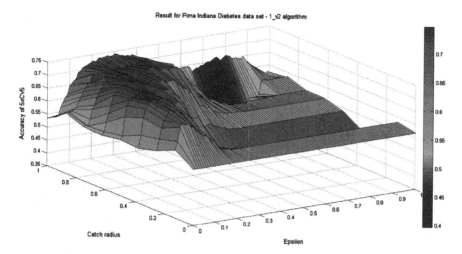

Fig. 9.8 1v2. Diabetes v2

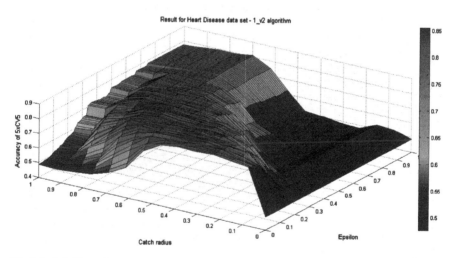

Fig. 9.9 1v2. Heart disease

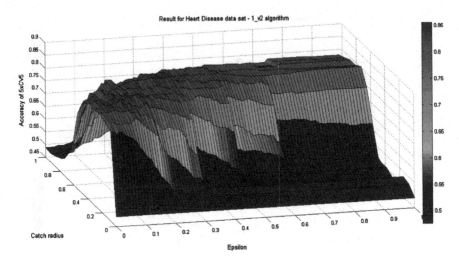

Fig. 9.10 1v2. Heart disease v2

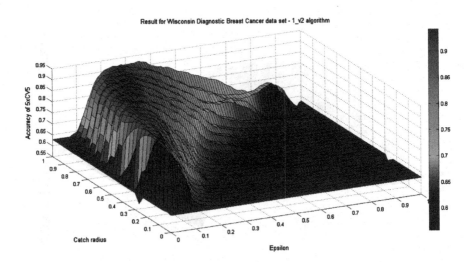

Fig. 9.11 1v2. WDBC

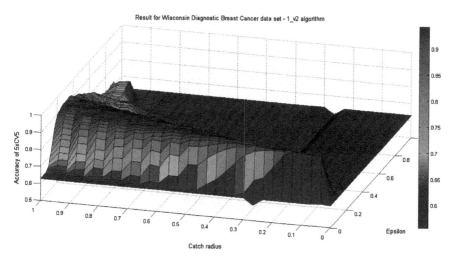

Fig. 9.12 1v2. WDBC v2

9.3 Results of Experiments with Classifiers Based on Parameters ε and r_{catch}

We now show results for 2_v2 classifier, cf. Sect. 4.6. We have separate results for all granulation radii, where the influence of r_{catch} and epsilon parameters is presented in cross-sections for particular granulation radii. Results for increasing granulation radii for Australian Credit data set are shown in Figs. 9.13, 9.14, 9.15, 9.16, 9.17, 9.18, 9.19, 9.20, 9.21, 9.22, 9.23, 9.24, 9.25, 9.26 and 9.27, for Pima Indians Diabetes, Heart Disease and Wisconsin Diagnostic Breast Cancer respectively in Figs. 9.28, 9.29, 9.30, 9.31, 9.32, 9.33, 9.34, 9.35, 9.36, 9.37, 9.38, 9.39, 9.40, 9.41, 9.42, 9.43, 9.44, 9.45, 9.46, 9.47, 9.48, 9.49, 9.50, 9.51, 9.52, 9.53, 9.54, 9.55, 9.56, 9.57, 9.58, 9.59, 9.60, 9.61, 9.62, 9.63, 9.64, 9.65, 9.66, 9.67, 9.68, 9.69, 9.70, 9.71, 9.72, 9.73, 9.74, 9.75, 9.76, 9.77, 9.78, 9.79, 9.80 and 9.81. There is visible increase in parameter influence with growing granulation radii.

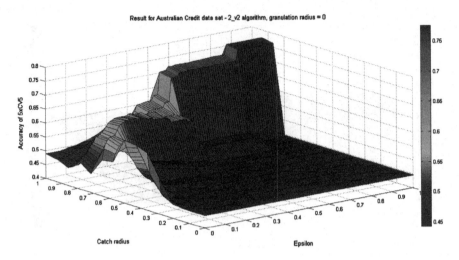

Fig. 9.13 2v2. Australian. Radius = 0

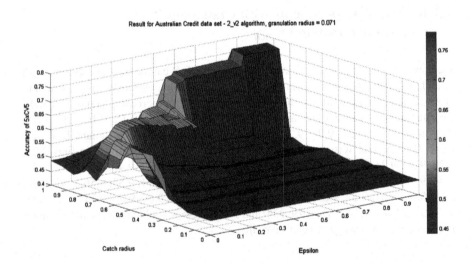

Fig. 9.14 2v2. Australian. Radius = 0.071

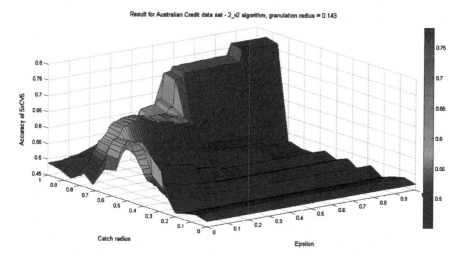

Fig. 9.15 2v2. Australian. Radius = 0.143

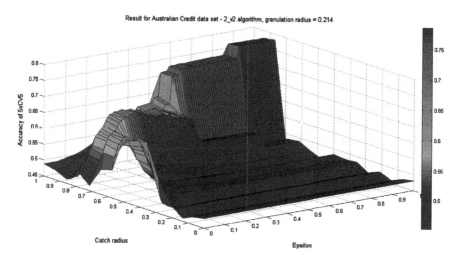

Fig. 9.16 2v2. Australian. Radius = 0.214

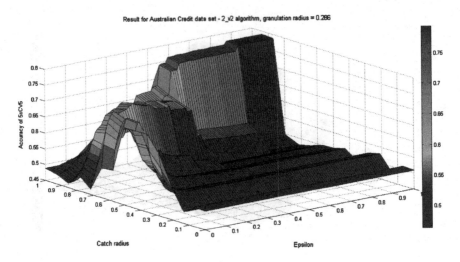

Fig. 9.17 2v2. Australian. Radius = 0.286

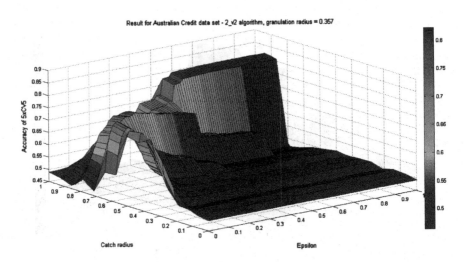

Fig. 9.18 2v2. Australian. Radius = 0.357

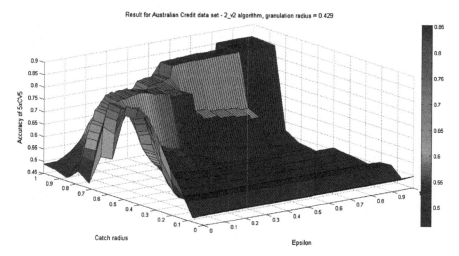

Fig. 9.19 2v2. Australian. Radius = 0.429

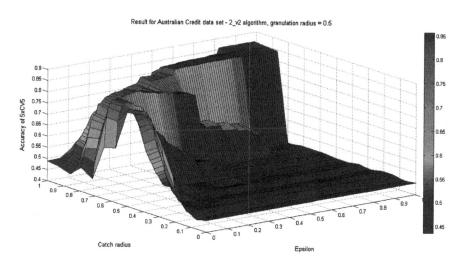

Fig. 9.20 2v2. Australian. Radius = 0.5

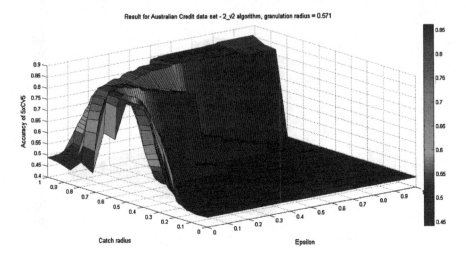

Fig. 9.21   2v2. Australian. Radius = 0.571

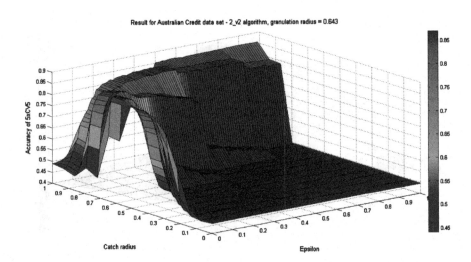

Fig. 9.22   2v2. Australian. Radius = 0.643

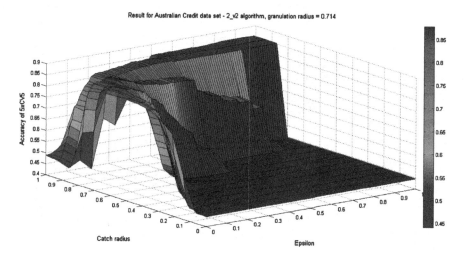

Fig. 9.23 2v2. Australian. Radius = 0.714

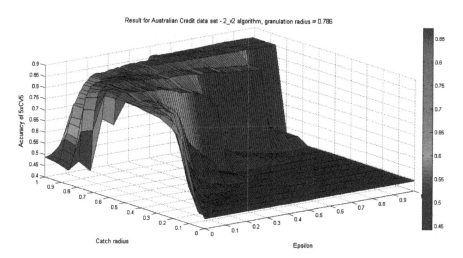

Fig. 9.24 2v2. Australian. Radius = 0.786

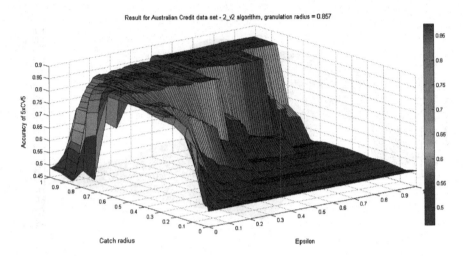

Fig. 9.25 2v2. Australian. Radius = 0.857

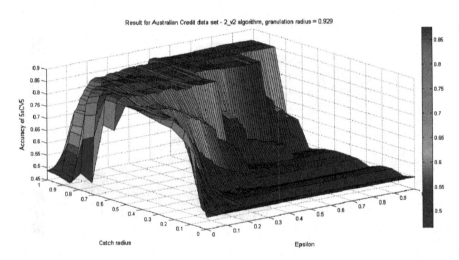

Fig. 9.26 2v2. Australian. Radius = 0.929

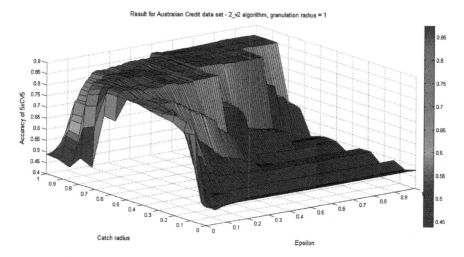

Fig. 9.27   2v2. Australian. Radius $= 1.0$

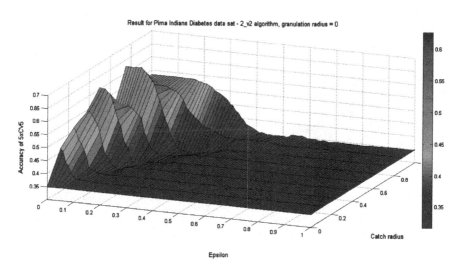

Fig. 9.28   2v2. Diabetes. Radius $= 0$

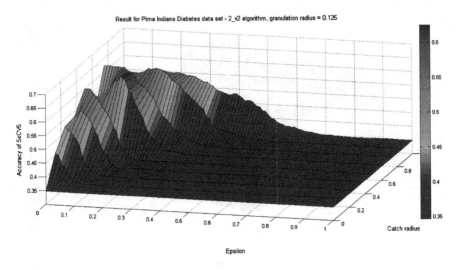

Fig. 9.29 2v2. Diabetes. Radius = 0.125

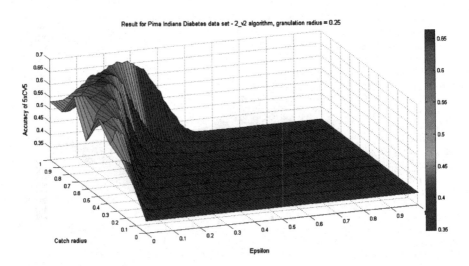

Fig. 9.30 2v2. Diabetes. Radius = 0.25

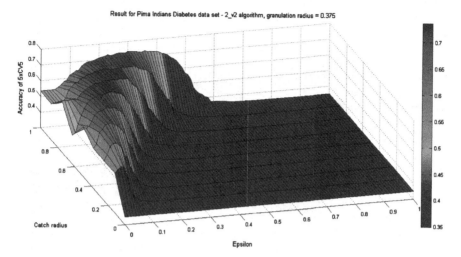

Fig. 9.31 2v2. Diabetes. Radius $= 0.375$

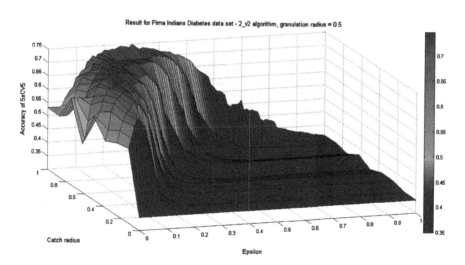

Fig. 9.32 2v2. Diabetes. Radius $= 0.5$

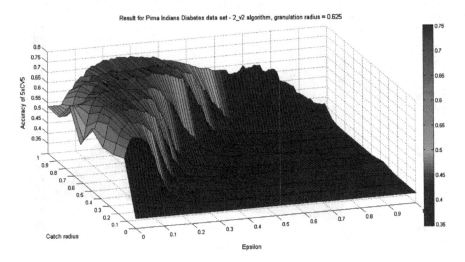

Fig. 9.33 2v2. Diabetes. Radius = 0.625

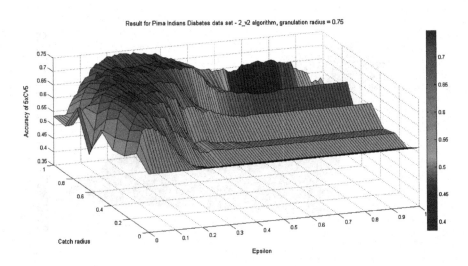

Fig. 9.34 2v2. Diabetes. Radius = 0.75

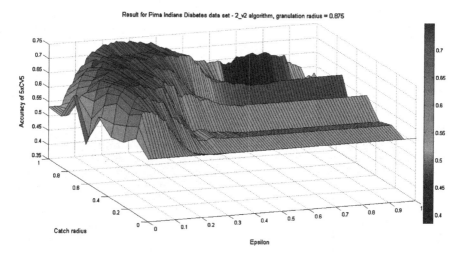

Fig. 9.35 2v2. Diabetes. Radius $= 0.875$

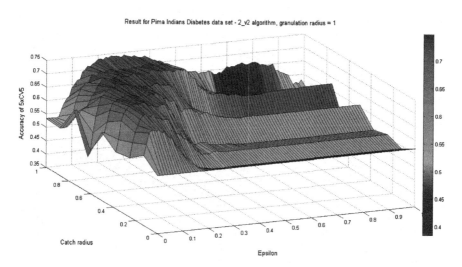

Fig. 9.36 2v2. Diabetes. Radius $= 1.0$

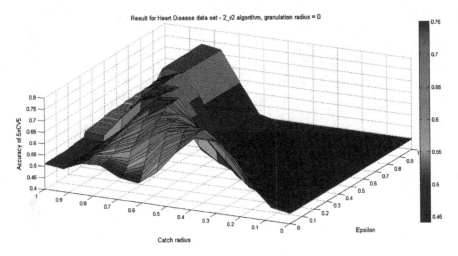

Fig. 9.37 2v2. Heart disease. Radius = 0

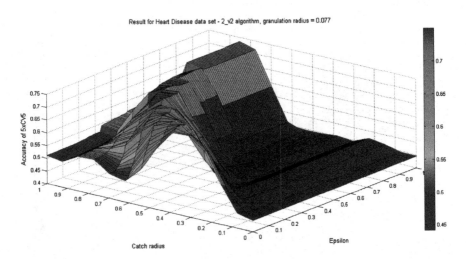

Fig. 9.38 2v2. Heart disease. Radius = 0.077

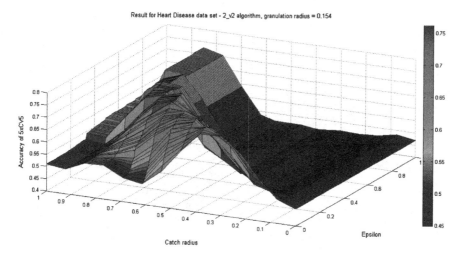

Fig. 9.39 2v2. Heart disease. Radius = 0.154

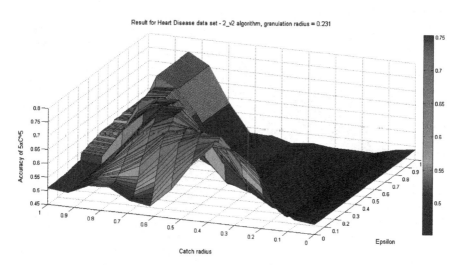

Fig. 9.40 2v2. Heart disease. Radius = 0.231

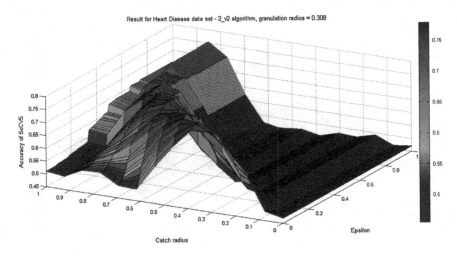

Fig. 9.41 2v2. Heart disease. Radius = 0.308

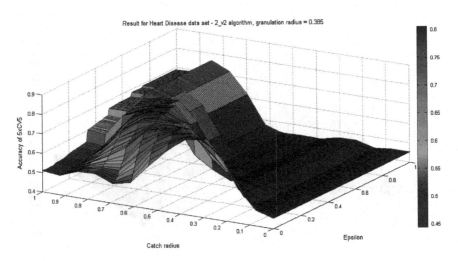

Fig. 9.42 2v2. Heart disease. Radius = 0.385

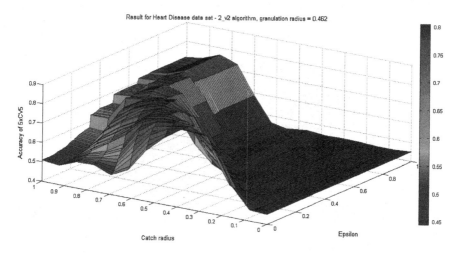

Fig. 9.43 2v2. Heart disease. Radius $= 0.462$

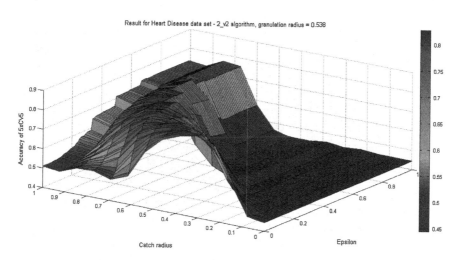

Fig. 9.44 2v2. Heart disease. Radius $= 0.538$

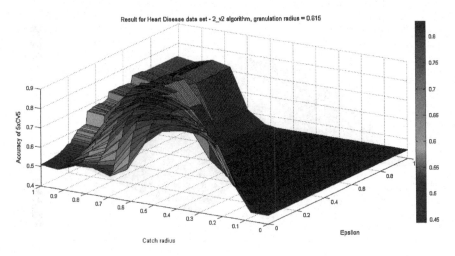

Fig. 9.45 2v2. Heart disease. Radius = 0.615

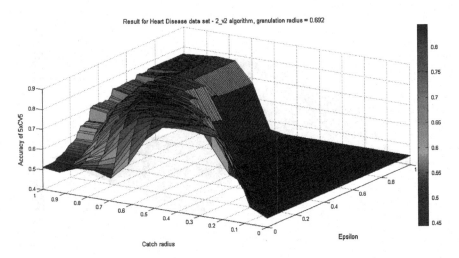

Fig. 9.46 2v2. Heart disease. Radius = 0.692

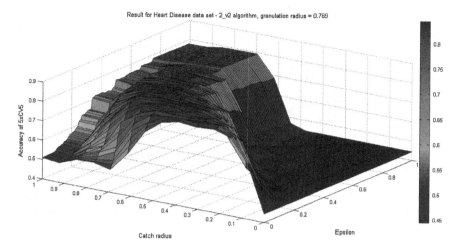

Fig. 9.47 2v2. Heart disease. Radius $= 0.769$

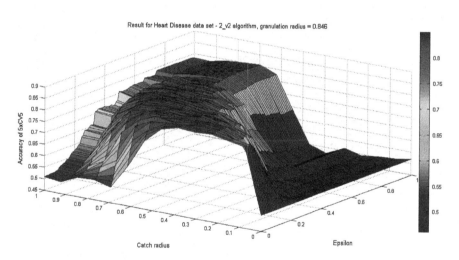

Fig. 9.48 2v2. Heart disease. Radius $= 0.846$

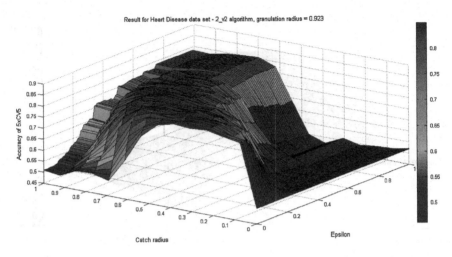

Fig. 9.49 2v2. Heart disease. Radius = 0.923

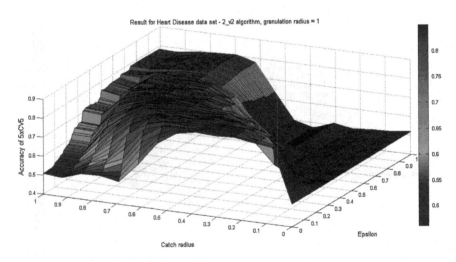

Fig. 9.50 2v2. Heart disease. Radius = 1.0

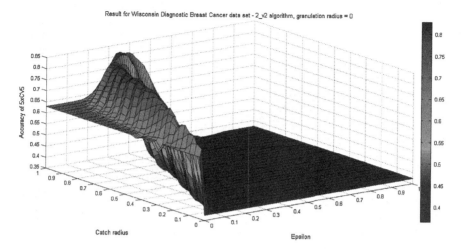

Fig. 9.51 2v2. WDBC. Radius $= 0$

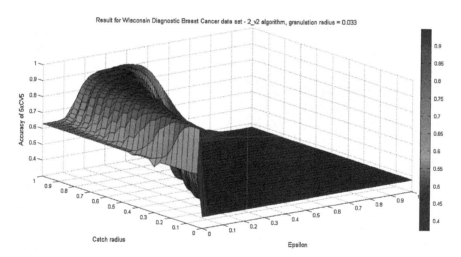

Fig. 9.52 2v2. WDBC. Radius $= 0.033$

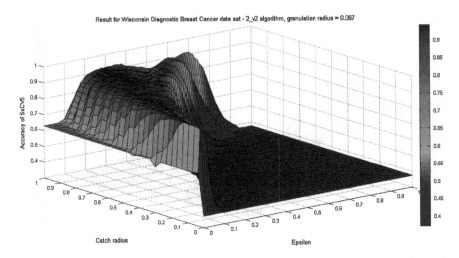

Fig. 9.53 2v2. WDBC. Radius = 0.067

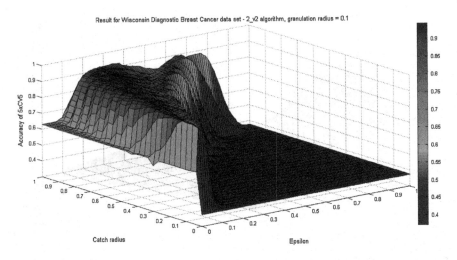

Fig. 9.54 2v2. WDBC. Radius = 0.01

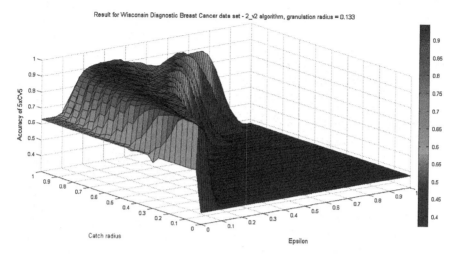

Fig. 9.55 2v2. WDBC. Radius $= 0.133$

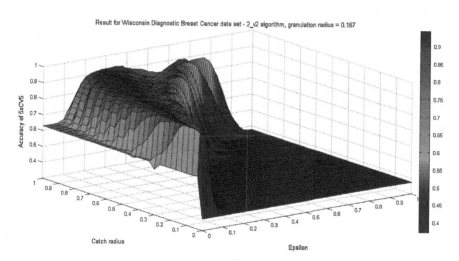

Fig. 9.56 2v2. WDBC. Radius $= 0.167$

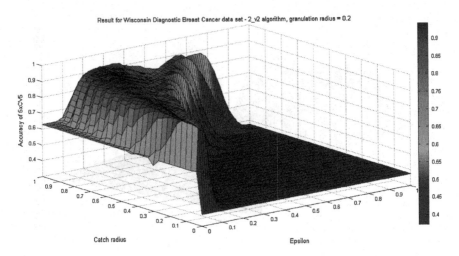

Fig. 9.57 2v2. WDBC. Radius = 0.2

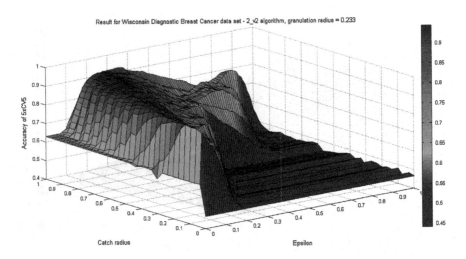

Fig. 9.58 2v2. WDBC. Radius = 0.233

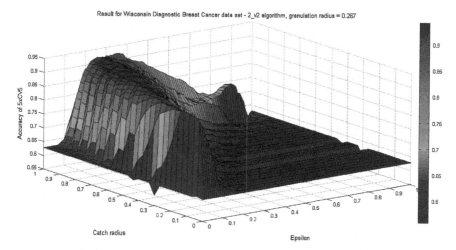

Fig. 9.59 2v2. WDBC. Radius = 0.267

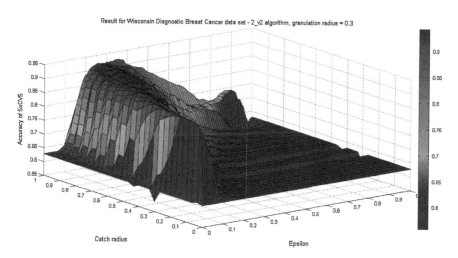

Fig. 9.60 2v2. WDBC. Radius = 0.3

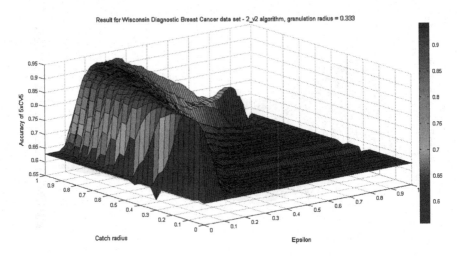

Fig. 9.61 2v2. WDBC. Radius = 0.333

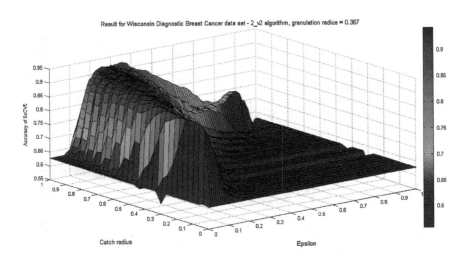

Fig. 9.62 2v2. WDBC. Radius = 0.367

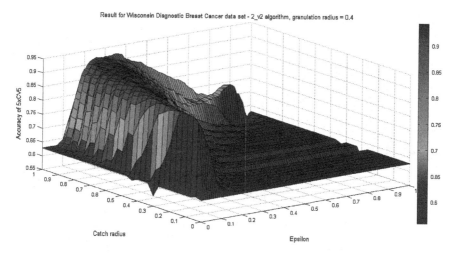

Fig. 9.63 2v2. WDBC. Radius = 0.4

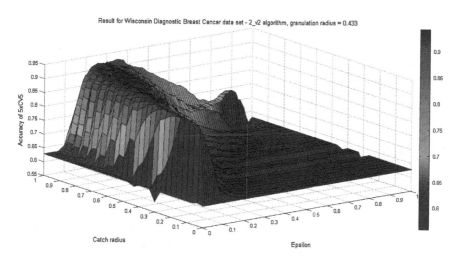

Fig. 9.64 2v2. WDBC. Radius = 0.433

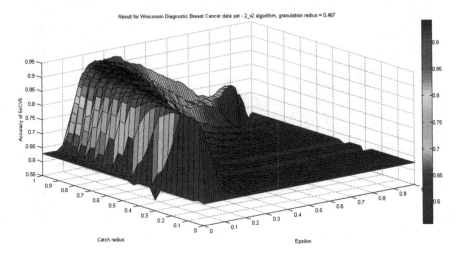

Fig. 9.65 2v2. WDBC. Radius = 0.467

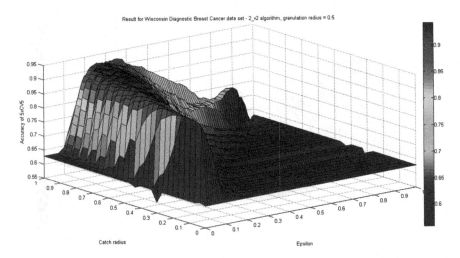

Fig. 9.66 2v2. WDBC. Radius = 0.5

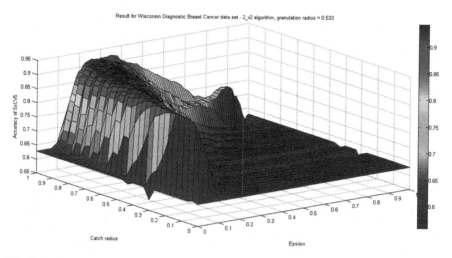

Fig. 9.67 2v2. WDBC. Radius $= 0.533$

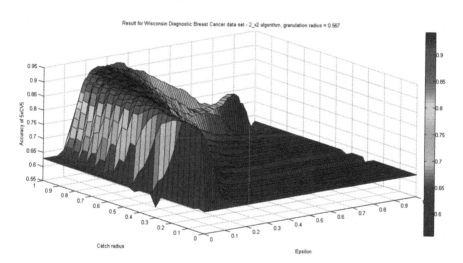

Fig. 9.68 2v2. WDBC. Radius $= 0.567$

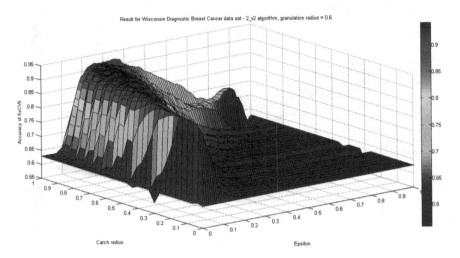

Fig. 9.69 2v2. WDBC. Radius = 0.6

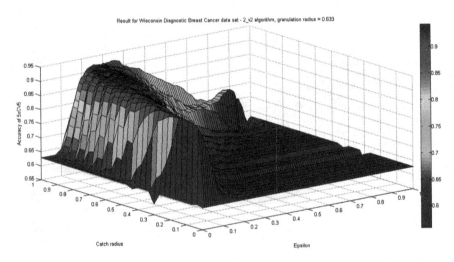

Fig. 9.70 2v2. WDBC. Radius = 0.633

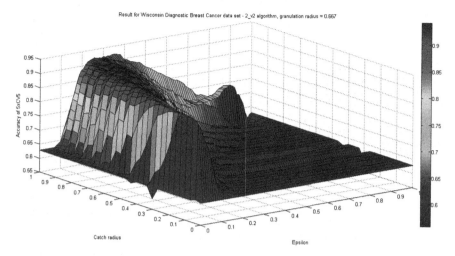

Fig. 9.71 2v2. WDBC. Radius $= 0.667$

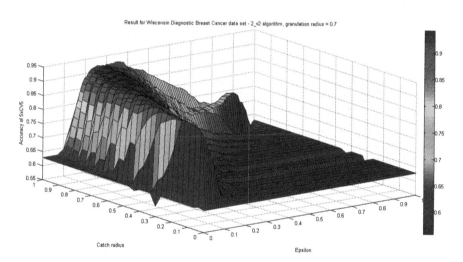

Fig. 9.72 2v2. WDBC. Radius $= 0.7$

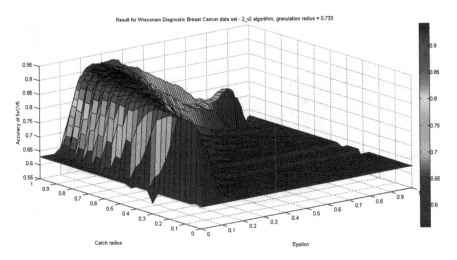

Fig. 9.73 2v2. WDBC. Radius = 0.733

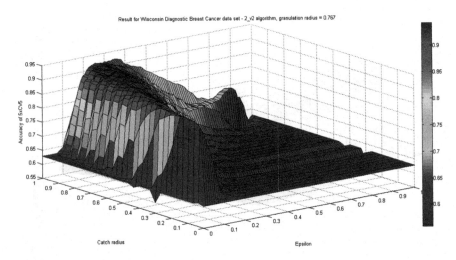

Fig. 9.74 2v2. WDBC. Radius = 0.767

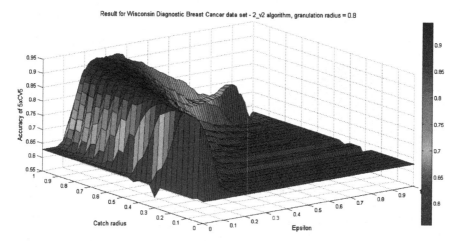

Fig. 9.75 2v2. WDBC. Radius $= 0.8$

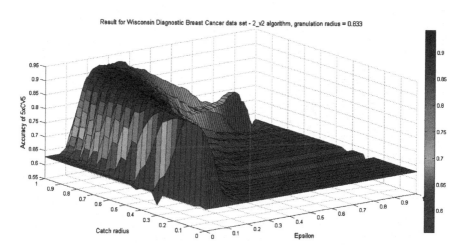

Fig. 9.76 2v2. WDBC. Radius $= 0.833$

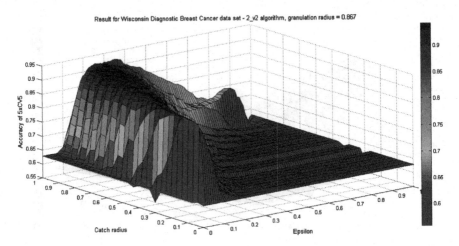

Fig. 9.77 2v2. WDBC. Radius = 0.867

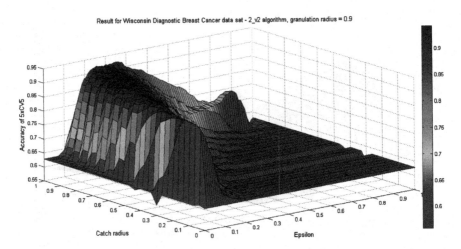

Fig. 9.78 2v2. WDBC. Radius = 0.9

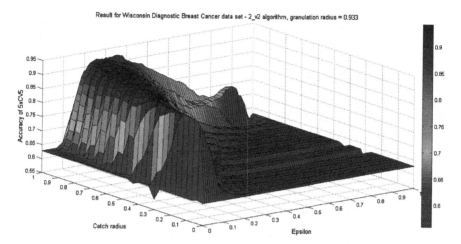

Fig. 9.79 2v2. WDBC. Radius $= 0.933$

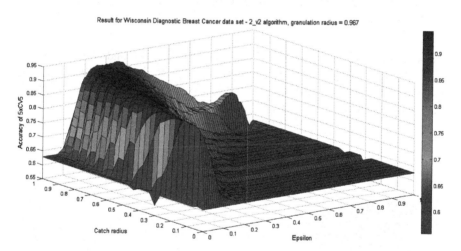

Fig. 9.80 2v2. WDBC. Radius $= 0.967$

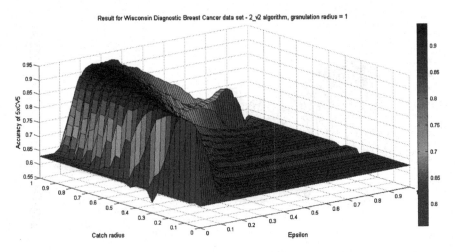

Fig. 9.81 2v2. WDBC. Radius = 1.0

9.4 Results of Experiments with Classifiers 5_v3, 6_v3, 7_v3 Based on the Parameter ε

In this Section, we show results for classifiers 5_v3, 6_v3, and, 7_v3 based on weighted voting by t-norms, cf. Sect. 4.6. In Figs. 9.82, 9.83, 9.84 and 9.85 we have results for classifier 5_v3, in Figs. 9.86, 9.87, 9.88, 9.89, 9.90, 9.91, 9.92 and 9.93, respectively, results are shown for 6_v3 and 7_v3. These results show regions of stable granulation—in the sense of optimal accuracy of classification.

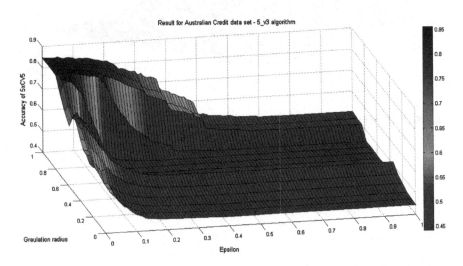

Fig. 9.82 5v3. Australian

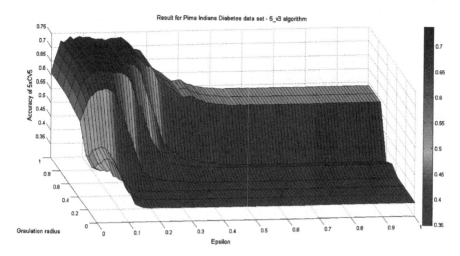

Fig. 9.83 5v3. Diabetes

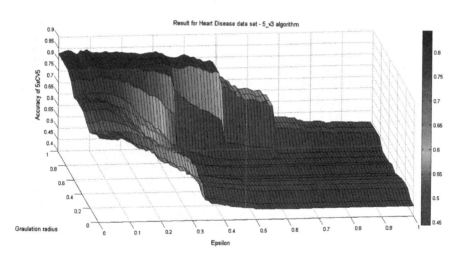

Fig. 9.84 5v3. Heart disease

We present a summary of results along with a visualization of results for studied in this chapter classifiers (Fig. 9.94).

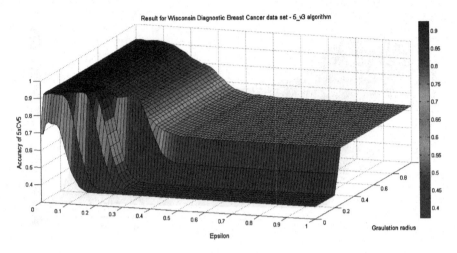

Fig. 9.85 5v3. WDBC

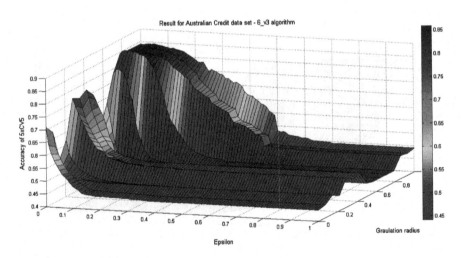

Fig. 9.86 6v3. Australian

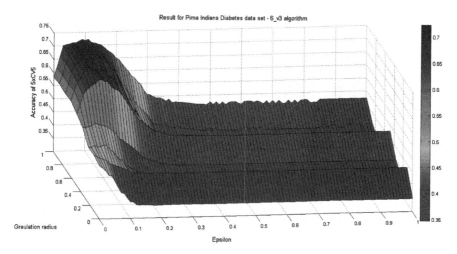

Fig. 9.87 6v3. Diabetes

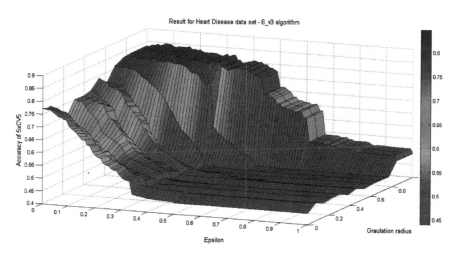

Fig. 9.88 6v3. Heart disease

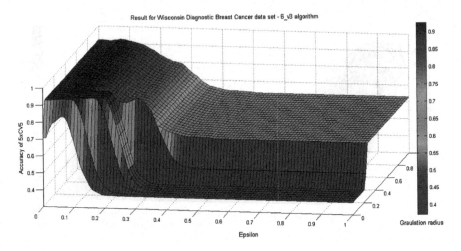

Fig. 9.89 6v3. WDBC

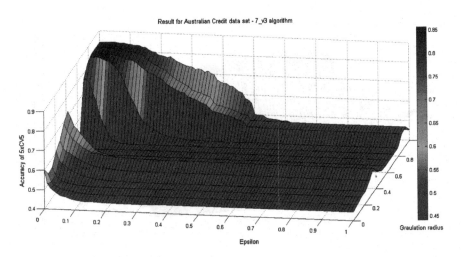

Fig. 9.90 7v3. Australian

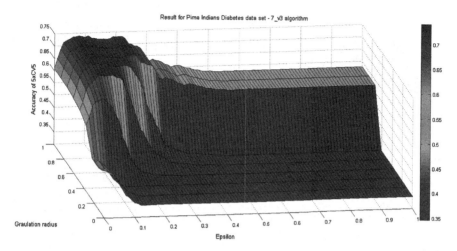

Fig. 9.91 7v3. Diabetes

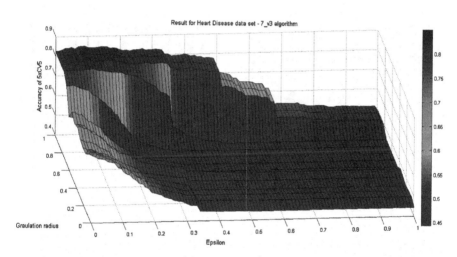

Fig. 9.92 7v3. Heart disease

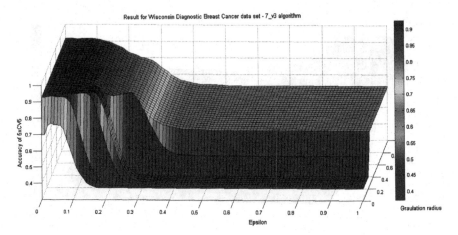

Fig. 9.93   7v3. WDBC

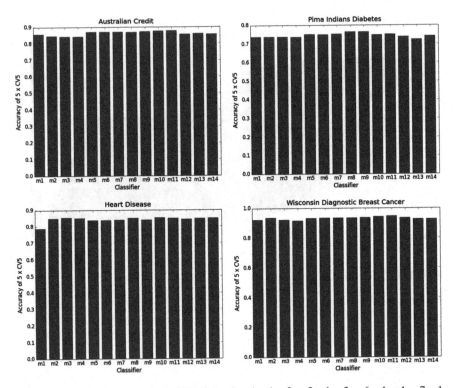

Fig. 9.94   Visualization of results for Table 9.1; $m1 = 1\_v1, m2 = 5\_v1, m3 = 6\_v1, m4 = 7\_v1,$
$m5 = 8\_v1, m6 = 8\_v2, m7 = 8\_v3, m8 = 8\_v4, m9 = 8\_v5, m10 = 1\_v2, m11 = 2\_v2,$
$m12 = 5\_v3, m13 = 6\_v3, m14 = 7\_v3$

Table 9.1 5xCV5; Selected best results for examined classifiers

	Australian	Diabetes	Heart disease	WisconsinDBC
1_v1	$acc = 0.857$	$acc = 0.738$	$acc = 0.787$	$acc = 0.922$
	$\varepsilon = 0.83–0.87$	$\varepsilon = 0.3$	$\varepsilon = 0.76–0.88$	$\varepsilon = 0.35$
5_v1	$acc = 0.846$	$acc = 0.738$	$acc = 0.847$	$acc = 0.935$
	$\varepsilon = 0.01, 0.04$	$\varepsilon = 0.08$	$\varepsilon = 0.08$	$\varepsilon = 0.06$
6_v1	$acc = 0.843$	$acc = 0.738$	$acc = 0853.$	$acc = 0.922$
	$\varepsilon = 0.04$	$\varepsilon = 0.07$	$\varepsilon = 0.08$	$\varepsilon = 0.06, 0.07$
7_v1	$acc = 0.843$	$acc = 0.736$	$acc = 0.85$	$acc = 0.915$
	$\varepsilon = 0.02, 0.03$	$\varepsilon = 0.1$	$\varepsilon = 0.08$	$\varepsilon = 0.06$
8_v1	$acc = 0.869$	$acc = 0.751$	$acc = 0.836$	$acc = 0.932$
8_v2	$acc = 0.869$	$acc = 0.748$	$acc = 0.837$	$acc = 0.933$
	$\varepsilon = 0.42$	$\varepsilon = 0.08–0.1$	$\varepsilon = 0.23–0.25$	$\varepsilon = 0–0.16$
	$\varepsilon = 0.49.0.5$	$\varepsilon = 0.16$	$\varepsilon = 0.3–0.33$	$\varepsilon = 0.97–1$
8_v3	$acc = 0.869$	$acc = 0.752$	$acc = 0839.$	$acc = 0.933$
	$\varepsilon = 0.33$	$\varepsilon = 0.95, 0.96$	$\varepsilon = 0.26$	$\varepsilon = 0.11–0.17, 0.19$
	$\varepsilon = 0.38–0.4$	$\varepsilon = 0.95, 0.96$	$\varepsilon = 0.28–0.32$	$\varepsilon = 0.2, 0.98, 0.99$
8_v4	$acc = 0.868$	$acc = 0.763$	$acc = 0.849$	$acc = 0.934$
	$\varepsilon = 0.92–1$	$\varepsilon = 0.56$	$\varepsilon = 0.68–0.7$	$\varepsilon = 0.39–0.42, 0.47$
8_v5	$acc = 0.872$	$acc = 0.763$	$acc = 0.839$	$acc = 0.936$
	$\varepsilon = 0.62–0.66$	$\varepsilon = 0.52, 0.57$	$\varepsilon = 0.36$	$\varepsilon = 0.45$
	$\varepsilon = 0.7–0.73$	$\varepsilon = 0.58$	$\varepsilon = 0.36$	$\varepsilon = 0.45$
1_v2	$acc = 0.875$	$acc = 0.747$	$acc = 0.854$	$acc = 0.942$
	$\varepsilon = 0.16$	$\varepsilon = 0.3$	$\varepsilon = 0.09$	$\varepsilon = 0.16$
	$r_{catch} = 0.714$	$r_{catch} = 0.75$	$r_{catch} = 0.308$	$r_{catch} = 0.7$
2_v2	$acc = 0.877$	$acc = 0.751$	$acc = 0.85$	$acc = 0.946$
	$\varepsilon = 0.04$	$\varepsilon = 0.3$	$\varepsilon = 0.16$	$\varepsilon = 0.23$
	$r_{gran} = 0.714$	$r_{gran} = 0.625$	$r_{gran} = 0.846$	$r_{gran} = 0.033$
	$r_{catch} = 0.571$	$r_{catch} = 0.75$	$r_{catch} = 0.385$	$r_{catch} = 0.733$
	$CombAGS = 0.514$	$CombAGS = 0.379$	$CombAGS = 0.425$	$CombAGS = 0.743$
5_v3	$acc = 0.855$	$acc = 0.737$	$acc = 0.843$	$acc = 0.933$
	$\varepsilon = 0.03$	$\varepsilon = 0.06$	$\varepsilon = 0.07$	$\varepsilon = 0.07$
	$r_{gran} = 0.786$	$r_{gran} = 0.75$	$r_{gran} = 0.769$	$r_{gran} = 0.267$
	$CombAGS = 0.432$	$CombAGS = 0.369$	$CombAGS = 0.433$	$CombAGS = 0.467$
6_v3	$acc = 0.858$	$acc = 0.722$	$acc = 0.847$	$acc = 0.925$
	$\varepsilon = 0.02$	$\varepsilon = 0.09$	$\varepsilon = 0.07$	$\varepsilon = 0.07$
	$r_{gran} = 0.786$	$r_{gran} = 1$	$r_{gran} = 0.846$	$r_{gran} = 0.033$
	$CombAGS = 0.433$	$CombAGS = 0.361$	$CombAGS = 0.424$	$CombAGS = 0.732$
7_v3	$acc = 0.855$	$acc = 0.74$	$acc = 0.85$	$acc = 0.926$
	$\varepsilon = 0.03$	$\varepsilon = 0.08$	$\varepsilon = 0.09$	$\varepsilon = 0.13$
	$r_{gran} = 0.786$	$r_{gran} = 625$	$r_{gran} = 0.846$	$r_{gran} = 0.067$
	$CombAGS = 0.432$	$CombAGS = 0.374$	$CombAGS = 0.425$	$CombAGS = 0.484$

References

1. Artiemjew, P.: Rough mereological classifiers obtained from weak rough set inclusions. In: Proceedings of International Conference on Rough Sets and Knowledge Technology RSKT'08, Chengdu China. LNAI, vol. 5009, pp. 229–236. Springer, Berlin (2008)
2. Polkowski, L.: Formal granular calculi based on rough inclusions (a feature talk). In: Proceedings of the 2006 IEEE International Conference on Granular Computing GrC'06, pp. 57–62. USA. IEEE Press, Atlanta (2006)
3. Polkowski, L.: The paradigm of granular rough computing. In: Proceedings of the 6th IEEE International Conference on Cognitive Informatics ICCI'07, Lake Tahoe NV USA, pp. 145–163. IEEE Computer Society, Los Alamitos (2007)
4. Polkowski, L.: A unified approach to granulation of knowledge and granular computing based on rough mereology: a survey. In: Pedrycz, W., Skowron, A., Kreinovich, V. (eds.) Handbook of Granular Computing, pp. 375–401. Wiley, New York (2008)
5. Polkowski, L.: Granulation of knowledge: similarity based approach in information and decision systems. In: Meyers, R.A. (ed.) Encyclopedia of Complexity and System Sciences. Springer, Berlin, Article 00788 (2009)

Chapter 10
Effects of Granulation on Entropy and Noise in Data

I thought of calling it 'information', but the word was overtly used, so I decided to call it 'uncertainty'... Von Neumann told me, 'You should call it entropy for two reasons. In the first place your uncertainty function has been used in statistical mechanics under that name, so it already has a name. In the second place, and more important, nobody knows what entropy really is, so in a debate you will always have the advantage'.

[Claude Shannon. In Tribus, McIrvine. Energy and information. Scientific American 224.]

10.1 On Entropy Behavior During Granulation

Assume that our data table does possess of M attributes in a set A, K objects, and N is the number of indiscernibility classes. We adopt a symmetric and transitive rough inclusion μ, see (2.40), induced by a t-norm L of Łukasiewicz, as the granulation tool, actually applied in computations in the preceding chapters. We denote with the symbol C_j the jth indiscernibility class for $j = 1, 2, \ldots, N$ and $|C_j|$ stands for the size of the jth class. Entropy of the partition into indiscernibility classes is

$$H_{ind} = -\sum_{j=1}^{N} \frac{|C_j|}{K} \cdot log_2 \frac{|C_j|}{K}. \tag{10.1}$$

Granules form a covering of the universe of objects, and, we modify the notion of entropy to that of the *covering entropy*:

$$H^*(r - Cov) = -\sum_{n=1}^{G} \frac{|g_n|}{S} \cdot log_2 \frac{|g_n|}{S}, \tag{10.2}$$

where $r - Cov$ is the covering by r-granules, G is the number of r-granules (under our assumption about the rough inclusion, $G = N$), and S is the sum of cardinalities of r-granules. Assuming that each class C_j produces a granule g_j of at least size $(1 + \gamma_j)$,

© Springer International Publishing Switzerland 2015
L. Polkowski and P. Artiemjew, *Granular Computing in Decision Approximation*,
Intelligent Systems Reference Library 77, DOI 10.1007/978-3-319-12880-1_10

for some positive value γ_j, so we may assume that S equals about $(1 + \sum_j \gamma_j) \cdot K$, the information gain is

$$[-\sum_j \frac{|C_j|}{K}] + log_2 K - [-\sum_j \frac{1 + \gamma_j)|C_j|}{(1 + \sum_j \gamma_j) \cdot K} \cdot log_2 (1 + \gamma_j)|C_j| + log_2 (K + \sum_j \gamma_j)].$$

In case each γ_j is slightly greater than 1, hence, approximately, $S = 2K$, one finds that the information gain is

$$-1 + 1 + \frac{1}{K} \cdot H - \frac{1}{K} \cdot H = 0,$$

which means that the entropy of data in this case is not increased in the process of granulation.

The same result is obtained when all γ_j's take a value about the common value γ, hence, a uniform granulation (a high entropy case) does not increase entropy of data.

In the opposite case when one granule dominates over all other in size, i.e., assume that at the extreme only one indiscernibility class induces a bigger granule, viz., $|g_{j_0}| = \gamma_0 \cdot |C_{j_0}|$, for some j_0, and, $|g_j| = |C_j|$ for $j \neq j_0$, the information gain is the sum of and the trade-off between the two factors, viz.,

$$\sum_{j \neq j_0} [\frac{|C_j|}{K + \gamma_0 \cdot |C_{j_0}|} \cdot log_2 (\frac{|C_j|}{K + \gamma_0 \cdot |C_{j_0}|}) - \frac{|C_j|}{K} \cdot log_2 (\frac{|C_j|}{K})],$$

which is positive as the function $\frac{a}{b+x} \cdot log_2 \frac{a}{b+x}$, where $b > a$, is increasing in x, and,

$$[\frac{(1 + \gamma_0) \cdot |C_{j_0}|}{K + \gamma_0 \cdot |C_{j_0}|} \cdot log_2 \frac{(1 + \gamma_0) \cdot |C_{j_0}|}{K + \gamma_0 \cdot |C_{j_0}|}] - \frac{|C_{j_0}|}{K} \cdot log_2 (\frac{|C_{j_0}|}{K}),$$

which is negative as the function $\frac{a+x}{b+x} \cdot log_2 \frac{a+x}{b+x}$, $b > a$ is decreasing in x.

A look at graphs of the two functions in Fig. 10.1, where the graph shows the behavior of two components over the interval $[0, 6]$ visualizes that the Component 1 is positive whereas the Component 2 is negative. Over the interval $[0, 2]$, the absolute value of Component 2 is greater than the value of Component 1, hence, the sum, i.e., the information gain, is negative.

10.2 On Noise in Data During Granulation

It is natural to define *noise* in a data table (a decision system) as the relative number of *decision-ambiguous objects*. An object u is decision-ambiguous in case when its indiscernibility class is covered by at least two decision classes. Formally, the

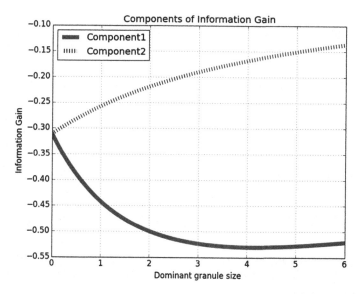

Fig. 10.1 Graphs of functions $\frac{a}{b+x} \cdot log_2 \frac{a}{b+x}$ (Component 1) and $\frac{a+x}{b+x} \cdot log_2 \frac{a+x}{b+x}$ (Component 2) for $a = 3, b = 4$

predicate $dec - amb(u)$ which does express this property is defined as

$$dec - amb(u) \text{ if and only if there is } v : \; Inf(u) = Inf(v), d(u) \neq d(v), \quad (10.3)$$

where d is the decision, $Inf(u)$ is the information vector of u.

The relative version of decision ambiguity, necessary in order to discuss noise in non-granulated data, begins with a *degree of indiscernibility* $r \in \{\frac{k}{|A|} : k = 0, 1, \ldots, |A|\}$, which in the granulated case is the granulation radius. The predicate $r - dec - amb$ is defined then as

$$r - dec - amb(u, v) \text{ if and only if } \frac{IND(u, v)}{|A|} = r. \quad (10.4)$$

The noise relative to r-indiscernibility is measured by the set $Noise_r$ defined in

$$u \in Noise_r \text{ if and only if there is } v : \; r - dec - amb(u, v), d(u) = d(v). \quad (10.5)$$

Noise in data relative to the r-indiscernibility (the granulation radius r in the granulated case) is expressed as the relative size of the set $Noise_r$,

$$noise_r = \frac{|Noise_r|}{|A|}. \quad (10.6)$$

In the following figures, the dotted line shows the noise in the above sense plotted for all fourteen data sets experimented with against the granulation radii in the granulated case and understood as degrees of indiscernibility in the non-granulated case. The results shown in Figs. 10.2, 10.3, 10.4, 10.5, 10.6, 10.7 and 10.8 demonstrate that as

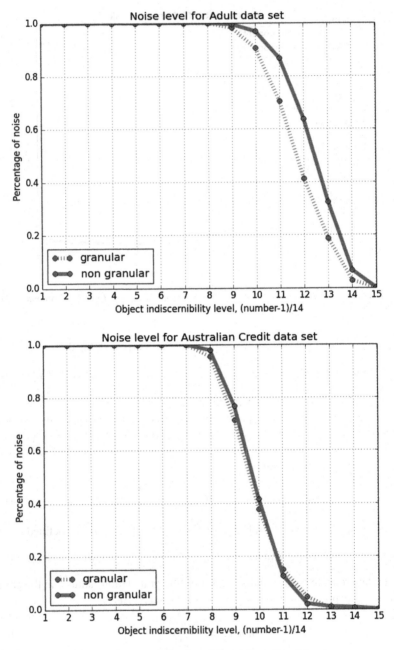

Fig. 10.2 Noise before and after granulation for adult and Australian data sets

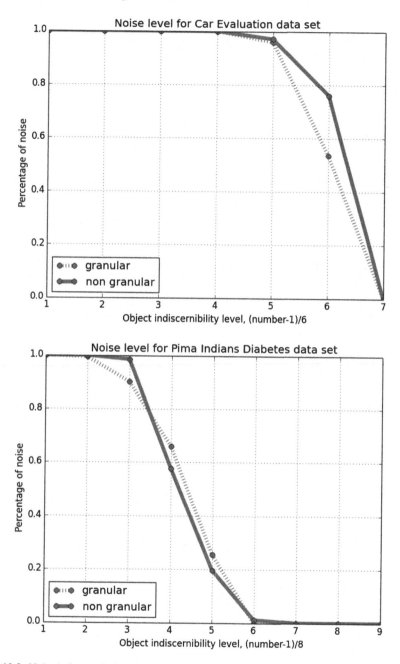

Fig. 10.3 Noise before and after granulation for Car and Diabetes data sets

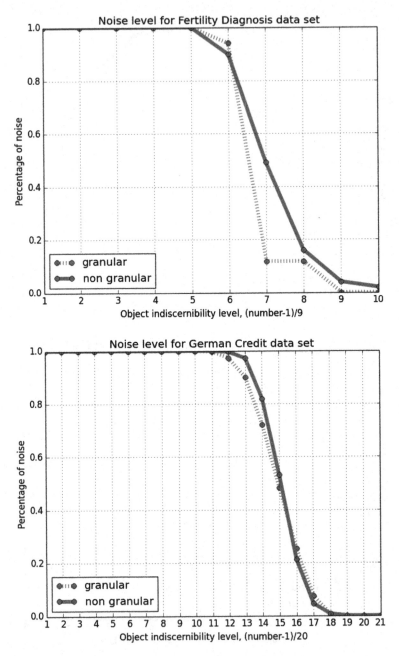

Fig. 10.4 Noise before and after granulation for Fertility and German data sets

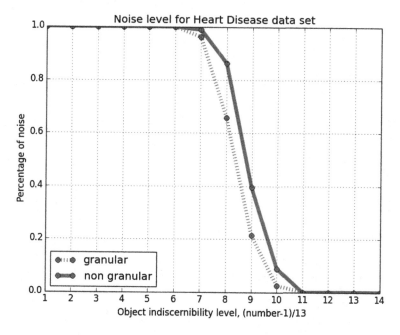

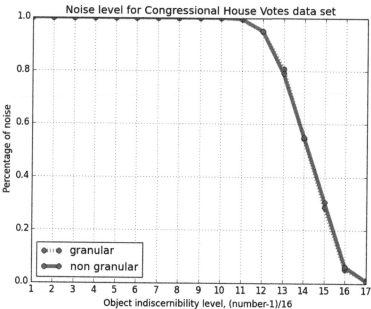

Fig. 10.5 Noise before and after granulation for Heart and Congressional Voting data sets

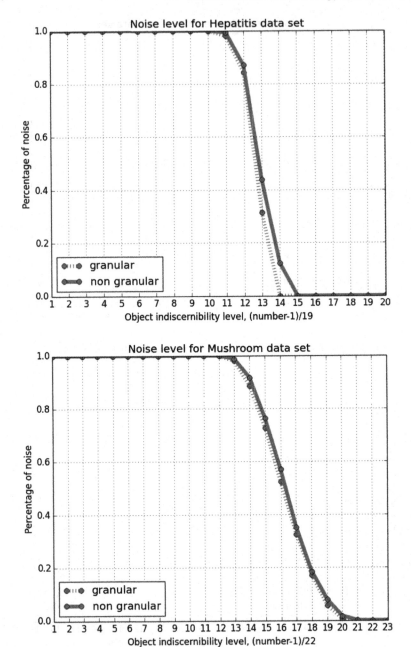

Fig. 10.6 Noise before and after granulation for Hepatitis and Mushroom data sets

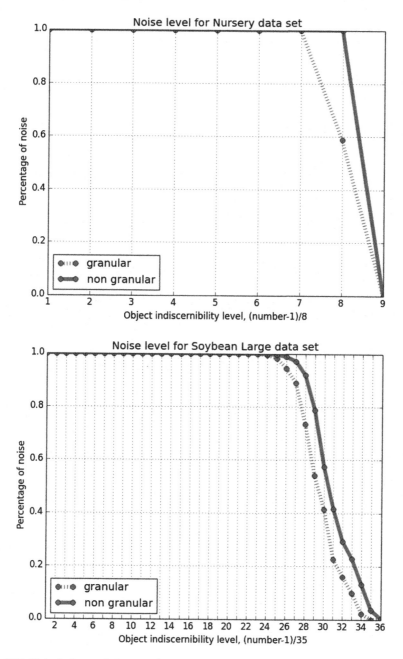

Fig. 10.7 Noise before and after granulation for Nursery and Soybean data sets

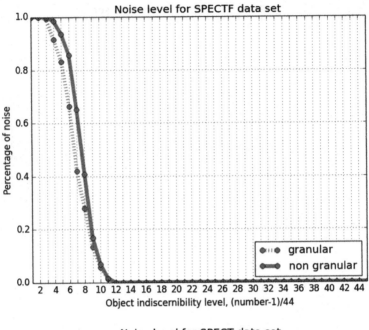

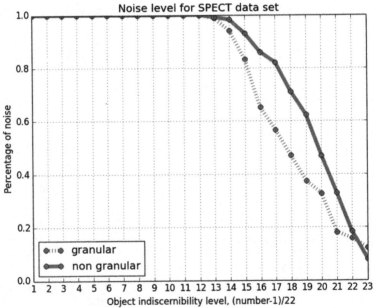

Fig. 10.8 Noise before and after granulation for SPECTF and SPECT data sets

a rule, granulation reduces noise in data. Results for granulated cases are obtained after two-layer granulation.

10.3 On Characteristics of Data Sets Bearing on Granulation

We continue our analysis in the context of Sect. 10.1, in particular the rough inclusion invoked here is that of (2.40).

We choose a granulation radius $r = \frac{k}{M}$ for some $k = 1, 2, \ldots, M - 1$. We denote by the symbol $\alpha_r(p, j)$ the cardinality of the set

$$A_r(p, j) = \{u \in C_j : \mu(u, x, r)\} \tag{10.7}$$

for some x in the class C_p (actually, every x in C_p). By symmetry and transitivity of μ, we have the following relations

$$\alpha_r(p, q) = \alpha_r(q, p) \tag{10.8}$$

for each pair $p \neq q$, and,

$$\text{If } x \in A_r(j, p), y \in A_r(j, q) \text{ then } x \in A_{T(r,r)}(q, p), y \in A_{T(r,r)}(p, q), \tag{10.9}$$

which shed some light on the dynamics of granulation.

The size of an r-granule about an indiscernibility class C_j is

$$\gamma(j, r) = |C_j| + \sum_{p \neq j} \alpha_r(j, p), \tag{10.10}$$

hence, the relative size of granules is

$$\beta_r(p, q) = \frac{|C_p| + \sum_{j \neq p} \alpha_r(p, j)}{|C_q| + \sum_{j \neq q} \alpha_r(q, j)}. \tag{10.11}$$

In view of (10.8), the relative size in (10.11) is determined by $|C_p|, |C_q|$,

$\sum_{j \neq p, q} \alpha_r(p, j), \sum_{j \neq p, q} \alpha_r(q, j)$ and one may assume that bigger indiscernibility classes induce bigger granules, if the mereological distance (see below) between classes C_p and C_q is sufficiently small.

With the rough inclusion (2.40), we can be more specific. In this case, a granule about an object does contain the whole indiscernibility class of the object, i.e., one may say that indiscernibility classes are inflated by granulation:

$$\text{If } u \in g(v, r, \mu) \text{ then } [u]_A \subseteq g(v, r, \mu). \tag{10.12}$$

For indiscernibility classes $C_p = [u]_A, C_q = [v]_A$, we define the *mereological distance* $dist_\mu$ as

$$dist_\mu(C_p, C_q) = min\{r : \text{ there are } x \in C_p, y \in C_q \text{ such that } \mu(u, v, r)\}. \quad (10.13)$$

The mereological distance $dist_\mu$ is 1 for identical classes (1), symmetric (2) and it does satisfy the triangle inequality (3):

- 1. $dist_\mu(C_p, C_p) = 1$.
- 2. $dist_\mu(C_p, C_q) = dist_\mu(C_q, C_p)$.
- 3. If $dist_\mu(C_p, C_q) = r$ and $dist_\mu(C_q, C_j) = s$, then $dist_\mu(C_p, C_j) \geq L(r, s) = max\{0, r + s - 1\}$.

Formula (10.12) can be rewritten with the indiscernibility class $C_p = [u]_A$ as the granule center

$$g(C_p, r, \mu) = \bigcup\{C_q : dist(C_p, C_q) \geq r\}. \quad (10.14)$$

Formulas for coefficients $\alpha_r(j, p), \beta_r(p, q)$ can be now rewritten as

- 1. $\alpha_r(p, j) = |\bigcup\{C_j : dist_\mu(C_j, C_p) \geq r\}|$.
- 2. $\beta_r(p, q) = \frac{|\bigcup\{C_j : dist_\mu(C_j, C_p) \geq r\}|}{|\bigcup\{C_j : dist_\mu(C_j, C_q) \geq r\}|}$.

It follows that important for analysis of granulation is the r-distribution of indiscernibility classes among granulation radii. It is shown, for data sets examined by us in the preceding chapters, in Figs. 10.9, 10.10, 10.11, 10.12 and 10.13.

It is to be seen from Figs. 10.9, 10.10, 10.11, 10.12 and 10.13 that, save for **SPECTF**, Congressional Voting Records and Mushroom, all remaining data sets demonstrate the r-distribution approximately normal with the mean about the median of the granulation radii.

It may be tempting to address the problem of the optimal granulation radius, cf., Table 6.49. Taking it as, approximately up to the 2-neighborhood, cf. Table 6.49, the value $r(1, 2)$, we find that, neglecting the outliers on the left and on the right of the histogram, having small values as compared to other, and counting the relative position of $r(1, 2)$ in the linear increasing order on other radii, all values of this position are between 0.715 and 0.88, i.e., in the 6th septile on the interval $[0, 1]$. We may thus express our rule of thumb as: *the optimal granulation radius is located approximately in the 6th septile of the linearly ordered in the increasing order set of granulation radii remaining after outliers on either side are rejected.*

Concerning noise reduction, when we consider intervals of granulation radii over which noise reduction is non-zero, and single out radii of the maximal noise reduction, then those values approximately coincide with values of radii $r(1, 2)$. A more detailed analysis of relations among those parameters will be given in Conclusion chapter.

Fig. 10.9 Distribution of indiscernibility classes among granulation radii for adult, Australian, Car data sets

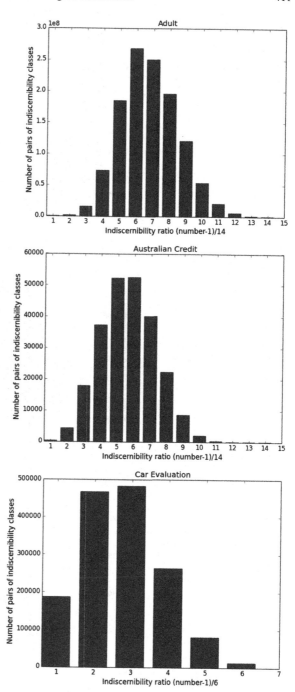

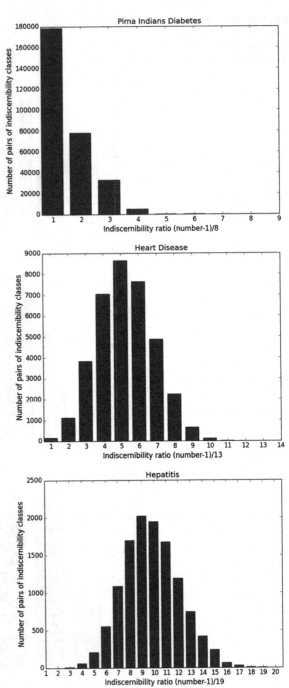

Fig. 10.10 Distribution of indiscernibility classes among granulation radii for Car, Diabetes, Hepatitis data sets

Fig. 10.11 Distribution of indiscernibility classes among granulation radii for Congr. Voting, Mushroom, Nursery data sets

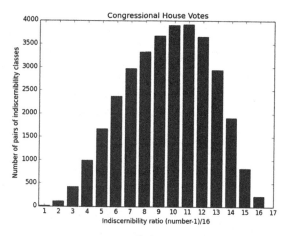

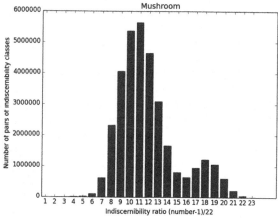

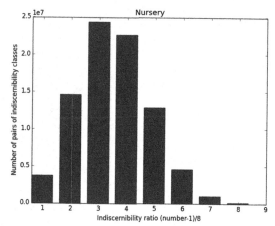

Fig. 10.12 Distribution of
indiscernibility classes among
granulation radii for Soybean,
SPECT, SPECTF data sets

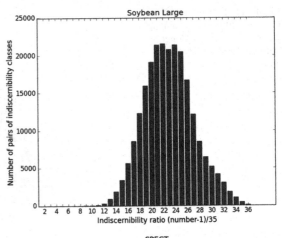

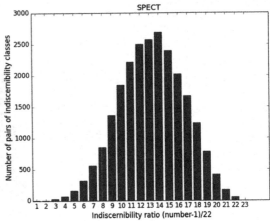

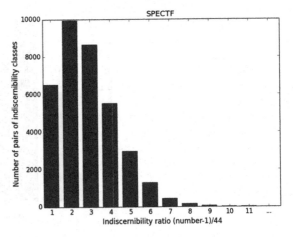

Fig. 10.13 Distribution of
indiscernibility classes among
granulation radii for Fertility
Diagnosis, German Credit
data sets

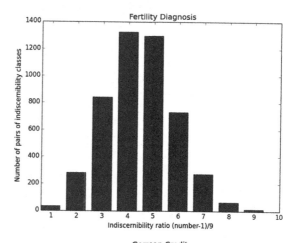

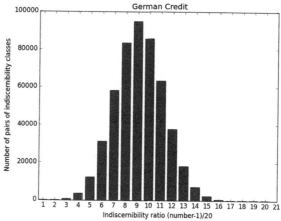

Chapter 11
Conclusions

I have done what I could; let those who can do better
[James Boswell. An Account of Corsica.]

In Tables 6.7, 6.8, 6.9, 6.10, 6.11, 6.12, 6.13, 6.14, 6.15, 6.16, 6.17, 6.18, 6.19, 6.20, 6.21, 6.22, 6.23, 6.24, 6.25, 6.26, 6.27, 6.28, 6.29, 6.30, 6.31, 6.35, 6.36, 6.37, 6.38, 6.39, 6.40, 6.41, 6.42, 6.43, 6.44, 6.45, 6.46, 6.47 and 6.48, Figs. 10.2, 10.3, 10.4, 10.5, 10.6, 10.7, 10.8, 10.9, 10.10, 10.11, 10.12 and 10.13, results are collected from which some parameters which may bear on the optimal granulation radius location are extracted. We single out the following values of granulation radii:

- $r(1, 2)$ = the granulation radius for which the decrease in size of the training set between the first and the second granulation layers is maximal.
- *rags* = the granulation radius for which maximal accuracy *combAGS* is achieved among all radii yielding accuracy in the interval $|acc_{r=1} - acc_r| \leq 0.01$, cf., Sect. 6.3.3.
- *rn* = the granulation radius for which the greatest reduction in noise is achieved.
- *rdec* = the granulation radius r_i in the increasing order of radii for which the decrease in number of pairs at a given distance between r_i and r_{i+1} is the greatest among positive decrease values.

We collect in Table 11.1 values of $r(1, 2)$, *rags*, *rn*, *rdec* from tables and figures listed above and in Table 11.2 we enclose values of absolute differences $|r(1, 2) - rags|$, $|rn - rags|$, and, $|rdec - rags|$ for all fourteen data sets examined by us.

In the following Table 11.3, we enclose values of a variant of the *experimental cumulative distribution functions* (ECDF's) for absolute differences $|r(1, 2) - rags$, $|rn - rags|$, and, $|rdec - rags|$. Data sets are numbered from 1 to 14, and, manifestly, for each ECDF, the numbering of data in increasing order is distinct. The value of each particular ECDF at each granulation radius r is the sum of values at radii less or equal to r (Figs. 11.1 and 11.2).

Figure 11.3 shows that ECDFs for $|rn - rags|$ and $|r(1, 2) - rags|$ are close to each other so we investigate distributions of these values. We apply to this end the

© Springer International Publishing Switzerland 2015
L. Polkowski and P. Artiemjew, *Granular Computing in Decision Approximation*,
Intelligent Systems Reference Library 77, DOI 10.1007/978-3-319-12880-1_11

Table 11.1 Values of granulation radii $r(1, 2)$, *rags*, *rn*, *rdec*

data set	$r(1, 2)$	rags	rn	rdec
adult	0.786	0.786	0.786	0.5
australian	0.571	0.571	0.571	0.429
car	0.833	0.833	0.833	0.333
diabetes	0.25	0.5	0.25	0
fertility	0.667	0.889	0.667	0.444
german	0.65	0.65	0.65	0.5
heart	0.538	0.462	0.538	0.385
hepatitis	0.632	0.632	0.632	0.526
congr. voting	0.938	0.625	0.875	0.813
mushroom	0.955	0.682	0.682	0.5
nursery	0.875	0.75	0.875	0.375
soybean	0.886	0.971	0.8	0.694
SPECT	0.818	0.682	0.727	0.773
SPECTF	0.114	0.091	0.136	0.045

Table 11.2 Values of absolute differences $|r(1, 2) - rags|$, $|rn - rags|$, and, $|rdec - rags|$

data set	$|r(1, 2) - rags|$	$|rn - rags|$	$|rdec - rags|$
adult	0.000	0.000	0.286
australian	0.000	0.000	0.142
car	0.000	0.000	0.500
diabetes	0.250	0.250	0.500
fertility	0.222	0.222	0.445
german	0.000	0.000	0.150
heart	0.076	0.076	0.077
hepatitis	0.000	0.000	0.106
congr. voting	0.313	0.250	0.188
mushroom	0.273	0.000	0.182
nursery	0.125	0.125	0.375
soybean	0.085	0.171	0.277
SPECT	0.136	0.045	0.091
SPECTF	0.023	0.045	0.046

empirical distribution function EDF known in the EDF tests for goodness-of-fit, cf., e.g., Allen [1], D'Agostino and Stephens [2]. *EDF* for a sample of data $x_1, x_2, \ldots, x_n$ ordered in the increasing order is defined as the function $S_n(x) = \frac{i(x)}{n}$, where $i(x)$ is the number of sample items x_i less or equal to x. In Table 11.4, we give values of EDFs for $|rn - rags|$ and $|r(1, 2) - rags|$, along with the *truncated EDFs*, $t - EDFs$, obtained by subtracting from each value of $S_n(x)$ the value $S_n(0)$. The shape of ECDFs suggests the choice of the *exponential distribution* $F(x, \alpha) = 1 - exp(-\alpha \cdot x)$. We evaluate the parameter α at the value $S_n(0.1)$ and with the value 2.41 of α, we obtain a satisfactory

Table 11.3 ECDFs for absolute differences $|r(1,2) - rags|$, $|rn - rags|$, and, $|rdec - rags|$

counter	ECDF for $|r(1,2) - rags|$	ECDF for $|rn - rags|$	ECDF for $|rdec - rags|$
1	0.000	0.000	0.046
2	0.000	0.000	0.123
3	0.000	0.000	0.214
4	0.000	0.000	0.320
5	0.000	0.000	0.462
6	0.023	0.000	0.612
7	0.099	0.045	0.794
8	0.184	0.090	0.982
9	0.309	0.166	1.259
10	0.445	0.291	1.545
11	0.667	0.462	1.920
12	0.917	0.684	2.365
13	1.190	0.934	2.865
14	1.503	1.184	3.365

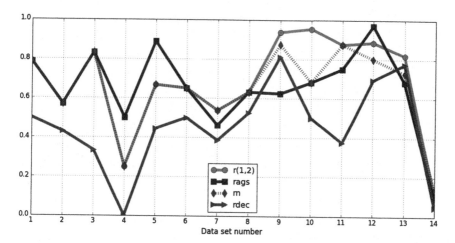

Fig. 11.1 Plot of parameters for Table 11.1

approximation to EDFs. We may conclude that both rn and $r(1,2)$ approximate fairly close the optimal granulation radius $rags$: the mean for $|r(1,2) - rags|$ is 0.107, the median is 0.076, and respective values for $|rn - rags|$ are 0.084 and 0.045. Henceforth, one is correct in assuming that values of rn and $r(1,2)$ are correlated with the value of $rags$ and can serve as estimates of it (Figs. 11.4 and 11.5).

We see from Table 11.4 that truncated EDFs are identical for $|rn - rags|$ and $|r(1,2) - rags|$ in the most important interval $[0, 0.3]$ in which all but one values are

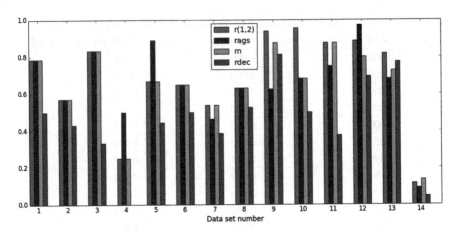

Fig. 11.2 Histogram of parameters for Table 11.1

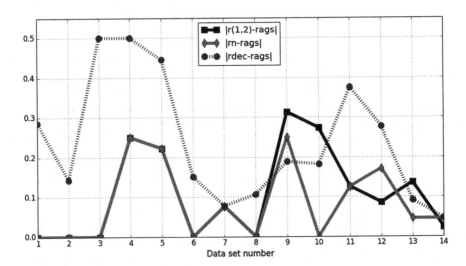

Fig. 11.3 Plot of radii differences for Table 11.2

Table 11.4 EDFs for absolute differences $|r(1, 2) - rags|$, $|rn - rags|$. Exponential approximation

x	EDF ($|r(1, 2) - rags|$)	$t - EDF$ ($|r(1, 2) - rags|$)	EDF ($|rn) - rags|$)	$t - EDF$ ($|rn - rags|$)	$F(x, 2.41)$
0.0	$\frac{5}{14}$	0.0	$\frac{6}{14}$	0.0	0.0
0.1	$\frac{8}{14}$	$\frac{3}{14}$	$\frac{9}{14}$	$\frac{3}{14} = 0.214$	0.214
0.2	$\frac{10}{14}$	$\frac{5}{14}$	$\frac{11}{14}$	$\frac{5}{14} = 0.357$	0.383
0.3	$\frac{13}{14}$	$\frac{8}{14}$	1.0	$\frac{8}{14} = 0.571$	0.515
0.4	1.0	$\frac{9}{14}$	1.0	$\frac{8}{14} = 0.571$	0.515

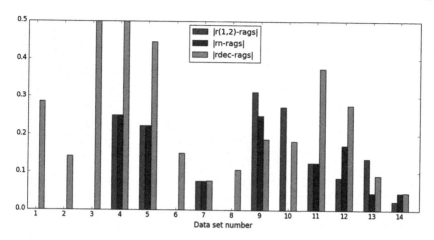

Fig. 11.4 Histogram of radii differences for Table 11.2

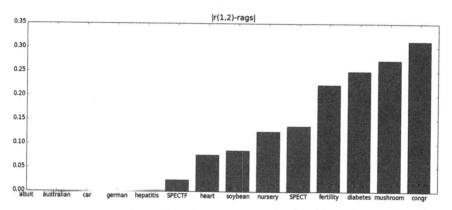

Fig. 11.5 Histogram for $|r(1, 2) - rags|$ for Table 11.2—sorted increasing

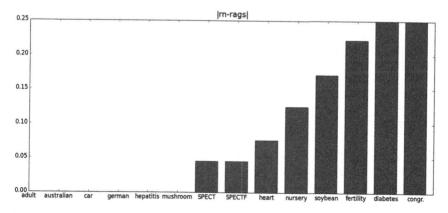

Fig. 11.6 Histogram for $|rn - rags|$ for Table 11.2—sorted increasing

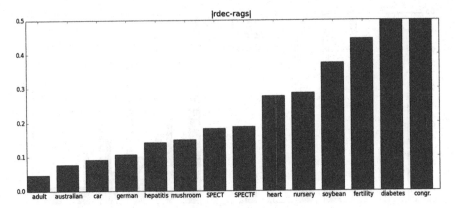

Fig. 11.7 Histogram for $|rdec - rags|$ for Table 11.2—sorted increasing

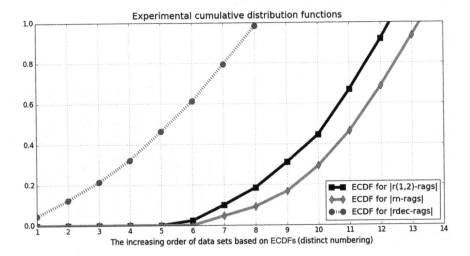

Fig. 11.8 Plot of ECDFs for Table 11.3

located. The exponential distribution with value 2.41 of the parameter α fits fairly closely EDFs. The error is $<10\%$ (Figs. 11.6, 11.7 and 11.8).

References

1. Allen, A.O.: Probability, Statistics and Queueing Theory with Computer Science Applications, 2nd edn. Academic Press, Boston (1990)
2. D'Agostino, R.B., Stephens, M.A. (eds.): Goodness-of-Fit Techniques. Marcel Dekker, New York (1986)

Appendix A
Data Characteristics Bearing on Classification

*I am not sure I should have dared to start, but I am sure I should
not have dared to stop*
[Winston Churchill. In Sked, Cook: Post-War Britain.
Penguin 1993.]

It is well-known that the structure of decision classes bears on classifier properties,
e.g., in case of two decision classes the Bayes classifier is a linear one whereas in
case of more than two decision classes it becomes polynomial of a higher degree, cf.,
[1]. Therefore, in Table A.1, we supply the main information about data sets, and, in
Tables A.2, A.3, A.4, A.5, A.6, A.7, A.8, A.9, A.10, A.11, A.12, A.13, A.14, A.15,
A.16, A.17, A.18, A.19, A.20, A.21, A.22, A.23, A.24, A.25, A.26, A.27, A.28 and
A.29, we include information about decision class distribution, the number of unique
attribute values in decision classes, and, the intersection of all central classes with
other classes. The intersection is defined as shown in Eq. A.1. The central class is a
class chosen as such among all decision classes (Table A.30).

The intersection of the central class with all other classes for particular attribute
a is defined as follows:

$$
\begin{aligned}
&Intersection_a(c_i, all\ other) \\
&= \frac{the\ number\ of\ all\ joint,\ unique\ attribute\ values_a}{the\ number\ of\ all\ unique\ attribute\ values\ in\ class\ c_{ia}}
\end{aligned}
\tag{A.1}
$$

© Springer International Publishing Switzerland 2015
L. Polkowski and P. Artiemjew, *Granular Computing in Decision Approximation*,
Intelligent Systems Reference Library 77, DOI 10.1007/978-3-319-12880-1

Table A.1 Data sets description—basic information

Name	Attr type	Attr no.	Obj no.	Class no.
Adult	categorical, integer	15	48842	2
Australian Credit	categorical, integer, real	15	690	2
Car Evaluation	categorical	7	1728	4
Diabetes	categorical, integer	9	768	2
Fertility Diagnosis	real	10	100	2
GermanCredit	categorical, integer	21	1000	2
Heart Disease	categorical, real	14	270	2
Hepatitis	categorical, integer, real	20	155	2
Congressional Voting Records	categorical	17	435	2
Mushroom	categorical	23	8124	2
Nursery	categorical	9	12960	5
soybean Large	categorical	36	307	19
SPECT Heart	categorical	23	267	2
SPECTF Heart	integer	45	267	2

Table A.2 Cardinality and percentage size of decision classes; data set: **adult**; *class_symbol* = label of decision class, *obj.no.* = size of decision class, *percentage* = percentage size of decision class in decision system

Class_symbol	Obj.no	Percentage
$c_1 = \leq 50K$	37155	76.1
$c_2 = > 50K$	11687	23.9

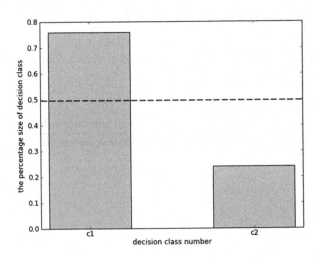

Table A.3 The number of unique values of attributes in decision classes and intersection of central decision class with all other classes (based on Eq. A.1); based on definition from Eq. A.1; data set **adult**

Attr.no.	Number of unique values		Class intersection	
	c_1	c_2	c_1-all	c_2-all
a_1	74	68	0.919	1
a_2	8	7	0.875	1
a_3	24560	8172	0.171	0.515
a_4	16	16	1	1
a_5	16	16	1	1
a_6	7	7	1	1
a_7	14	14	1	1
a_8	6	6	1	1
a_9	5	5	1	1
a_{10}	2	2	1	1
a_{11}	92	35	0.043	0.114
a_{12}	79	28	0.101	0.286
a_{13}	95	86	0.895	0.988
a_{14}	41	40	0.976	1

Table A.4 Cardinality and percentage size of decision classes; data set **australian credit**; *class_symbol* = label of decision class, *obj.no.* = size of decision class, *percentage* = percentage size of decision class in decision system

Class_symbol	Obj.no	Percentage
$c_1 = 0$	383	55.5
$c_2 = 1$	307	44.5

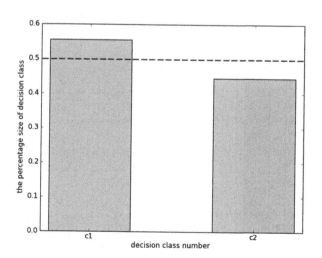

Table A.5 The number of unique values of attributes in decision classes and intersection of central decision class with all other classes (based on Eq. A.1); data set: **australian credit**

Attr.no.	Number of unique values		Class intersection	
	c_1	c_2	c_1-all	c_2-all
a_1	3	2	0.667	1
a_2	232	226	0.466	0.478
a_3	148	145	0.547	0.559
a_4	2	3	1	0.667
a_5	14	14	1	1
a_6	8	8	1	1
a_7	70	118	0.8	0.475
a_8	2	2	1	1
a_9	2	2	1	1
a_{10}	12	23	1	0.522
a_{11}	2	2	1	1
a_{12}	3	3	1	1
a_{13}	108	110	0.435	0.427
a_{14}	117	150	0.231	0.18

Table A.6 Cardinality and percentage size of decision classes; data set **car evaluation**; *class_symbol* = label of decision class, *obj.no.* = size of decision class, *percentage* = percentage size of decision class in decision system

Class_symbol	Obj.no	Percentage
$c_1 = unacc$	1210	70
$c_2 = acc$	384	22.2
$c_3 = vgood$	65	3.8
$c_4 = good$	69	4

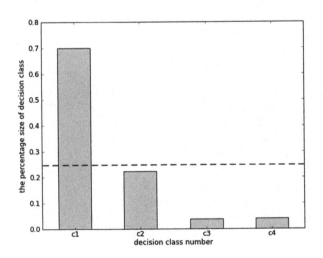

Table A.7 The number of unique values of attributes in decision classes and intersection of central decision class with all other classes (based on Eq. A.1); data set **car evaluation**

Attr.no.	Number of unique values				Class intersection			
	c_1	c_2	c_3	c_4	c_1-all	c_2-all	c_3-all	c_4-all
a_1	4	4	2	2	1	1	1	1
a_2	4	4	3	2	1	1	1	1
a_3	4	4	4	4	1	1	1	1
a_4	3	2	2	2	0.667	1	1	1
a_5	3	3	2	3	1	1	1	1
a_6	3	2	1	2	0.667	1	1	1

Table A.8 Cardinality and percentage size of decision classes; data set **diabetes**; *class_symbol =* label of decision class, *obj.no.* = size of decision class, *percentage* = percentage size of decision class in decision system

Class_symbol	Obj.no	Percentage
$c_1 = 1$	268	34.9
$c_2 = 0$	500	65.1

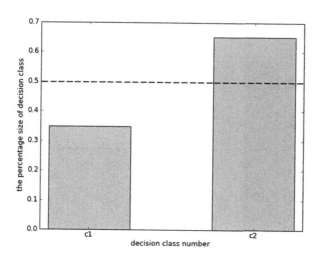

Table A.9 The number of unique values of attributes in decision classes and intersection of central decision class with all other classes (based on Eq. A.1); data set **diabetes**

Attr.no.	Number of unique values		Class intersection	
	c_1	c_2	c_1-all	c_2-all
a_1	17	14	0.824	1
a_2	104	111	0.76	0.712
a_3	39	43	0.897	0.814
a_4	43	46	0.884	0.826
a_5	93	137	0.473	0.321
a_6	148	210	0.743	0.524
a_7	231	372	0.372	0.231
a_8	45	51	0.978	0.863

Table A.10 Cardinality and percentage size of decision classes; data set **fertility diagnosis**; *class_symbol* = label of decision class, *obj.no.* = size of decision class, *percentage* = percentage size of decision class in decision system

Class_symbol	Obj.no	Percentage
$c_1 = N$	88	88
$c_2 = O$	12	12

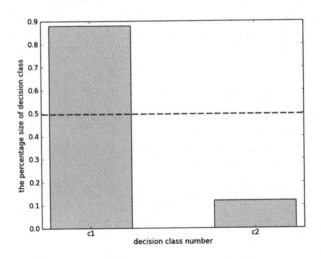

Table A.11 The number of unique values of attributes in decision classes and intersection of central decision class with all other classes (based on Eq. A.1); data set **fertility diagnosis**

Attr.no.	Number of unique values		Class intersection	
	c_1	c_2	c_1-all	c_2-all
a_1	4	4	1	1
a_2	18	6	0.333	1
a_3	2	2	1	1
a_4	2	2	1	1
a_5	2	2	1	1
a_6	3	3	1	1
a_7	5	3	0.6	1
a_8	3	3	1	1
a_9	14	7	0.5	1

Table A.12 Cardinality and percentage size of decision classes; data set **german credit**; *class_symbol* = label of decision class, *obj.no.* = size of decision class, *percentage* = percentage size of decision class in decision system

Class_symbol	Obj.no	Percentage
$c_1 = 1$	700	70
$c_2 = 2$	300	30

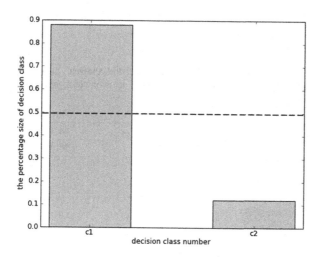

Table A.13 The number of unique values of attributes in decision classes and intersection of central decision class with all other classes (based on Eq. A.1); data set **germancredit**

Attr.no.	Number of unique values		Class intersection	
	c_1	c_2	c_1-all	c_2-all
a_1	4	4	1	1
a_2	31	25	0.742	0.92
a_3	5	5	1	1
a_4	10	10	1	1
a_5	655	294	0.043	0.095
a_6	5	5	1	1
a_7	5	5	1	1
a_8	4	4	1	1
a_9	4	4	1	1
a_{10}	3	3	1	1
a_{11}	4	4	1	1
a_{12}	4	4	1	1
a_{13}	53	47	0.887	1
a_{14}	3	3	1	1
a_{15}	3	3	1	1
a_{16}	4	4	1	1
a_{17}	4	4	1	1
a_{18}	2	2	1	1
a_{19}	2	2	1	1
a_{20}	2	2	1	1

Table A.14 Cardinality and percentage size of decision classes; data set **heart disease**; *class_symbol* = label of decision class, *obj.no.* = size of decision class, *percentage* = percentage size of decision class in decision system

Class_symbol	Obj.no	Percentage
$c_1 = 2$	120	44.4
$c_2 = 1$	150	55.6

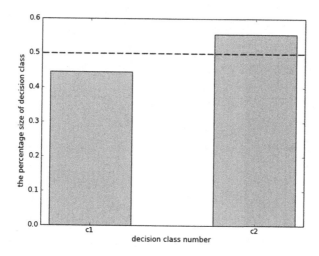

Table A.15 The number of unique values of attributes in decision classes and intersection of central decision class with all other classes (based on Eq. A.1); data set **heart disease**

Attr.no.	Number of unique values		Class intersection	
	c_1	c_2	c_1-all	c_2-all
a_1	35	39	0.943	0.846
a_2	2	2	1	1
a_3	4	4	1	1
a_4	35	37	0.714	0.676
a_5	84	99	0.464	0.394
a_6	2	2	1	1
a_7	3	3	1	1
a_8	67	67	0.657	0.657
a_9	2	2	1	1
a_{10}	34	26	0.618	0.808
a_{11}	3	3	1	1
a_{12}	4	4	1	1
a_{13}	3	3	1	1

Table A.16 Cardinality and percentage size of decision classes; data set **hepatitis**; *class_symbol* = label of decision class, *obj.no.* = size of decision class, *percentage* = percentage size of decision class in decision system

Class_symbol	Obj.no	Percentage
$c_1 = 2$	123	79.4
$c_2 = 1$	32	20.6

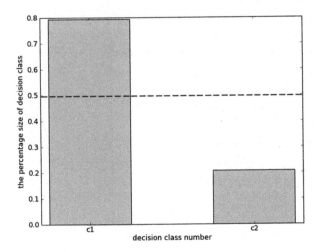

Table A.17 The number of unique values of attributes in decision classes and intersection of central decision class with all other classes (based on Eq. A.1); data set **hepatitis**

Attr.no.	Number of unique values		Class intersection	
	c_1	c_2	c_1-all	c_2-all
a_1	2	2	1	1
a_2	44	27	0.5	0.815
a_3	2	1	0.5	1
a_4	2	2	1	1
a_5	2	2	1	1
a_6	2	2	1	1
a_7	2	2	1	1
a_8	2	2	1	1
a_9	2	2	1	1
a_{10}	2	2	1	1
a_{11}	2	2	1	1
a_{12}	2	2	1	1
a_{13}	2	2	1	1
a_{14}	2	2	1	1
a_{15}	24	23	0.542	0.565
a_{16}	66	24	0.106	0.292
a_{17}	74	25	0.203	0.6
a_{18}	24	15	0.417	0.667
a_{19}	36	15	0.194	0.467

Table A.18 Cardinality and percentage size of decision classes; data set **congressional voting records**; *class_symbol* = label of decision class, *obj.no.* = size of decision class, *percentage* = percentage size of decision class in decision system

Class_symbol	Obj.no	Percentage
c_1 = republican	168	38.6
c_2 = democrat	267	61.4

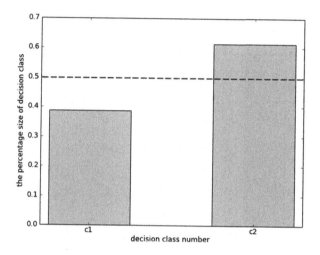

Table A.19 The number of unique values of attributes in decision classes and Intersection of central decision class with all other classes (based on Eq. A.1); data set **congressional voting records**

Attr.no.	Number of unique values		Class intersection	
	c_1	c_2	c_1-all	c_2-all
a_1	2	2	1	1
a_2	2	2	1	1
a_3	2	2	1	1
a_4	2	2	1	1
a_5	2	2	1	1
a_6	2	2	1	1
a_7	2	2	1	1
a_8	2	2	1	1
a_9	2	2	1	1
a_{10}	2	2	1	1
a_{11}	2	2	1	1
a_{12}	2	2	1	1
a_{13}	2	2	1	1
a_{14}	2	2	1	1
a_{15}	2	2	1	1
a_{16}	2	2	1	1

Table A.20 Cardinality and percentage size of decision classes; data set **mushroom**; *class_symbol* = label of decision class, *obj.no.* = size of decision class, *percentage* = percentage size of decision class in decision system

Class_symbol	Obj.no	Percentage
$c_1 = p$	3916	48.2
$c_2 = e$	4208	51.8

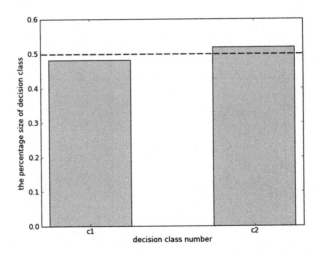

Table A.21 The number of unique values of attributes in decision classes and intersection of central decision class with all other classes (based on Eq. A.1); data set **mushroom**

Attr.no.	Number of unique values		Class intersection	
	c_1	c_2	c_1-all	c_2-all
a_1	6	7	1	0.857
a_2	5	5	0.8	0.8
a_3	4	3	0.75	1
a_4	8	10	1	0.8
a_5	2	2	1	1
a_6	7	3	0.143	0.333
a_7	2	2	1	1
a_8	2	2	1	1
a_9	2	2	1	1
a_{10}	10	10	0.8	0.8
a_{11}	2	2	1	1
a_{12}	3	4	1	0.75
a_{13}	4	4	1	1
a_{14}	4	4	1	1
a_{15}	6	6	0.5	0.5

(continued)

Table A.21 (continued)

Attr.no.	Number of unique values		Class intersection	
	c_1	c_2	c_1-all	c_2-all
a_{16}	6	6	0.5	0.5
a_{17}	1	1	1	1
a_{18}	2	3	0.5	0.333
a_{19}	3	2	0.667	1
a_{20}	4	3	0.5	0.667
a_{21}	5	8	0.8	0.5
a_{22}	4	6	1	0.667

Table A.22 Cardinality and percentage size of decision classes; data set **nursery**; *class_symbol* = label of decision class, *obj.no.* = size of decision class, *percentage* = percentage size of decision class in decision system

Class_symbol	Obj.no	Percentage
c_1 = recommend	2	0
c_2 = priority	4266	32.9
c_3 = not_recom	4320	33.3
c_4 = very_recom	328	2.5
c_5 = spec_prior	4044	31.2

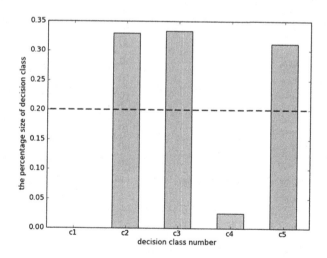

Table A.23 The number of unique values of attributes in decision classes and intersection of central decision class with all other classes (based on Eq. A.1); data set **nursery**

Attr.no.	Number of unique values					Class intersection				
	c_1	c_2	c_3	c_4	c_5	c_1-all	c_2-all	c_3-all	c_4-all	c_5-all
a_1	1	3	3	2	3	1	1	1	1	1
a_2	1	5	5	3	5	1	1	1	1	1
a_3	1	4	4	4	4	1	1	1	1	1
a_4	1	4	4	4	4	1	1	1	1	1
a_5	1	3	3	3	3	1	1	1	1	1
a_6	1	2	2	2	2	1	1	1	1	1
a_7	2	3	3	2	3	1	1	1	1	1
a_8	1	2	1	1	2	1	1	0	1	1

Table A.24 Cardinality and percentage size of decision classes; data set **soybean large**; *class_symbol* = label of decision class, *obj.no.* = size of decision class, *percentage* = percentage size of decision class in decision system

Class_symbol	Obj.no	Percentage
$c_1 = diaporthe - stem - canker$	10	3.3
$c_2 = charcoal - rot$	10	3.3
$c_3 = rhizoctonia - root - rot$	10	3.3
$c_4 = phytophthora - rot$	40	13
$c_5 = brown - stem - rot$	20	6.5
$c_6 = powdery - mildew$	10	3.3
$c_7 = downy - mildew$	10	3.3
$c_8 = brown - spot$	40	13
$c_9 = bacterial - blight$	10	3.3
$c_{10} = bacterial - pustule$	10	3.3
$c_{11} = purple - seed - stain$	10	3.3
$c_{12} = anthracnose$	20	6.5
$c_{13} = phyllosticta - leaf - spot$	10	3.3
$c_{14} = alternarialeaf - spot$	40	13
$c_{15} = frog - eye - leaf - spot$	40	13
$c_{16} = diaporthe - pod - \& - stem - blight$	6	2
$c_{17} = cyst - nematode$	6	2
$c_{18} = 2 - 4 - d - injury$	1	0.3
$c_{19} = herbicide - injury$	4	1.3

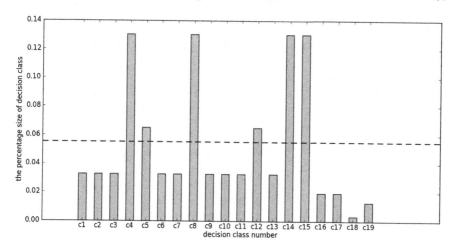

Table A.25 The number of unique values of attributes in decision classes; data set **soybean large**

Attr.no.	c_1	c_2	c_3	c_4	c_5	c_6	c_7	c_8	c_9	c_{10}	c_{11}	c_{12}	c_{13}	c_{14}	c_{15}	c_{16}	c_{17}	c_{18}	c_{19}
a_1	1	1	1	2	1	1	1	1	1	3	1	1	1	1	1	1	1	1	1
a_2	4	4	4	4	4	5	5	6	4	4	4	7	3	4	4	3	3	1	2
a_3	1	1	2	1	2	2	2	2	2	2	1	2	2	2	2	2	1	1	1
a_4	1	1	1	2	2	2	1	2	2	2	1	1	2	2	2	2	1	1	1
a_5	1	2	1	3	3	2	3	2	2	3	3	2	2	2	2	1	1	1	1
a_6	2	2	2	2	2	2	2	2	2	2	2	2	2	2	1	1	1	1	1
a_7	3	4	4	4	3	4	4	4	4	4	4	4	3	4	4	2	2	1	2
a_8	2	2	1	1	3	4	3	4	4	4	4	4	3	4	4	2	2	1	2
a_9	2	1	2	2	2	2	2	3	2	2	1	2	2	2	2	1	1	1	1
a_{10}	2	2	2	2	2	3	2	3	2	2	2	3	3	3	3	1	1	1	1
a_{11}	3	3	2	3	3	3	2	3	3	3	3	3	3	3	3	3	1	1	1
a_{12}	1	1	1	1	2	1	1	2	1	2	1	2	2	1	2	1	1	1	1
a_{13}	1	1	1	1	2	1	1	1	1	1	2	2	1	1	1	1	1	1	1
a_{14}	1	1	1	2	2	1	2	1	2	2	2	1	1	1	1	1	1	1	2
a_{15}	1	1	1	2	2	1	1	1	1	2	2	1	1	1	1	1	1	1	2
a_{16}	1	1	1	2	2	1	1	1	1	1	2	1	1	1	1	1	1	1	2
a_{17}	1	1	1	1	1	1	1	2	2	2	1	1	2	2	1	1	1	1	1
a_{18}	1	1	1	1	1	1	2	1	2	2	1	1	2	1	1	1	1	1	1
a_{19}	1	1	1	1	1	1	1	1	1	1	1	1	1	1	1	1	1	1	1
a_{20}	1	1	1	1	1	1	1	2	1	1	2	1	1	1	2	1	1	1	1
a_{21}	2	2	2	2	2	1	1	1	1	1	2	2	1	1	1	1	1	1	1
a_{22}	1	1	1	3	1	1	1	2	1	1	1	2	1	1	2	1	1	1	1
a_{23}	2	1	1	1	2	1	1	3	1	1	1	2	1	1	3	1	1	1	1
a_{24}	1	1	1	1	1	1	1	2	1	1	1	2	1	1	2	1	1	1	1

(continued)

Table A.25 (continued)

Attr.no.	c_1	c_2	c_3	c_4	c_5	c_6	c_7	c_8	c_9	c_{10}	c_{11}	c_{12}	c_{13}	c_{14}	c_{15}	c_{16}	c_{17}	c_{18}	c_{19}
a_{25}	1	1	1	2	1	1	1	2	1	1	1	2	1	1	2	1	1	1	1
a_{26}	1	1	2	1	1	1	1	1	1	1	1	1	1	1	1	1	1	1	1
a_{27}	1	1	1	1	1	1	1	1	1	1	1	1	1	1	1	1	1	1	1
a_{28}	1	1	1	1	1	1	1	1	1	1	1	1	1	1	1	1	1	1	1
a_{29}	1	1	1	2	1	1	1	1	1	1	2	2	1	1	2	1	1	1	1
a_{30}	1	1	1	2	2	1	1	2	1	1	2	2	1	1	3	1	1	1	1
a_{31}	1	1	1	1	1	1	1	1	1	2	1	2	1	2	2	2	1	1	1
a_{32}	1	1	1	1	1	1	1	1	1	2	1	2	1	1	1	1	1	1	1
a_{33}	1	1	1	1	1	1	1	1	1	2	1	2	1	2	2	1	1	1	1
a_{34}	1	1	1	1	1	1	1	1	1	2	1	2	1	1	2	1	1	1	1
a_{35}	1	1	1	1	1	1	1	1	1	1	1	2	1	1	2	1	1	1	1

Table A.26 Intersection of central decision class with all other classes (based on Eq. A.1); data set **soybean − large − all − filled**; based on definition from Eq. A.1

Attr.no.	c_1	c_2	c_3	c_4	c_5	c_6	c_7	c_8	c_9	c_{10}	c_{11}	c_{12}	c_{13}	c_{14}	c_{15}	c_{16}	c_{17}	c_{18}	c_{19}
a_1	1	1	1	1	1	1	1	1	1	1	1	1	1	1	1	1	1	1	1
a_2	1	1	1	1	1	1	1	1	1	1	1	1	1	1	1	1	1	1	1
a_3	1	1	1	1	1	1	1	1	1	1	1	1	1	1	1	1	1	1	1
a_4	1	1	1	1	1	1	1	1	1	1	1	1	1	1	1	1	1	1	1
a_5	1	1	1	1	1	1	1	1	1	1	1	1	1	1	1	1	1	1	1
a_6	1	1	1	1	1	1	1	1	1	1	1	1	1	1	1	1	1	1	1
a_7	1	1	1	1	1	1	1	1	1	1	1	1	1	1	1	1	1	1	1
a_8	1	1	1	1	1	1	1	1	1	1	1	1	1	1	1	1	1	1	1
a_9	1	1	1	1	1	1	1	1	1	1	1	1	1	1	1	1	1	1	1
a_{10}	1	1	1	1	1	1	1	1	1	1	1	1	1	1	1	1	1	1	1
a_{11}	1	1	1	1	1	1	1	1	1	1	1	1	1	1	1	1	1	1	1
a_{12}	1	1	1	1	1	1	1	1	1	1	1	1	1	1	1	1	1	1	1
a_{13}	1	1	1	1	1	1	1	1	1	1	1	1	1	1	1	1	1	1	1
a_{14}	1	1	1	1	1	1	1	1	1	1	1	1	1	1	1	1	1	1	1
a_{15}	1	1	1	1	1	1	1	1	1	1	1	1	1	1	1	1	1	1	1
a_{16}	1	1	1	1	1	1	1	1	1	1	1	1	1	1	1	1	1	1	1
a_{17}	1	1	1	1	1	1	1	1	1	1	1	1	1	1	1	1	1	1	1
a_{18}	1	1	1	1	1	1	1	1	1	1	1	1	1	1	1	1	1	1	1
a_{19}	1	1	1	1	1	0	0	1	1	1	1	1	1	1	1	1	1	1	1
a_{20}	1	1	1	1	1	1	1	1	1	1	1	1	1	1	1	1	1	1	1
a_{21}	1	1	1	1	1	1	1	1	1	1	1	1	1	1	1	1	1	1	1
a_{22}	1	1	1	1	1	1	1	1	1	1	1	1	1	1	1	1	1	1	1
a_{23}	1	1	1	1	1	1	1	1	1	1	1	1	1	1	1	1	1	1	1

(continued)

Table A.26 (continued)

Attr.no.	c_1	c_2	c_3	c_4	c_5	c_6	c_7	c_8	c_9	c_{10}	c_{11}	c_{12}	c_{13}	c_{14}	c_{15}	c_{16}	c_{17}	c_{18}	c_{19}
a_{24}	1	1	1	1	1	1	1	1	1	1	1	1	1	1	1	1	1	1	1
a_{25}	1	1	1	1	1	1	1	1	1	1	1	1	1	1	1	1	1	1	1
a_{26}	1	1	0.5	1	1	1	1	1	1	1	1	1	1	1	1	1	1	1	1
a_{27}	1	0	1	1	0	1	1	1	1	1	1	1	1	1	1	1	1	1	1
a_{28}	1	0	1	1	1	1	1	1	1	1	1	1	1	1	1	1	1	1	1
a_{29}	1	1	1	1	1	1	1	1	1	1	1	1	1	1	1	1	0	1	1
a_{30}	1	1	1	1	1	1	1	1	1	1	1	1	1	1	1	1	1	1	1
a_{31}	1	1	1	1	1	1	1	1	1	1	1	1	1	1	1	1	1	1	1
a_{32}	1	1	1	1	1	1	1	1	1	1	1	1	1	1	1	1	1	1	1
a_{33}	1	1	1	1	1	1	1	1	1	1	1	1	1	1	1	1	1	1	1
a_{34}	1	1	1	1	1	1	1	1	1	1	1	1	1	1	1	1	1	1	1
a_{35}	1	1	1	1	1	1	1	1	1	1	1	1	1	1	1	1	1	1	1

Table A.27 Cardinality and percentage size of decision classes; data set **SPECT**; $class\_symbol =$ label of decision class, $obj.no. =$ size of decision class, $percentage =$ percentage size of decision class in decision system

Class_symbol	Obj.no	Percentage
$c_1 = 1$	212	79.4
$c_2 = 0$	55	20.6

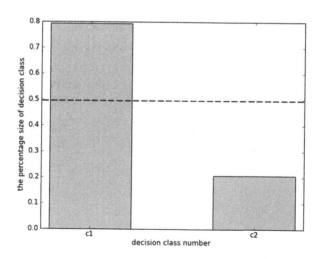

Table A.28 The number of unique values of attributes in decision classes and intersection of central decision class with all other classes (based on Eq. A.1); data set **SPECT**

Attr.no.	Number of unique values		Class intersection	
	c_1	c_2	c_1-all	c_2-all
a_1	2	2	1	1
a_2	2	2	1	1
a_3	2	2	1	1
a_4	2	2	1	1
a_5	2	2	1	1
a_6	2	2	1	1
a_7	2	2	1	1
a_8	2	2	1	1
a_9	2	2	1	1
a_{10}	2	2	1	1
a_{11}	2	2	1	1
a_{12}	2	2	1	1
a_{13}	2	2	1	1
a_{14}	2	2	1	1
a_{15}	2	2	1	1
a_{16}	2	2	1	1
a_{17}	2	2	1	1
a_{18}	2	1	0.5	1
a_{19}	2	1	0.5	1
a_{20}	2	2	1	1
a_{21}	2	2	1	1
a_{22}	2	2	1	1

Table A.29 Cardinality and percentage size of decision classes; data set **SPECTF**; *class_symbol* = label of decision class, *obj.no.* = size of decision class, *percentage* = percentage size of decision class in decision system

Class_symbol	Obj.no	Percentage
$c_1 = 1$	212	79.4
$c_2 = 0$	55	20.6

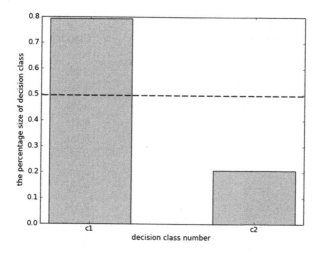

Table A.30 The number of unique values of attributes in decision classes and intersection of central decision class with all other classes (based on Eq. A.1); data set **SPECTF**

Attr.no.	Number of unique values		Class intersection	
	c_1	c_2	c_1-all	c_2-all
a_1	58	23	0.379	0.957
a_2	39	21	0.538	1
a_3	43	25	0.581	1
a_4	33	19	0.545	0.947
a_5	40	15	0.375	1
a_6	42	20	0.429	0.9
a_7	44	19	0.432	1
a_8	37	14	0.378	1
a_9	41	17	0.39	0.941
a_{10}	36	18	0.472	0.944
a_{11}	43	20	0.442	0.95
a_{12}	32	18	0.531	0.944
a_{13}	37	18	0.486	1
a_{14}	38	19	0.474	0.947
a_{15}	36	17	0.472	1
a_{16}	46	17	0.37	1
a_{17}	47	17	0.362	1
a_{18}	37	15	0.405	1
a_{19}	42	19	0.452	1
a_{20}	40	22	0.525	0.955
a_{21}	39	23	0.59	1
a_{22}	32	18	0.563	1

(continued)

Table A.30 (continued)

Attr.no.	Number of unique values		Class intersection	
	c_1	c_2	c_1-all	c_2-all
a_{23}	37	18	0.486	1
a_{24}	39	22	0.538	0.955
a_{25}	41	23	0.512	0.913
a_{26}	56	21	0.375	1
a_{27}	59	22	0.356	0.955
a_{28}	40	24	0.55	0.917
a_{29}	49	22	0.408	0.909
a_{30}	48	20	0.396	0.95
a_{31}	48	20	0.354	0.85
a_{32}	29	15	0.517	1
a_{33}	35	14	0.371	0.929
a_{34}	34	16	0.412	0.875
a_{35}	38	18	0.474	1
a_{36}	36	16	0.444	1
a_{37}	44	15	0.341	1
a_{38}	37	24	0.649	1
a_{39}	40	23	0.575	1
a_{40}	50	20	0.38	0.95
a_{41}	50	18	0.36	1
a_{42}	54	18	0.296	0.889
a_{43}	58	20	0.293	0.85
a_{44}	50	22	0.4	0.909

Reference

1. Rish, I., Hellerstein, J., Thathachar, J.: An analysis of data characteristics that affect naive Bayes performance. IBM Technical Report RC 21993 (2001)

Author Index

© Springer International Publishing Switzerland 2015
L. Polkowski and P. Artiemjew, *Granular Computing in Decision Approximation*,
Intelligent Systems Reference Library 77, DOI 10.1007/978-3-319-12880-1

General Index

© Springer International Publishing Switzerland 2015
L. Polkowski and P. Artiemjew, *Granular Computing in Decision Approximation*,
Intelligent Systems Reference Library 77, DOI 10.1007/978-3-319-12880-1

Symbols

Printed in the United States
By Bookmasters